SAFETY MANAGEMENT

A COMPREHENSIVE APPROACH TO DEVELOPING A SUSTAINABLE SYSTEM

SAFETY MANAGEMENT

A COMPREHENSIVE APPROACH TO DEVELOPING A SUSTAINABLE SYSTEM

Dr. Chitram Lutchman
Dr. Rohanie Maharaj
Eng. Waddah Ghanem

CRC Press
Taylor & Francis Group
Boca Raton London New York

CRC Press is an imprint of the
Taylor & Francis Group, an **informa** business

CRC Press
Taylor & Francis Group
6000 Broken Sound Parkway NW, Suite 300
Boca Raton, FL 33487-2742

Version Date: 20111107

International Standard Book Number: 9781439862612 (Hardback)

International Standard Book Number: 9781041220978 (Paperback)

Library of Congress Cataloging-in-Publication Data

Lutchman, Chitram.
Safety management : a comprehensive approach to developing a sustainable system / authors, Chitram Lutchman, Rohanie Maharaj, and Waddah Ghanem.
p. cm.
Includes bibliographical references and index.
ISBN 978-1-4398-6261-2 (alk. paper)
1. Industrial safety. I. Maharaj, Rohanie. II. Ghanem, Waddah. III. Title.

T55.L88 2012
658.4'08--dc23 2011041848

Visit the Taylor & Francis Web site at
http://www.taylorandfrancis.com

and the CRC Press Web site at
http://www.crcpress.com

For Product Safety Concerns and Information please contact our EU representative:
GPSR@taylorandfrancis.com
Taylor & Francis Verlag GmbH,
Kaufingerstraße 24,
80331 München, Germany.

We dedicate the knowledge provided in this book to the families of the more than two million workers who lose their lives in the workplace every year. This book is also dedicated to the families of and the estimated 250 million men and women who are injured and suffer illnesses from adverse working conditions and ongoing uncontrolled work hazards exposures every year. With increasing global competition, organizations are faced with greater challenges to satisfy shareholder needs. We applaud the many organizations that have transcended the focus of profit maximization to value maximization, in which the health and safety of our most precious assets—the lives of our workers—are foremost in the decision-making process.

Contents

Foreword

The corporate world has undergone much change in recent years. Society's heightened expectations and increased scrutiny that require companies to conduct safe and environmentally responsible operations have much more impact on a company's financial bottom line and license to operate than ever before. In today's business environment, a book like this is needed more than ever. I am struck by the comprehensive nature of this book that serves as an encyclopedia to identify and define many of the key and critical concepts to effect significant safety improvements in the operating work environment. I am impressed by the historical compendium of information as well as the various charts and templates provided that can be directly used and applied to improve any company's safety program and performance.

The book has an exceptional chapter on leadership, which is the foundation and true driver for safe operation. The reality is that many of those in leadership positions from supervisors to CEOs continue to underestimate the quantity and quality of their safety communications and direct leadership necessary to guide their organizations through an ever-changing business landscape that is becoming less and less tolerant to accidents and injuries.

This book is written for the serious student, safety professional, business leader, and executive who have a passion to learn the inner secrets of how businesses create a safe operation in their industry. Although this is written from an oil and gas business perspective for the most part, the principles and concepts can be applied to any business or facility operation.

I believe the tools, concepts, and insights in this book will provide readers and leaders seeking answers and direction with the inspiration to drive world-class results into their organizations and inspire others to value their safety and the safety of their workforce above all else.

Bob Batch
President, Operations Executive Advisors LLC

Preface

This book provides many recommendations and practical solutions for improving health and safety in the workplace. The authors recognize and sell workplace health and safety as essential for sustained long-term profitability of all organizations, regardless of the industry in which they operate. The authors also emphasize the values derived from the following sustained improvements in workplace health and safety:

1. Business environment trends—an understanding helps to guide our approaches to managing health and safety in the workplace
2. Safety management system (SMS)—a must
3. Process safety management (PSM)—must be integrated into business practices
4. Leadership commitment and shared learning in health and safety
5. Contractor safety management—must be leveraged to generate real improvements in workplace safety
6. Gaps in organizational SMS—proactively identified and addressed by using audits as a collaborative process

Although the information provided in this book is neither new nor groundbreaking, compiling this wealth of knowledge into a single book provides organizations, safety professionals, and teaching and educational institutions a holistic approach to upgrading the way in which health and safety are managed in the workplace.

Acknowledgments

We humbly acknowledge the help, support, and input from the many organizations that contributed to the content of this book. Our sincere thanks to those who have performed much of the early research work to generate new knowledge in the field of health and safety. Sincere thanks are also extended to Kevan Lutchman and Bill Pilon for their editorial and graphics design contributions. Similar gratitude is also extended to Sita, Megan, and Alexa Lutchman for contributions in editorial and formatting of the manuscript. The authors acknowledge the contributions of Steve Stewart (Director, Personal and Process Safety Management, Suncor Energy Inc.) and Ramesh Sharma (MOC Specialist, Suncor Energy Inc.) who reviewed the content of this book. The authors would also like to thank Mike Doyle (Silverstar Well Servicing Ltd.) and Martin Mudryk for their value-added contributions in the form of practical tools and processes for improving health and safety at the field level. Special thanks to Kavita Kissoon for contributions that improved comprehension and the flow of the knowledge provided. Finally, the authors thank Carlos V. Tan (EHS advisor, Group EHSQ Directorate, ENOC) and Bona Pinto (personal assistant to the director of EHSQ Compliance Directorate, ENOC) for their peer review and editing efforts.

Authors

Dr. Chitram Lutchman, a recipient of the McMaster Alumni Arch Award 2011, is an experienced safety professional with extensive frontline and leadership experience in the energy industry. Having sustained some disability early in his career from an industrial work accident, Dr. Lutchman developed a passion for improving health and safety in the workplace. From his experience in the international oil and gas industry, Dr. Lutchman has experienced the two extremes of organizational health and safety practices. As an employee of Canada's largest oil and gas producer, he pioneered work within his organization aimed at improving contractor health and safety management. In prior roles, Dr. Lutchman has functioned as a corporate leader in loss management and emergency response. He also functioned in project management leadership roles in the commissioning and start-up of a commercial SAGD and cogeneration facility in Fort McMurray, Alberta, Canada. Dr. Lutchman published his first book—*Project Execution: A Practical Approach to Industrial and Commercial Project Management*—in 2010. Dr. Lutchman is the principal author of this book and has coordinated the contributions of his coauthors to provide readers with this practical package of knowledge that can contribute significantly to improving the health and safety skills and capabilities of new graduates, safety professionals, and business leaders.

Dr. Rohanie Maharaj is a graduate of the University of the West Indies (UWI) in natural sciences (BSc Honours) and food technology (MSc and MPhil). She holds a PhD in food science and technology, specializing in food processing/postharvest technology, from the Université Laval in Québec, Canada. She is a senior member of the American Society for Quality (ASQ) and the Institute of Food Technologists (IFT). She is also trained in six sigma methodology. Her current research interests are in the areas of food safety, hazard analysis and critical control points (HACCP), and the use of ultraviolet radiation to preserve horticultural crops. She has been an associate professor at the University of Trinidad and Tobago (UTT) since 2009, teaching in the areas of food safety, human nutrition, and veterinary public health in the Biosciences, Agriculture and Food Technology Unit. Dr. Maharaj also lectures in the areas of quality assurance, environment, and health and safety in the Petroleum Engineering Unit. Prior to working at UTT, she was a regional director of quality assurance and regulatory affairs at Johnson & Johnson Caribbean and a board member for more than 7 years. She has extensive industrial experience, working in various capacities as operations director, process excellence director, compliance director, and quality assurance manager over the last 12 years at Johnson & Johnson in Trinidad, Jamaica, Dominican Republic, and Puerto Rico. She received the Award of

Excellence in 2005 at Johnson & Johnson for achieving exceptional business results. She was also instrumental in obtaining the ISO 14001 environmental certification for the Trinidad manufacturing facility in 2004, which was recertified in 2008. Apart from the many environmental and health and safety trainings offered at Johnson & Johnson, Dr. Maharaj has completed a number of international trainings, including behind the wheel (BTW), international health care compliance, process excellence examiner, marketing companies compliance, green belt, and crisis management trainings.

Waddah Ghanem is the chief EHSQ compliance officer for the ENOC Group. He is the chairman of various committees within the organization, including the Wellness & Social Activities Program that serves over 6000 employees. Waddah is an environmental engineer who graduated from the University of Wales, College Cardiff. He began his career working as a consultant for Hyder Consulting Middle East and later transitioned into the first oil refinery in Dubai, UAE, ENOC Processing Company LLC (EPCL) during the construction and precommissioning phases. Waddah holds a distinction-level MSc in environmental science from the University of the UAE. He has obtained an Executive MBA from Bradford University in the United Kingdom in which he specialized in organizational safety behavior. With extensive hands-on and frontline experience, Waddah has broad knowledge of environmental management systems and pollution control, fire and safety compliance and design reviews, development and administration of occupational health management systems, EHS management systems auditing, job safety task analysis, total quality management, and other HSE capabilities. His contribution to the EHS field includes more than 50 presentations and technical papers at various local, regional, and international conferences and forums. Waddah has been a member of the executive committee of the Emirates Safety Group, and has served as an advisor on the HCT Environmental Sciences Program Committee. He was a member of the Petroleum and Lube Specification Committee for the Government of the UAE, and more recently became a member of the Dubai Corporate Social Responsibility (CSR) Committee. He is also the chairman of the ENOC Group's Wellness & Social Activities Program Committee. He is passionate about EHSQ issues and his current doctoral research work with the Bradford School of Management, Bradford University, UK is focused on operational corporate governance, EHS leadership, and compliance systems. Waddah has also recently been appointed the vice-chairman of the board of the Dubai Centre for Carbon Excellence (DCCE).

1

Introduction

What does workplace safety mean to you? In some instances, scholars have identified workplace safety from a human perspective as the discipline of ensuring worker welfare in the workplace. Others have extended safety to include process safety incidents and management. In this book, the term *workplace safety* refers to both personnel and process safety management. Accordingly, workplace safety, when discussed in this book, deals with the protection of the health and welfare of all workers, the environment, company assets and facilities, and ultimately the company's image and sustainability. Essentially, the long-term viability of the organization is at risk when health and safety are not well managed.

Occupational health and safety (OH&S); health, safety, and environment (HSE); and environmental health and safety (EH&S) are all different umbrellas under which the safety of workers and assets is managed in the workplace. In some organizations, organizational security and quality assurance are also managed under these umbrellas. Safety in the workplace has evolved from an obscure, irrelevant, and often overlooked area of business to its present all-important, critical, corporate-defining business support requirement. Today, the mantra in every workplace is *safety first*. This message is irrespective of industry, and, regardless of where you may work, leaders encourage workers to place safety ahead of other business deliverables. Organizations are continually differentiating themselves from others by leveraging differences in safety performance.

Since the earliest time, workplace safety has always been an issue with which business leaders have had to contend with and struggle. Early history shows varying relationships between business and proponents of safety information. Today, business and stakeholders address workplace safety in cooperative and collaborative working relationships. Improvements in organizational safety have evolved over time from confrontational relationships to today's current collaborative work environments whereby all stakeholders work together to create a safer work environment that improves the health and safety of workers. In addition, key stakeholders have also evolved to include all direct and indirect participants in the workplace. Figure 1.1 shows the historical evolution of working relationships in the evolution of safety improvements in the workplace.

In the early 1900s, stakeholders were primarily unions, workers, and employers. In addition to these, today's work environment also includes

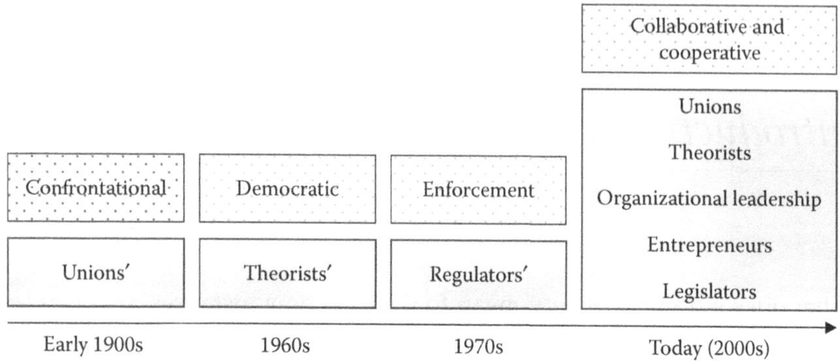

FIGURE 1.1
Evolving workplace relationships for improvements in organizational safety.

regulators, shareholders, governments, and the global community as all necessary components of stakeholder group. Essentially, everyone who uses the workplace is interested in and concerned about the safety of the workplace. History has shown that unions have helped transform safety in the workplace both directly and indirectly. History also points to severe criticism of unions for their focus on pursuing financial awards for their members as opposed to demanding safer and healthier workplaces for their members. The confrontational approaches of the early 1900s between unions and mine owners in the United Kingdom led to improvements in the ways anthracite coal miners were compensated for silicosis lung diseases, resulting from exposure to silica dust (Bufton and Melling, 2005).

Today, strong safety performance means not only exceptional organizational performance from worker productivity but also the ability to attract and retain excellent talent from the workforce. Indeed, Generation Y workers (born after 1970) have infiltrated the workplace, placing higher demands on organizations, not only for ensuring a safer workplace but also for creating business processes that are less damaging to the environment, while optimizing the use of assets.

A safe workplace is a strong marketing platform for attracting the best and brightest in the workforce. Senior leaders are slowly shifting behaviors and attitudes to embrace and recognize expenditures designed to enhance the organizational safety culture as an investment in the future of the organization as opposed to expenses. Data also support that organizational effort to improve safety processes is not an expense with little or no return on the investment but rather quite the opposite. Improved safety performance contributes to improved reliability, production run time, and therefore profitability. Improvements in safety have not been limited to any particular industry. Data show tremendous improvements in the oil and gas industry, the mining industry, and shipping and transportation industries, to name a few.

In today's work environments, unions and leadership have fostered more collaborative relationships toward improving workplace safety. The United Steel Workers, Auto Workers Unions (CAW/UAW), and Communication, Energy and Paperwork (CEP) are all organized labor unions that strive to remove workplace hazards for the health and safety of workers. The early work of Frederick Herzberg, which focused on hygiene and motivating factors in the workplace, suggested that when hygiene factors were absent in the workplace, workers became dissatisfied. Research clearly shows that dissatisfied workers will turnover very quickly or in the worst-case scenario *quit and stay*. Herzberg identified hygiene factors that impact turnover to include the following:

Working conditions (including safety and workplace hazards)
Level and quality of supervision
Company policy and administration
Interpersonal relations

Ministries of Labor, Occupational Safety and Health Administration (OSHA), worker compensation boards, and many other governing bodies have developed safety regulations and standards by which businesses must operate. Over time, regulations have evolved further to enhance the safety of the workplace, and businesses are responsive to these improved standards. Many safety-conscious organizations seek to exceed these regulatory requirements through their own health and safety management systems (SMS). Regulations have brought about lasting impact on the way safety is managed in the workplace. Nevertheless, while regulations have evolved and brought success, such success has been limited since the oversight and enforcement branches of governments have not been adequately resourced to keep on top of violators. At the root of this is the underlining issue that politicians do not see the real benefits of workplace safety.

Figure 1.2 shows the historical evolution of safety focus in the workplace. Today, leadership focuses on changing the behavior of the worker using principles of behavior-based safety and modern social sciences.

The early 1900s was characterized as a period for improving welfare. Toward the mid-1900s, with the evolution of the Factories Act, health and safety became a focal point. Today's focus on leadership as the means for improving safety in the workplace places huge emphasis on the social skills, capabilities, and competence of the leader. The involvement of every worker in this process creates a sense of ownership and clearly sends the message to all workers that the employer cares for the well-being of its workers. Leadership skills are therefore crucial for continuous improvements in safety in today's fast-paced work environment.

Improvements in organizational safety have not come without cost. During the period 1969–2000, a total of 81,258 fatalities were recorded globally from

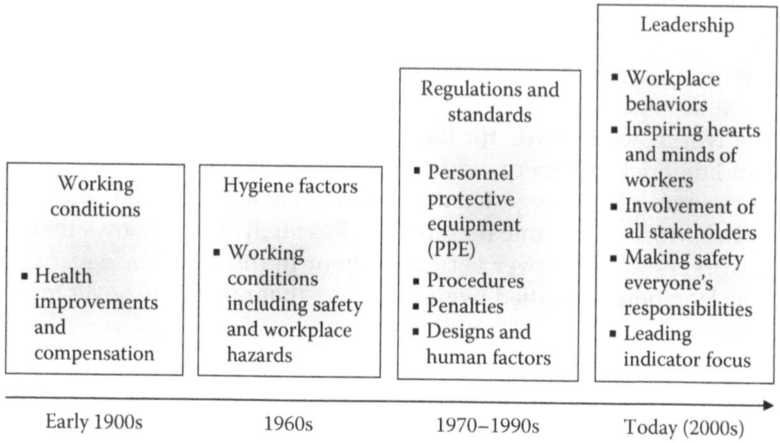

FIGURE 1.2
Evolving workplace focus for improvements in organizational safety.

1870 accidents in the energy-related industries (Burgherr and Hirschberg, 2008). These data do not include accidents with less than five deaths per accident for the same period. Table 1.1 shows the distribution of these accidents within the energy sector and the number of deaths associated with each sector of the energy industry.

Historical landmark events such as the Bhopal incident of 1984 led to 3787 deaths within the first 24 hours, after a gas release at a chemical plant (Krishnan, 2009). Studies show that more than 560,000 residents were affected by this incident; today, almost 100,000 are still chronically ill (Krishnan, 2009). The public reaction to the fallouts from this incident led to major

TABLE 1.1
Significant Energy Industry Accidents and Fatalities (1969–2000)

Energy Industry	Number of Accidents	% of Accidents	Number of Fatalities	% of Fatalities
Coal	1221	65	25,107	31
Oil	397	21	20,218	25
Natural gas	135	7	2043	3
LPG	105	6	3921	5
Hydro	11	1	29,938	37
Nuclear	1	0	31	0
Total	1870	100	81,258	100

Source: Adapted from Burgherr, P. and Hirschberg, S. 2008. *Human & Ecological Risk Assessment, 14*(5), 947–973. Retrieved February 3, 2010 from EBSCOHost database.

changes in the way multinational corporations view safety in developing countries.

Similarly, the Piper Alpha North Sea Oil Platform destruction by fire in 1988, which killed 167 workers (Hibbert, 2008), led to major improvements in the offshore industry. Despite these improvements, offshore operations continue to be risky and organizations are challenged to be continuously vigilant for improvement opportunities. The 1989 Exxon Valdez spill led to the release of 11 million gallons of hydrocarbons into the Alaskan coastlines. Harrald et al. (1990) pointed to a direct cleanup cost of >$2 billion dollars and according to Aspen Publishers Inc. (2009), this event led to punitive charges to the employer (including interests) of ~$1 billion dollars. Fallouts from the ecological impact continue to haunt us even today—20 years later. This single event and the lessons learned from the Exxon Valdez spill of 1989 helped to dramatically reduce the numbers and volumes of spills experienced annually for many years to follow. Figure 1.3 shows reducing trends in the numbers of U.S. coastal spill volumes and spill numbers.

The Exxon Valdez incident led to an outpouring of public support for changes in the accountabilities of businesses for environmental accidents. According to Ramseur (2009), this single event and ensuing public outrage—from the media-generated attention coupled with the Oil Pollution Act of 1990—led to significant declines in the numbers of spills in U.S. coastal waters. More recently, the Texas City BP incident that occurred in March

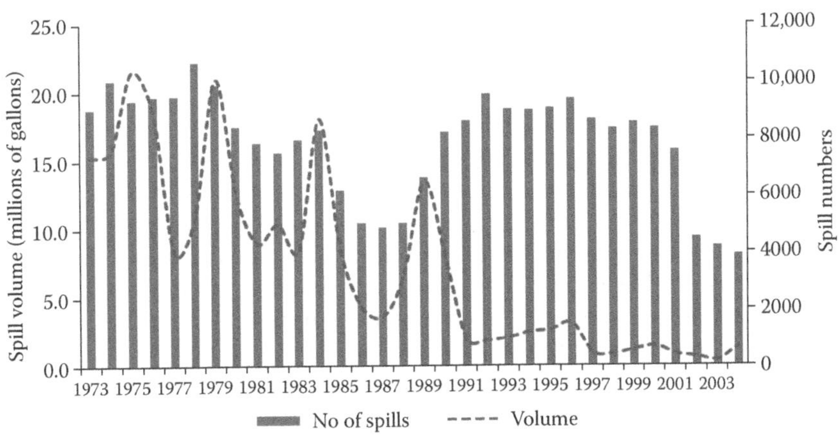

FIGURE 1.3
Spill volume and spill numbers—U.S. coastal waters 1973–2004. (Compiled using data from the U.S. Department of Homeland Security, 2009. *Oil Spills Compendium, 1973–2004*. United States Coastguard. Retrieved February 7, 2010 from http://homeport.uscg.mil/mycg/portal/ep/contentView.do?channelId=-18374&contentId=120051&programId=91343&programPage=%2Fep%2Fprogram%2Feditorial.jsp&pageTypeId=13489&contentType=EDITORIAL&BV_SessionID=@@@@1864989697.1265590118@@@@&BV_EngineID=cccdadejigldhfmcfjgcfgfdffhdghm.0.)

2005 claimed the lives of 15 workers and injured 500 and was regarded as the worst industrial accident in the last 15 years (Chemical Safety Board, 2007a). This event continues to enhance the way in which process safety is managed in organizations through regulatory changes and genuine desire by business leaders to avoid similar incidents. Many organizations have adopted proactive measures to enhance process safety management within their operations to protect workers and assets from a similar repeat incident.

No one will forget the April 20, 2010 explosion of the Deepwater Horizon offshore oil drilling about 40 miles southeast of Louisiana in the Gulf of Mexico, which claimed the lives of 11 and injured 98 of which 17 were seriously injured. This incident resulted in the largest oil spill and environmental disaster ever experienced by the United States. This spill has been regarded as the worst the world has ever seen with estimated volumes between 60,000 and 80,000 barrels of oil released into the Gulf of Mexico daily while BP worked to fix the damaged wellhead. When this event was finally resolved, this single event cost BP more than $20 billion dollars for cleanup and claims settlements alone. Experts estimated that four million barrels of oil were released into the environment and if the experiences of Exxon Valdez and Bhopal are any indicators, the fallout of this single event will affect this region for decades into the future.

It is important to place this event in context to fully understand the basis of the disaster. With declining easy-to-access natural resources, industries are forced to access resources that are riskier to locate. This deepwater drilling incident occurred more than a mile under the sea and was executed with technology about which the then BP CEO suggested the organization may not have been fully knowledgeable. This situation is no different from the challenges experienced by the whale oil industry in the 1800s, which resulted in whale oil factories on ships. Access to whales forced whaling ships to venture into the Arctic and Antarctic regions in search of the dwindling whale supplies from which many of the ships and crews never returned.

As we proceed into the future and as industries venture into unchartered territories with new technology, new and unknown risks and challenges will arise. There will be an increase in new safety incidents and hazards to address. History has shown that man-made disasters prompted changes in the way businesses operated, much like the manner in which the Texas City BP incident transformed the oil and gas industry to reexamine and improve its approaches to process safety management. The Gulf of Mexico drilling disaster will be no different and will serve as a great opportunity for regulators and stakeholders to ensure that industry improvements are generated, to improve access to dwindling supplies of oil and gas reserves.

The way we manage safety in the workplace must similarly evolve to meet the needs of changing industry requirements. Changes in technology, operating environments, financial allocations, training needs, stakeholder demands, and environmental conditions will indeed influence the application and administration of workplace safety. While best practices require effective risk assessment and mitigation strategies, as the above conditions are imposed on organizations by the insatiable demands of our nations, mistakes will be made. We must work collaboratively to minimize these mistakes and to learn from them. For many incidents, you may also find that the concept of *you do not know what you do not know* rings true where safety is concerned, and as leaders and safety practitioners into whose hands the workforce places its safety, we must do more to preserve and ensure their health and safety.

Safety is a costly business for many stakeholders. With hundreds of billions of dollars expended each year to address the direct and indirect impacts of workplace injuries and diseases, there is growing emphasis on creating a zero-harm work environment. Here, the focus is on the welfare of the worker, preservation of the environment, and protection of the assets of the organization.

Organizations place great emphasis on asset management, process efficiency, resources management, image and goodwill preservation, and the ability to attract and retain the best and brightest workforce through a caring work environment. Other stakeholders are focused on producing solutions to today's workplace hazards all aimed primarily at the welfare of the worker. The current focus on safety is on *cooperation*, *collaboration*, and *generating and sharing new knowledge* to enhance workplace safety.

This book is intended to provide an overview to all workers on the benefits of a safe work environment and how we can contribute to this goal. The book focuses on providing practical approaches to enhancing the welfare of workers and guides business leaders on the fundamental requirements for creating a safer workplace. The authors share practical knowledge from field experience and leverage research knowledge to create a step-by-step model for creating a safer workplace.

The authors place tremendous responsibilities on leadership for creating a compelling sense of direction and desire among workers for a safer workplace. In addition, they explore key attributes of an effective SMS to enhance safety in the workplace. Further, the authors recognize prevention of personnel injuries and prevention of damage to the environment and assets as key elements of organizational safety programs. Subsequent discussions in this book are intended to build upon prior knowledge generated in the field of safety and introduce current industry focus in a structured format for enhancing workplace safety.

2

Trends in Safety

This chapter defines major trends in occupational health and safety by regions and where possible, globally. Some of the trends examined include fatalities, workplace injuries, labor force changes, labor force demographics, and other trendable information that can demonstrate the evolution of safety in the workplace. A brief discussion is provided on key useful trends for industries and businesses.

Some of the key safety trends focused on in industry include injury frequencies, fatalities, illnesses, and demographic-related information. These trends at best are directional indicators since data management for trending injuries and fatalities in the workplace varies across regions and countries. Reporting of incidents also affects the quality and availability of global data on fatalities, illnesses, and injuries in the workplace. Useful trends are derived from good-quality data that can be manipulated to provide credible information.

Where workplace illnesses, injuries, and fatalities are concerned, useful information is best derived from annual trends. Where illnesses and injuries are concerned, frequencies that reflect normalized data trended over some fixed period may be the best approach to determine progress made on these variables. On the other hand, where fatalities are concerned, the absolute numbers of deaths experienced may be the best method for trending.

OSHA requires injuries to be grounded to 200,000 work hours. Using this base, some standard measures for injury trends may include, but are not limited to, the examples provided below. In addition, these measures may vary across companies and industries (benchmarking purposes) in the definitions of what may be considered a recordable injury.

1. Total recordable injury frequency (TRIF)—This frequency is derived from this equation: TRIF = (total number of recordable injuries for all workers × 200,000) divided by the total number of hours worked. For example, if eight recordable injuries were recorded from 1,000,000 h worked at a worksite, then TRIF = (8 × 200,000) ÷ 1,000,000. This calculation yields a TRIF of 1.6.
2. Employee recordable injury frequency (ERIF)—This frequency is derived from this equation: ERIF = (number of employee recordable injuries × 200,000) divided by the total number of hours worked. Therefore, in the above example, if two of the eight recordable injuries

were employees being hurt, then ERIF = (2 × 200,000) ÷ 1,000,000. This calculation yields an ERIF of 0.4.

3. Contractor recordable injury frequency (CRIF)—This frequency is derived from this equation: CRIF = (number of contractor recordable injuries × 200,000) divided by the total number of hours worked. Therefore, in the above examples, of the eight recordable injuries, two were employee recordable and the remaining six were contractor recordable injuries. CRIF = (6 × 200,000) ÷ 1,000,000. This calculation yields a CRIF of 1.2.
4. TRIF = ERIF + CRIF. In the examples above, TRIF = 0.4 + 1.2 = 1.6. In some cases, TRIF may be called injury severity rate.
5. Other measures may include variations of the above frequencies. Some companies may measure the frequency of all injuries (first aids and recordable).
6. Other useful measures may include lost work day frequency (LWDF). This is calculated similarly as follows: LWDF = (number of work days lost from injuries × 200,000) divided by the total number of hours worked at the worksite.

The use of frequency rates in environment, health and safety (EH&S) key performance indicators (KPIs) is very useful and effective in risk assessment (RA) and risk management (RM) studies since they are related to the function of probability. Within an organization, these measures are very important in determining the success of an organization's safety program. TRIFs, ERIFs, and CRIFs that trend to zero indicate that success is being derived and recordable injuries are becoming fewer and fewer over time. Other meaningful injury frequencies that may be measured by organizations are disabling injury frequencies, in which case, the number of disabling injuries replaces the number of recordable injuries in the formulae above.

Global and Regional Trends in Employment

Figure 2.1 shows the changing employment numbers in the United States, Canada, France, and Germany over the period 1985–2007. The data show a growing employment trend in the United States, while the labor force growth has remained relatively flat in Canada, France, and Germany. For the United States, this growth has tremendous implications for organizational safety management strategies.

The introduction of Generation Y workers into the workplace, foreign workers from different cultures, and growing number of women in the workforce, in nontraditional roles, will influence the way in which safety is managed in

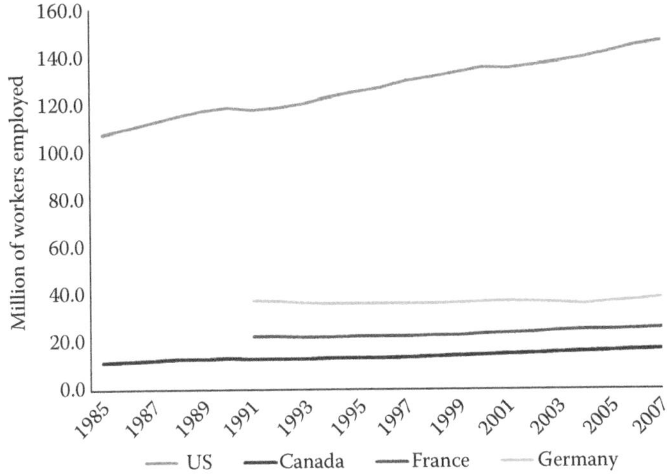

FIGURE 2.1
Employment growth in the United States, Canada, France, and Germany, 1985–2007. (Compiled using data from ILO, http://laborsta.ilo.org/STP/guest.)

the workplace. Indeed, physical differences between men and women will necessarily affect workplace design and human/machinery interface.

Figure 2.2 demonstrates changes in growth in the workforce according to gender. Notably, there are increasing numbers of women entering the workforce in the United States. These changes have implications for human factor

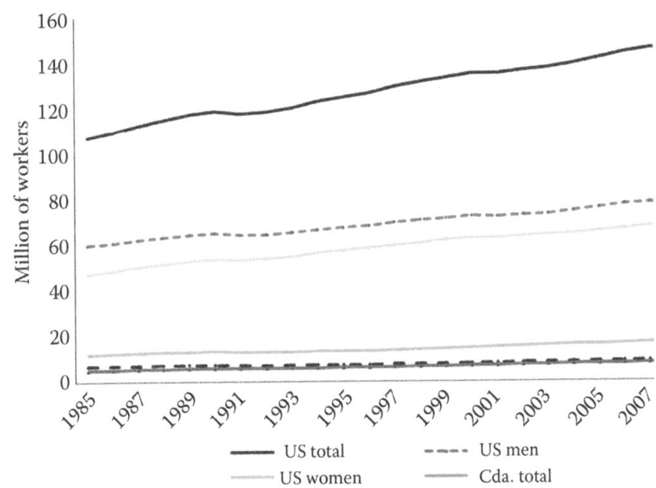

FIGURE 2.2
Employment growth by gender in the United States and Canada, 1985–2007. (Compiled using data from ILO, http://laborsta.ilo.org/STP/guest.)

considerations in the design of the workplace. While data were not available for high-growth regions such as China and India, it is plausible that similar changes in the workforce have been occurring in these regions. Failure to evolve the workplaces to accommodate a growing presence of women in the workforce can have disastrous long-term workplace health and safety consequences. Similarly, the introduction of younger Generation Y workers into the workplace strongly influences the way in which safety is delivered in the workplace. These challenges will be investigated and discussed later.

Trends in Injuries and Fatalities

Figure 2.3 demonstrates trends in the number of recordable injuries in the United States, Canada, France, and Germany. Comparative statistics for other developing nations such as China and India were not available. In all cases, since the early 1990s, there appears to be a declining trend in the number of injuries that result in temporary and permanent disabilities. Nevertheless, within the United States alone, there were almost 2.5 million injuries in 2001 and when the United States trend line is extrapolated to 2007, the four countries considered in this figure pointed to ~4.0 million injuries during 2007. As one can imagine, such staggering injury numbers can create havoc within our health care systems and have a significant negative impact on cost associated with required treatments. Moreover, the impact on productivity and lost production can be overwhelming.

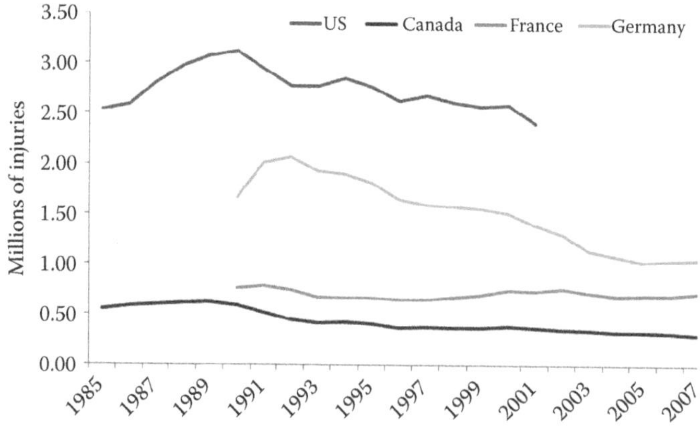

FIGURE 2.3
Nonfatal injuries with temporary and permanent disabilities, 1985–2007. (Compiled using data from ILO, http://laborsta.ilo.org/STP/guest.)

When more detailed analyses are conducted, more meaningful information can be derived. Such detailed analyses can help in developing the right response plan at a country, industry, organizational, or business unit/facility level. In Canada and the United States, OH&S and OSHA, the U.S. Bureau of Labor Statistics, and the U.S. Department of Labor are excellent sources of work-related information relating to safety.

For global data, the International Labor Organization (ILO) is an excellent source of information. Concerns with international data include source reliability, consistency, and frequency of reporting. As a consequence, there may be gaps in the data and it is difficult to compare safety performance on a regional or country basis.

Nevertheless, global trends show strong improvements in workplace safety as demonstrated by trends that reflect reducing numbers of disabling injuries and fatalities. Clearly, the impact of historical measures to address workplace safety issues has worked. While these successes should be lauded, we have collectively harvested the low-hanging fruits where managing occupational health and safety is concerned. Moving forward, improvements to health and safety in the workplace will become even more difficult as low-hanging fruits are fully exploited. We must now focus on organizational safety cultures, leadership, process safety management, and other innovative means, yet to be determined for major improvements in workplace health and safety.

Injuries and Fatality Trends: Canada

Figures 2.4 through 2.9 show injuries and fatality trends across Canada and across its provinces. The most disconcerting figure in this range is Figure 2.7, which shows an increasing trend in the number of fatalities in the workplace. When considered with a relatively flat growth in employment in Canada as demonstrated in Figure 2.2, Canada is obviously trending in the wrong direction as far as workplace fatalities are concerned.

Can this trend be the outcome of an aging workforce in Canada? Coles (2009) advises that an aging workforce is placing additional cost pressures on employers in Canada. According to Cole, Canada has an aging population that resulting in retaining workers in the workforce longer. In addition, employers are continually seeking cost-cutting strategies to better manage cost. Can these actions be indirectly contributing to an increase in workplace fatalities? More detailed analyses may be required to better understand why this is happening.

Some industries may also be more vulnerable than others. Hopkins (2008) pointed to major skill shortages in the oil and gas industry. He advised, "average age is about 50 and their average retirement age is 55" (Hopkins, 2008, p. 150). He concluded, "the industry faces a major skills crisis in the next 5 to 10 years, as more than half the experienced workforce will leave the industry" (Hopkins, 2008, p. 150). Can this skill shortage lead to increasing

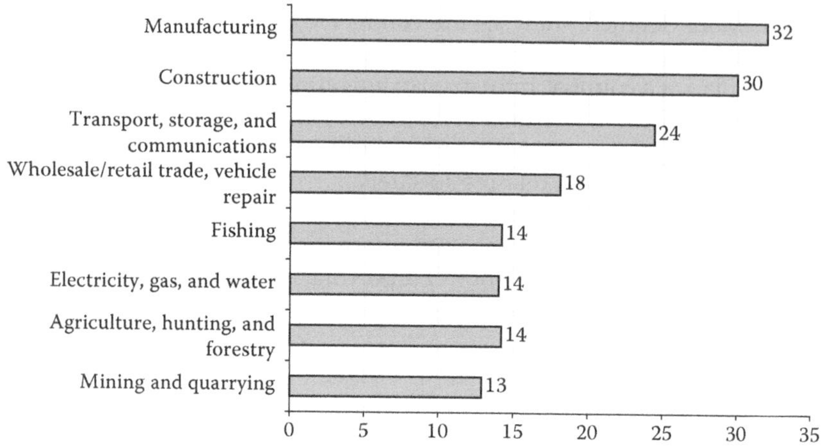

FIGURE 2.4
Work injuries in Canada by industries per 1000 employed, 2007. (Adapted from HRDC Canada, http://www4.hrsdc.gc.ca/.3nd.3c.1t.4r@-eng.jsp?iid=20#M_3.)

fatalities and injuries in the workplace? The authors argue that an aging workforce, coupled with tighter cost controls that can impact on training budgets, as well as skill shortages in some industries, can certainly impact on the number of workplace fatalities and injuries. Furthermore, as employers seek to augment these skill shortages with foreign workers, differences in culture, practices, education, and language will introduce many additional health and safety challenges to employers. The impact of these variables is

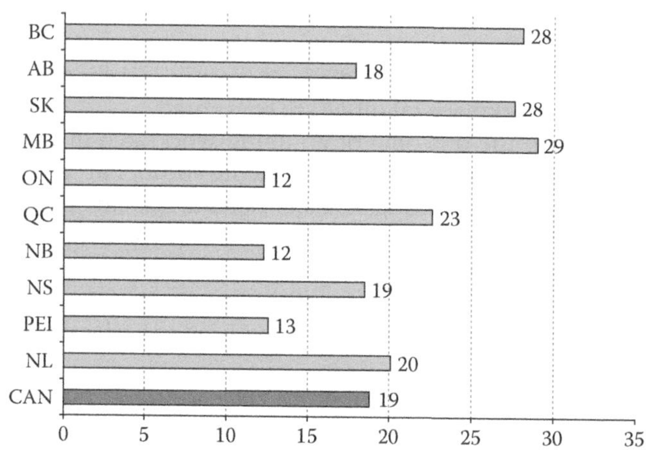

FIGURE 2.5
Work-related injuries in Canada by provinces per 1000 employed, 2007. (Adapted from HRDC Canada, 2011, http://www4.hrsdc.gc.ca/.3nd.3c.1t.4r@-eng.jsp?iid=20#M_3.)

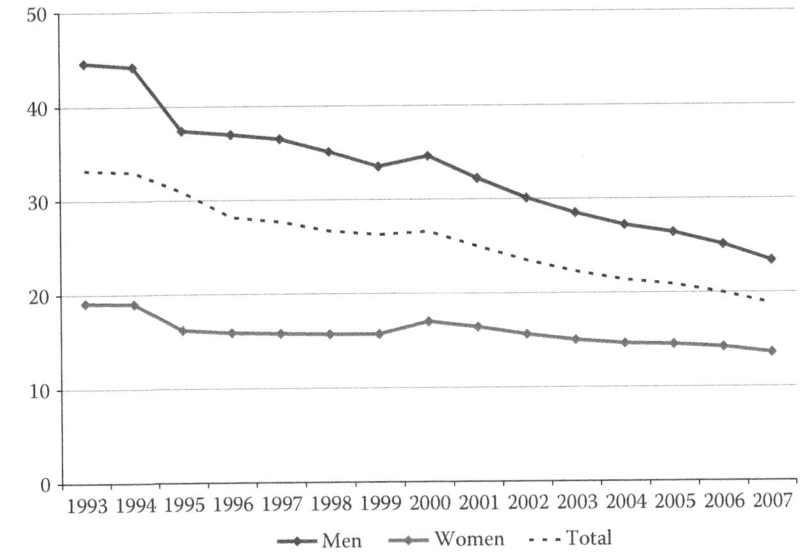

FIGURE 2.6
Work-related injuries by gender, Canada, 1993–2007. (Adapted from HRDC Canada, 2011, http://www4.hrsdc.gc.ca/.3nd.3c.1t.4r@-eng.jsp?iid=20#M_3.)

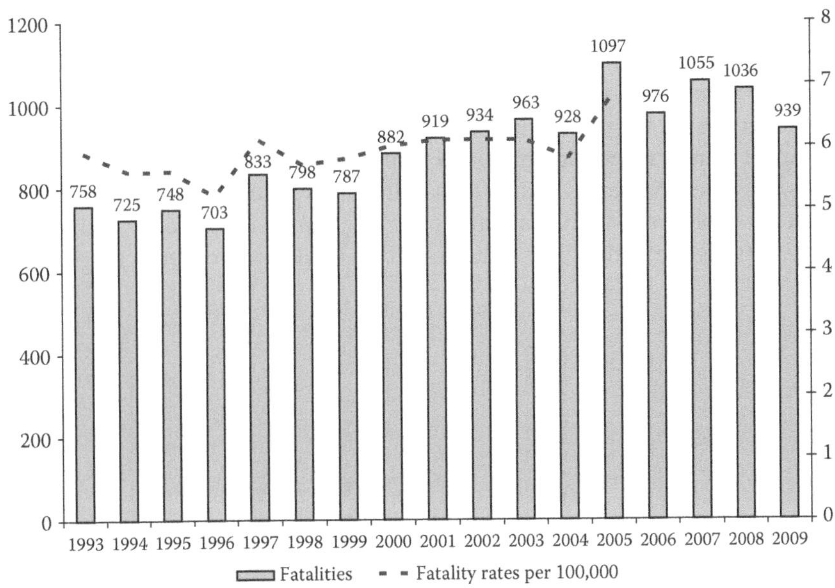

FIGURE 2.7
Work-related fatalities and fatality rates per 100,000 employed in Canada, 1993–2009. (Adapted from WCB of Canada, 2011, http://www.awcbc.org/common/assets/nwisptables/fat_summary_jurisdiction.pdf.)

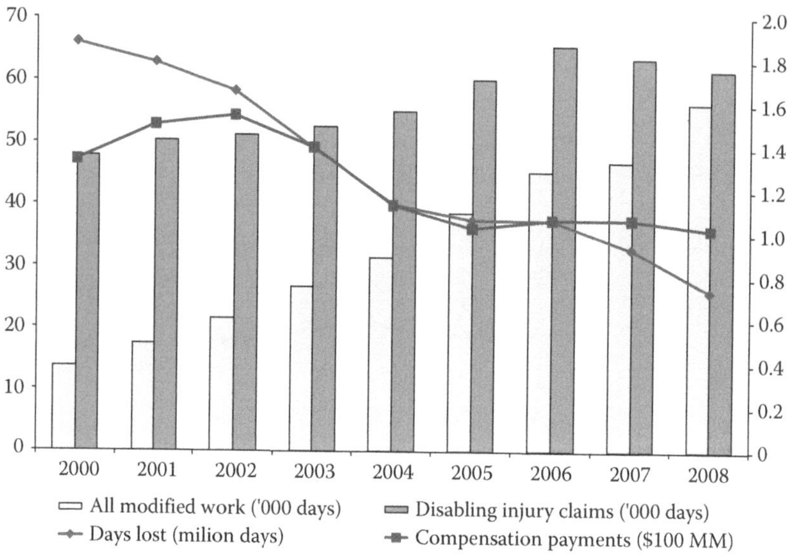

FIGURE 2.8
Worker compensation and injury trends Alberta, 2000–2009. (Adapted from WCB Alberta, OH&S Canada, 2009.)

greater in high-risk industries such as the oil and gas, manufacturing, and power-generating industries.

Figure 2.8 is quite interesting for the province of Alberta, Canada. As shown in this figure, there is an increasing number of disabling injury claims (in thousands of person days) coupled with decreasing claims cost and loss days. The most salient trend, however, is a sharp increase in the incidence of modified work days. What this means is that ill workers are returning to work much faster than before, which may suggest a decline in the severity of incidents. Some experts are of the view that this is due to higher workers' compensation

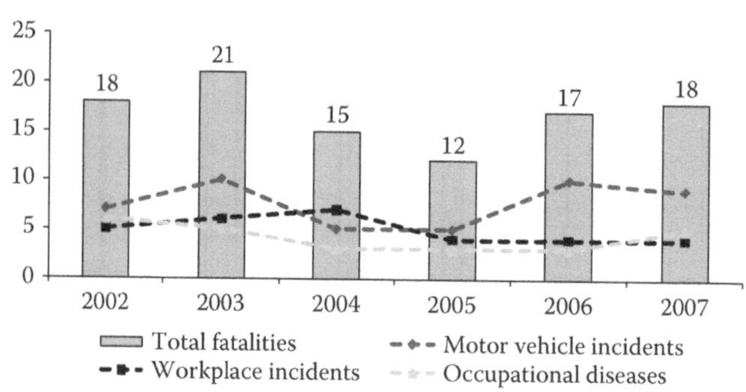

FIGURE 2.9
Upstream oil and gas fatalities in Alberta, 2002–2007. (Adapted from WCB Alberta, OH&S Canada, 2008.)

board (WCB) costs and better claims management and pressures from senior leadership for better safety indicators. By bringing workers back to work early through aggressive claims management, employers save money through WCB premiums. While this method of management has been effective in bringing workers back to the workplace faster, there are unintended consequences of worker harassment and intimidation. In addition, it also means that employers are now bearing the cost of workers injured in the workplace, from reduced productivity, since workers returning to work on modified work will generally perform at less than optimum capacity and often in light-duty jobs. One can therefore argue that we have taken the cost from one pocket and placed it into another, and we are no further ahead than before where workplace injuries are concerned in Alberta. This phenomenon may also not be unique to Alberta.

Figure 2.9, which examines upstream oil and gas fatalities (exploration and production) in Alberta, shows strong progress in reducing work-related fatalities during the period 2003–2005. However, when examined relative to a base year of 2002, the province was no further ahead in 2007 with 18 work-related fatalities.

Injuries and Fatality Trends: United States

Figures 2.10 through 2.13 show trends in work-related fatalities in the United States. Interestingly, Figure 2.10, when coupled with rising employment shown in Figure 2.2, is in stark contrast to Canada's performance. The United States shows continuous decline in work-related fatalities and fatality rate per 100,000 employed since 1992. This performance against a backdrop of

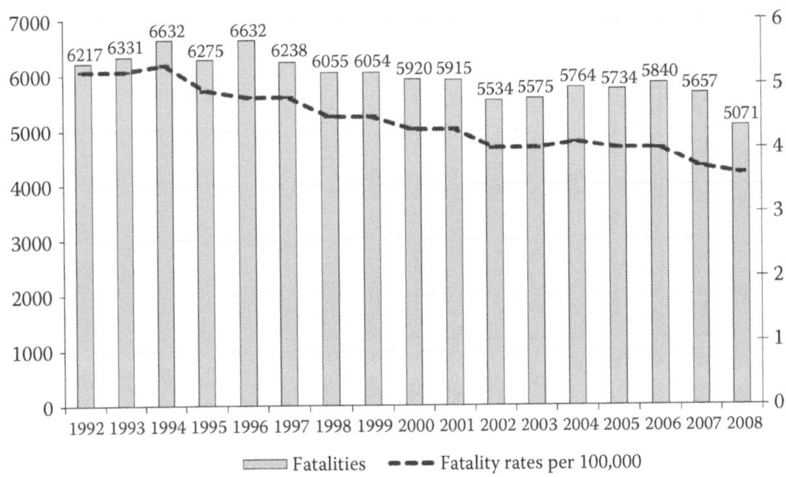

FIGURE 2.10
Work-related fatalities and fatality rates per 100,000 employed in the United States, 1992–2008. (Adapted from U.S. Bureau of Labor Statistics, U.S. Department of Labor, 2008.)

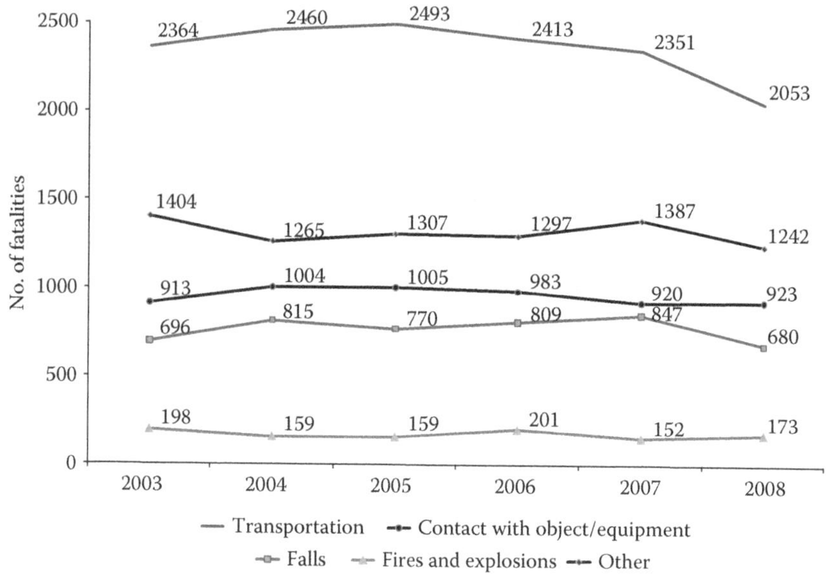

FIGURE 2.11
Principal causes of work fatalities in the United States, 2003–2008. (Adapted from U.S. Bureau of Labor Statistics, 2009.)

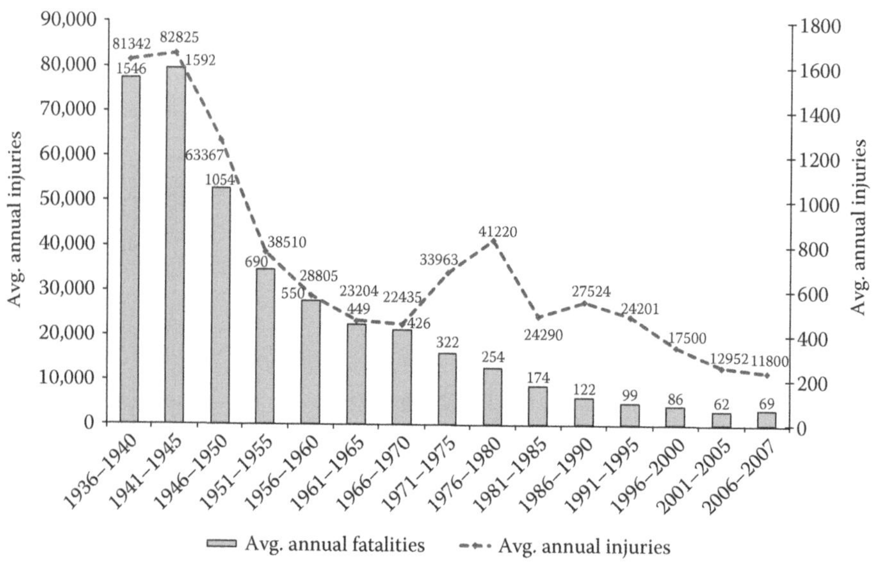

FIGURE 2.12
Historical mining fatalities and injuries in the United States, 1936–2007. (Adapted from U.S. Department of Labor, 2009.)

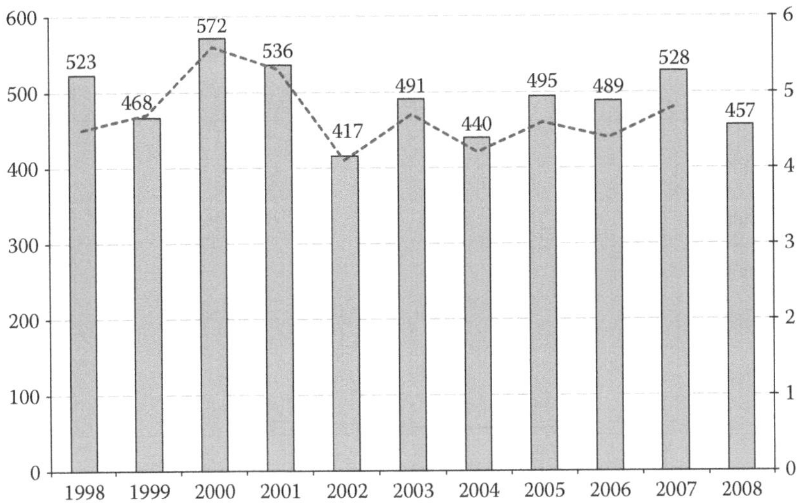

FIGURE 2.13
Work-related fatalities and fatality rates per 100,000 employed in the State of Texas, 1998–2008. (Adapted from Texas Department of Insurance, Division of Workers' Compensation (TDI-DWC), 2009.)

increasing trends in the number of workers employed is very encouraging to the United States.

Figure 2.12 shows tremendous improvement in both fatalities and injuries in the mining industry in the United States. As shown in Figure 2.13, which highlights fatality trends in the state of Texas, full control of fatality in the workplace continues to be an elusive target with no definite downward trend shown over a 10-year period. Figure 2.14 shows a reducing trend in workplace

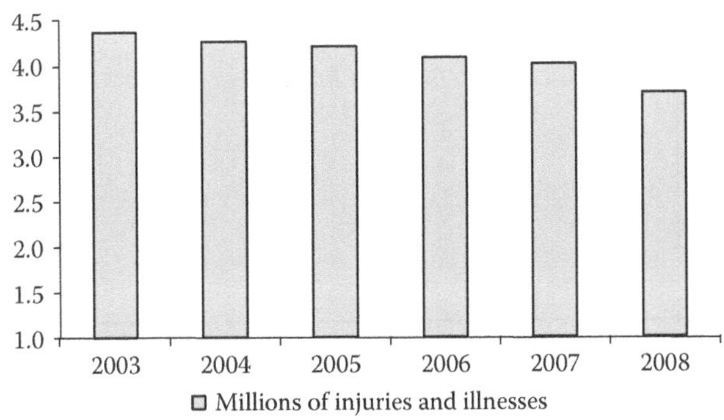

FIGURE 2.14
Declining trend in workplace injuries and illnesses in the United States. (Adapted from U.S. Department of Labor, 2007–2009.)

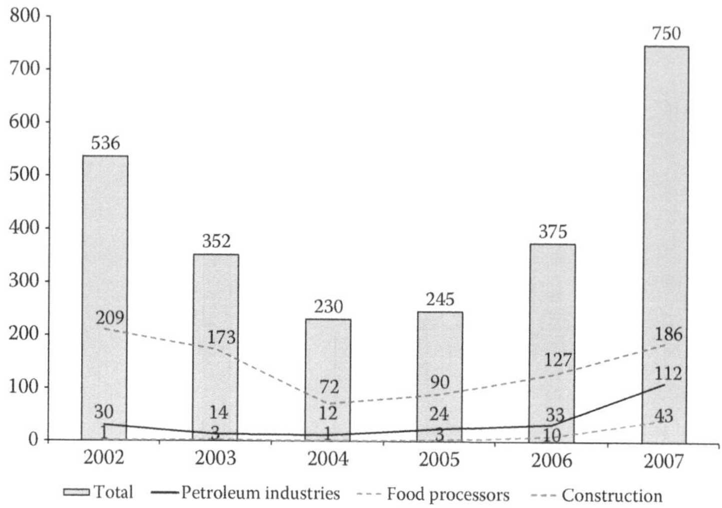

FIGURE 2.15
Work-related disabling injuries in Trinidad and Tobago, 2002–2007. (Adapted from Central Statistical Office (CSO), Trinidad and Tobago, 2009.)

injuries in the United States. Despite strong and continuous improvements, however, in 2008, almost 3.7 million workplace injuries and illnesses were experienced by the U.S. workforce.

When similar numbers were trended (Figures 2.15 through 2.17) for a developing industrialized country such as Trinidad and Tobago, an increasing

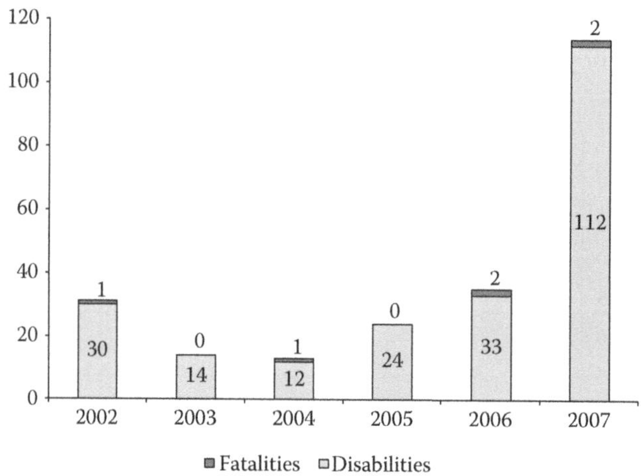

FIGURE 2.16
Work-related disabling injuries and fatalities in the petroleum industry of Trinidad and Tobago, 2002–2007. (Adapted from Central Statistical Office (CSO), Trinidad and Tobago, 2009.)

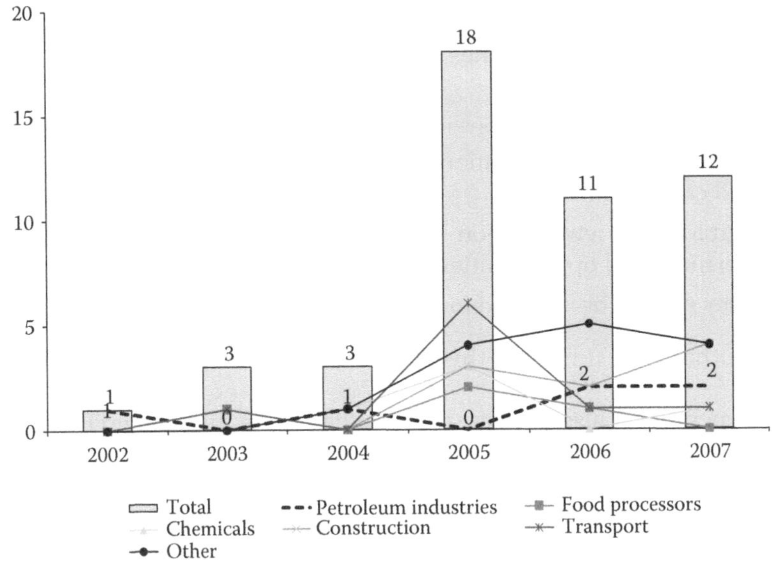

FIGURE 2.17
Work-related fatalities in Trinidad and Tobago by industries, 2002–2007. (Adapted from Central Statistical Office (CSO), Trinidad and Tobago, 2009.)

trend in both workplace fatalities and injuries and illnesses since 2004 is present. Prior to 2004, some improvements were made in occupational health and safety which appeared to have deteriorated as a result of growing economic activities in the country.

Generation Y and the Workforce

The technological revolution that began in the 1980s created a significant impact on the workplace with the influx of a growing group of Generation Y workers within the workforce. According to Cable (2005), Generation Y workers include workers who are characterized as follows:

- Workers entering the workforce who are generally <30 years old and born during the mid-1970s through the early 2000s
- Computer literate and eager to please
- Want a safe workplace and eager for safety training in the workplace
- Strong ability to multitask
- Willing to embrace change
- Strong ability to absorb information—and willing to ask questions when not sure
- Demand more from trainers

- Demand training from multimedia, one-on-one, and print materials
- Will change jobs 4–5 times in the first 10 years

Austin (2005) added that

- They have high expectations of employers and they must be known and catered for.
- Employers viewed upon as a resource pool for relationships, learning, and opportunities to excel and to be rewarded.
- They need to be involved in finding solutions to their safety concerns.

The authors add further to this list to include the following:

- Generation Y also utilizes social media such as Twitter and Facebook. This is very different from Generations X who sees this type of technology as frivolous in the workplace and often seeks to quash its use at the risk of disengaging workers.
- Generation Y dislikes authoritarian leadership and management style.

Generation Y poses different challenges to the workplace where safety is concerned. The learning behaviors, work methods, expectations, and growing influence in the workplace require innovative solutions from leaders to create a work environment that caters to their needs, while meeting the needs of a traditional Generation X workforce. In addition, the general tolerance to risk and the risk-taking attitudes is different between Gen X and Gen Y, which impacts directly on the perception of consequences (how likely they are) and thus on behavioral-based safety (BBS).

Some of the challenges presented to leadership of the new workplace as dictated by an ever-increasing Generation Y workgroup are highlighted subsequently (Cable, 2005).

Challenges for Workplace Safety

- Need to cater for both Generation Y and traditional Generation X (older workers).
    - Different learning styles (older workers prefer print versus Generation Y workers who are more effective with computers and information technology learning).
    - Generation X possesses greater field experience and may therefore be more set in ways, due to their experiences in safety.
    - Generation Y's lack of field experience encourages them to trust leaders by placing their safety into the hands of leaders.
- Generation Y depends on *engage-me* strategies and want to be a part of workplace safety solutions.

- Generation Y has shorter attention spans and requires clear and precise messaging. This can be challenging, since some workplace safety measures may require detailed procedures and full compliance to ensure the safety of workers.
- Generation Y seeks personal protective equipment (PPE) that is trendy and stylish. Often, there may be a trade-off between PPE that is cool versus PPE that is safe.
- Generation Y likes change and continuous improvements. Leaders must keep ideas fresh to retain the attention of Generation Y workers.
- Safety trainers challenged to meet training needs for a group of workers that think differently and will often question the expertise and credibility of trainers. Such Generation Y behaviors are often information-seeking as opposed to deliberately seeking to discredit the trainer. As such, trainers must be tolerant and open to any discussion that is aimed at gaining their trust. On most occasions, trainers are Generation X workers who can be easily offended and intimidated by such behaviors. Trainers must therefore be fully knowledgeable and well read on the abundance of materials available to the keen Generation Y worker, whose principal source of information may be the Internet, which as we all know, is readily accessible to all and carries an abundance of peer-reviewed as well as generic information.
- Today's work environment requires leaders to leverage technology in training such as the use of computers and the Internet. Such actions can potentially alienate some Generation X workers who are not computer savvy and are often intimidated by the use of technology.
- Considerations for e-learning include
  - Assess your organization's needs and readiness.
  - Choose the right partner or vendor when buying safety services.
  - Be strategic when attaining buy-in within your e-learning program.
  - Embrace change.
  - Identify champions or coaches, who will motivate others to value and defend your e-learning initiatives.
  - Engage a cross section of management to ensure corporate buy-in and build trust with workers.
  - Create a pilot program to demonstrate value to the entire organization.
  - Pilot the program in an area of the organization which you believe has the greatest chance for success.

How We Retain Them

Cable (2005) suggested the following ways to retain Generation Y workers:

- Keep them safe and motivated.
- Involve and engage them in finding solutions to meet their safety needs.
- Leverage information technology to cater to their learning needs (computers/videos).
- Training providers must be dynamic and knowledgeable.
- Be flexible in training methods.
- Demonstrate and replicate—use role play to simulate situations.

Women in the Workforce

Today, there are increasing numbers of women in the workforce. More importantly, women are increasingly more abundant in nontraditional work areas. For example, women can now be found in many areas of trade careers such as construction, power engineering, firefighting, forestry, welding, light- and heavy-duty mechanic, and offshore oil and gas operations. These are areas traditionally dominated by a male workforce. The presence of an increasing number of women in these areas has changed the way in which safety is administered and will continue to do so as women continue to become more comfortable in these areas. In addition, women are generally more empathetic and caring to coworkers. They also take greater care in assessing workplace hazards and can often be more thorough in workplace hazards assessments.

According to Wesley et al. (2009), in India alone, since the 1980s, women became more active in the workforce and at present stands at 28% of the paid workforce. Figure 2.18 highlights a dramatic shift in workforce participation by women in the United States since the 1970s. In 2005, women workforce participation rates very closely resembled that of the male workforce participation rates.

Today, increasing number of women in the workforce poses a real challenge for employers as it relates to the prevention of muscular injuries and industrial diseases that may affect the reproductive system. Furthermore, limited research in the areas of chronic diseases and illnesses related to female workers in today's chemical-filled workplaces limits the corrective and preventative measures that organizations may proactively adopt to protect this group of workers in the workplace. These unknowns and inadequacy of current SMSs to adequately address the health and safety needs of women in the workplace will have an impact on society for years to come.

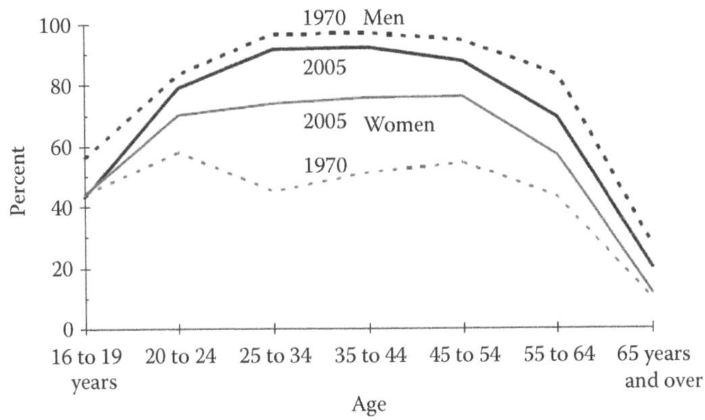

FIGURE 2.18
Labor force participation men and women, 1970 and 2005. (Adapted from rom U.S. Bureau of Labor Statistics, 2009.)

Roles of Unions and Union Membership

History points to highly confrontational relationships between unions and employers in addressing workplace safety and working conditions. In the early 1900s, confrontational approaches between unions and mine owners in the United Kingdom led to improvements in the ways anthracite coal miners were compensated for silicosis lung diseases resulting from exposure to silica dust (Bufton and Melling, 2005). Research also suggests that unions reduce the competitiveness of organizations (Gill, 2009). Gill also advised that unions help facilitate effective communication among its membership. This capability can be extremely beneficial in communicating organizational safety strategies and practices among its membership.

Survival of unions in today's business environment requires unions and organizational leadership to foster and maintain collaborative relationships toward improving workplace safety. The United Steel Workers, CAW/UAW, and CEP are all examples of organized labor unions that strive to work collaboratively to remove workplace hazards for the health and safety of workers. Figure 2.19 highlights declining union membership among both men and women in the United States.

Shorter Work Tenure among Men

As shown in Figure 2.20, men remain working for the same employer for shorter periods. Several possible reasons can account for this behavior, among which is an increasingly more mobile workforce. In addition, the drive for success pushes men to gain experience and knowledge in one organization and transfer these skills to another organization where generally better recognition and acknowledgement is provided. Generally too, when workers

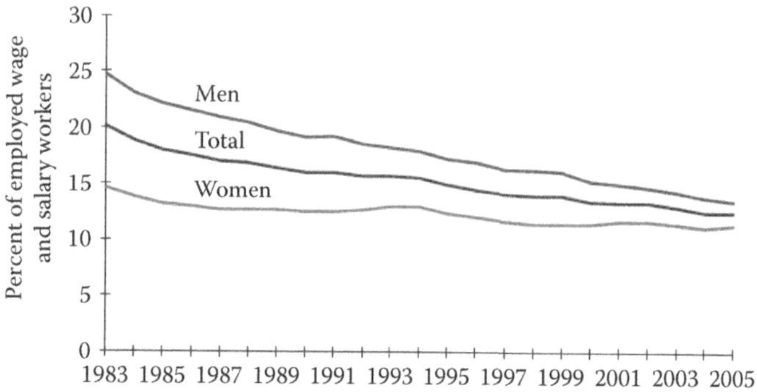

FIGURE 2.19
Declining union membership in the United States. (Adapted from U.S. Bureau of Labor Statistics, 2009.)

obtain a particular level of competence and expertise and are unable to progress up the corporate ladder, they are no longer afraid to change jobs and even employers to seek greater access to climbing the corporate ladder. Another contributory factor is that in most instances, household income is determined by both spouses in the family. This allows one of the family members to take the risk of changing jobs much more frequently than was possible in the past where single incomes were more prevalent among households.

This problem is likely to worsen as Generation Y males continue to increase their presence in the workforce, particularly as organizations fail to address their specific needs regarding corporate social responsibility and workplace safety. Ultimately, such mobility results in increased cost

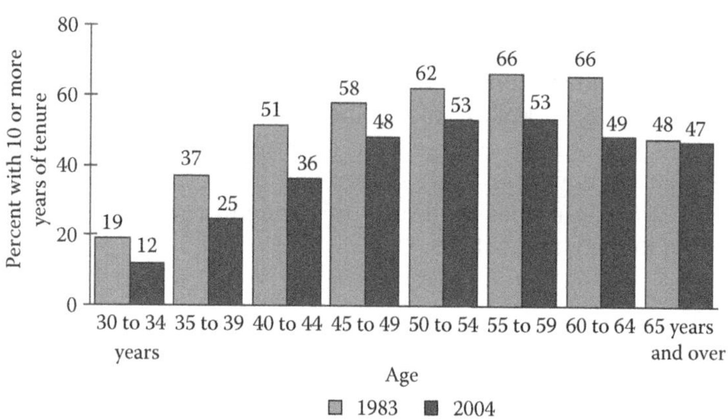

FIGURE 2.20
Fewer men are working for the same employer for ≥10 years. (Adapted from U.S. Bureau of Labor Statistics, 2009.)

from recruitment and retention to the organization, which can affect its competitiveness.

A More Educated and Knowledgeable Workforce

As shown in Figure 2.21, the education level of the workforce has changed dramatically since the 1970s. Since the 1970s, we have seen a continuous shift in the educational levels of workers toward degree level and higher education. In 2003, ~60% of the workforce was trained to a first-degree level compared to ~25% in the 1970s.

While it is difficult to obtain empirical studies linking the educational levels of workers with safety performance and the demand for a safer workplace, there is an abundance of studies linking education with higher workplace safety performance. Truxillo et al. (1998), in a 10-year study of the relationship between education and job performance, confirmed that education makes a difference, in on-the-job performance. In addition, human resource management clearly links workplace performance with education and training. By extension, therefore, we can predict that a more educated workforce should result in a safer workplace.

It also follows that a more educated workforce will also demand a safer workplace from employers. We must hasten to point out that a common theme among many major workplace incidents is a lack of training and competency assurance, or inadequate training and ultimately education, which have generally contributed to the incident. The major issue still remains however, as to how much, or better still how little, there is in terms

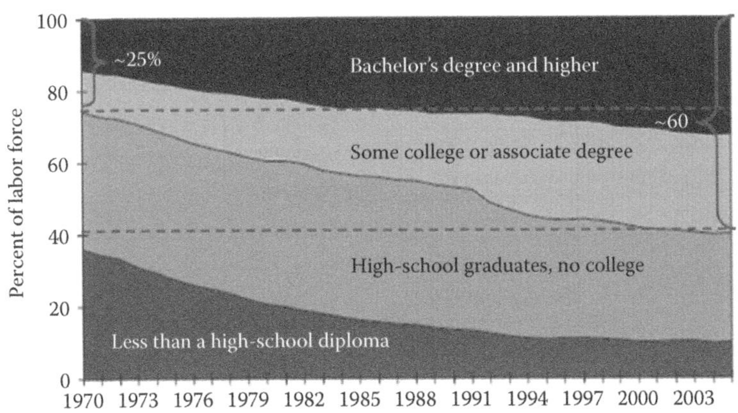

FIGURE 2.21
Changing educational levels of the U.S. labor force. Note: Data are from the march 1970–2005 current population survey and are for person age 25–64. Beginning in 1992, data are based on highest diploma or degree received; prior to this time, data were based on years of school completed.

of health and safety in the curriculums of vocational and academic qualifications.

Changing Leadership Environment

The leadership environment has changed dramatically over time and continues to evolve daily as the workforce transitions across generations. Today's leadership environment is characterized by high expectations from workers and high leadership efforts necessary to meet the expectations of the workforce. Figure 2.22 demonstrates the evolving leadership environments. In addition, the leadership skills required to lead and inspire the hearts and minds of a growing Generation Y worker population (workers born after 1970) are very different from the leadership skills two decades ago. As a consequence, leaders who are resistant to change and fail to evolve with the changing workplace requirements are destined to fail.

Leaders must master modern leadership skills and inspire Generation Y workers. At the same time, they must be able to cater to the needs of Generation X workers in the same workplace. Meeting the different needs of both workgroups must be seamless and efficient. This is particularly important in the development needs of both workgroups, and in the way we train these workers. Research suggests that there may be a disconnection

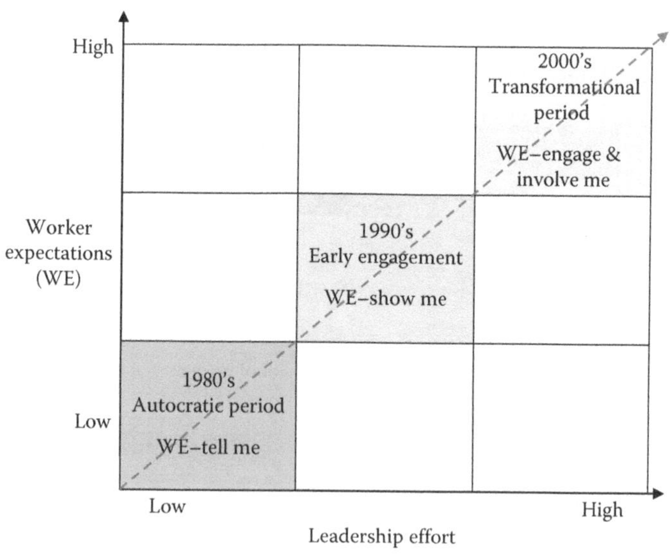

FIGURE 2.22
Recent evolution of the leadership environment. (Adapted from Lutchman, C. 2010. *Project Execution: A Practical Approach to Industrial and Commercial Project Management*. Boca Raton, FL, CRC Press.)

between the perceptions Generation Y workers have of the field working environment versus the reality. As a consequence, leaders are faced with the ever-increasing challenge of keeping this valuable workgroup safe and engaged in the field.

Compounding the workplace challenge is the rapid introduction of women into nontraditional and skilled trades' jobs. Increasingly, women are finding increasing opportunities in the construction, power engineering, mechanic, and mining industries. Consistent with their influx into these work environments is the need for leaders to further adjust behaviors to accommodate for a workgroup with whom they are not familiar. For Generation X leaders, this is indeed a challenge, as personal biases and failure to evolve with the changing workforce can significantly impact on leadership decisions.

Creating and sharing the safety vision of an organization are critical to developing the safety culture of the organization. If a leader is unable to adjust to the current leadership demands of the work environment, overall financial and corporate success will be compromised in addition to safety. Creating a shared vision requires leaders to demonstrate commitment to safety in a visible manner. They must have the ability to clearly articulate this by making the often painful decision of placing safety before profits, for example, by shutting down operations in a facility to perform required maintenance on an unsafe system. For most leaders, this can be a very difficult decision for a site leader where the CEO is remote to the site and does not fully comprehend the need for such actions, particularly when shareholders are demanding more from the organization.

Some organizations have been successful in connecting the dots, which suggests that a safe and healthy work environment ultimately leads to a more motivated workforce and greater operational performance and profitability. However, to get here comes at great short-term cost that many senior leaders are not prepared to accept until a crisis occurs in which case it is too late. Keeping workers motivated is a continuous challenge for leadership. A safe work environment goes a long way in motivating the workforce, when the worker knows that his or her safety is the highest priority of the organization.

Motivation is high when organizational leadership demonstrates behaviors such as

1. A willingness to forgo profits while making the workplace safer.
2. A willingness to strike the right balance between profits and a safe workplace.
3. Continuous training to ensure workers are competent to perform work that they are assigned.
4. Creating a nonpunitive environment to bring forward mistakes or errors where unsafe actions or conditions may exist, or are created.

5. Creating a workplace where accountability at all levels is understood and demonstrated to create trust and consistency that improves safety performance and reduces risks.
6. Acting in a timely manner when unsafe actions or conditions are brought to the organization's attention.
7. Creating a culture where employees automatically recognize unsafe acts and conditions and are empowered to correct such unsafe acts and conditions.
8. Proactive actions are taken to enhance the health and safety of workers.

Subsequent interviews by CNN of workers involved in the April 27, 2010 oil spill in the Gulf of Mexico showed a significant disconnect between the way frontline workers and leadership interpreted BP's focus on safety. While leaders felt they were doing a great job on improving safety, workers on the other hand saw leaders compromising safety in the pursuit of profitability.

To address this concern, leaders must seek to promote full engagement and involvement in looking after the health and safety of all workers. Leaders must be seen as trustworthy. Trust is earned from transparent and consistent ethical behaviors, the ability to do the right thing, the ability to recognize and acknowledge errors and mistakes, and the ability to demonstrate genuine empathy and care for employees. Demonstrated honesty and integrity by organizational leaders are absolute requirements to create a motivated workforce. From experience, followers always look toward leaders for guidance, as they try to understand how decisions are made, especially in their quest to emulate strong and effective leadership behaviors. In the absence of honest leaders with integrity, workplace safety becomes challenging and genuinely committed workers with integrity will always seek an alternative organizational environment that cares about their values. Another way to ensure that this type of disconnect is avoided is by ensuring that all personnel are trained on safety, and the organization's expectations regarding safety are clearly understood by all workers. There is also further discussion on the critical importance of trust and the development of a safety culture (safety culture maturity model) in Chapter 17.

3

World Class in Safety

Many organizations boast of having achieved world-class levels of performance in safety. What do they mean by world class in safety? This chapter seeks to examine what organizations mean when the claim is made of having achieved world-class safety system performance or status. For most organizations, the aspiration is to achieve world-class safety performance. To better understand this goal, we need to understand what defines a world-class safety system, performance, or status. When used loosely, these terms can mean different things to different stakeholders and are therefore subject to interpretation.

Defining World-Class Safety Performance or Status

The authors of this book contend that when an organization achieves its vision, and has excelled in operational excellence derived from focused attention to the interrelationships among people, facilities, and technology, it is approaching world-class performance in safety. World-class safety system performance or status may be defined as an organization with demonstrated exceptional and industry-leading safety performance that is supported with the right personnel, in the right roles, for stewarding safety. Such personnel must possess the right attitude and aptitude for doing the job. They must be able to inspire the hearts and minds of all workers to be engaged in looking after the health and safety of both themselves and coworkers. Such an organization is characterized by a strong safety culture with demonstrated support and commitment to superior safety performance from all levels of the organization. Such organizations recognize safety as a core value and provide systems, processes, and technologically advanced solutions to deliver on exceptional safety performance. Finally, the safety performance of the organization exceeds the performance of peers involved in the same business.

There are many definitions for world-class performance in safety. According to Geller (2006), a world-class safety system is a culture in which people maintain safety as a value, hold themselves personally responsible for the safety of themselves and coworkers, and are prepared to act on this feeling of responsibility. Leemann (2007) summarized the thoughts of a group of career safety professionals enrolled in postgraduate work in occupational safety and health management at Tulane University, Arizona, about their perspectives of world-class safety performance. The following two definitions were captured from their discussions (Leemann, 2007). First, a world-class safety system was defined as a health and safety program that "accomplishes the safety and health mission of the EHS department and the organization and it incorporates operational and administrative efforts into the safety process so workers are always thinking of safety" (Leemann, 2007, p. 32). Students argue that leadership ensures that safety is not an afterthought. Second, a world-class safety and health system "occurs when safety and health are an integral part of every component in an organization" (Leemann, 2007, p. 32). Leadership commitment is strong and filters down through successive organizational levels. In such systems, incidents are considered unacceptable and behaviors are such that safety is at the forefront when work is done.

Leemann's students highlighted the flaws of applying world-class terminology to safety as follows:

1. It is more of a vision perspective—driven primarily by the lowest numbers as opposed to having a self-sustaining process to continually improve and enhance safety performance.
2. There are no metrics for measurement—there are no defined and established metrics to determine what is considered world-class safety performance. In addition, such standards may vary by industry.
3. Best in class is better—the focus should be on developing the skills and capabilities of all workers while placing safety ahead of production. In addition, the focus should also be on demonstrated supporting leadership behaviors, employee involvement, and the use of best practices to ensure superior safety performance. Best in class provides measurable and tangible targets to aspire for relative to world-class performance where metrics are not generally available.
4. A perception of a marketing cliché—focus on numbers without concerns for supporting structures to sustain strong safety performance.
5. It is easier to define what world-class safety is not—essentially, it is easier to define a world-class safety system or performance as one that is *not* characterized by disengaged management. It is also *not* characterized by weak or unavailable safety training, disinterested and disengaged employees, high injury rates, lack of leading indicator focus, and repeat incidents.
6. A perception of being too elitist—world-class status requires commitment and behaviors for which many organizations aspire but are unwilling to commit the time and resources to truly achieve this elite status.

7. An idealized state—the end goal to which many organizations aspire but are required to take small progressive steps toward this aspired-for state. The culture of the organization supports safety; however, few true world-class safety organizations exist.

Are World-Class Performance and Status Achievable?

Is it wrong, therefore, for organizations to aspire for world-class safety performance and standards? The short answer is absolutely not. However, there must be clear guidelines around what this means for the organization—its leadership, workers, stakeholders, and culture. When an organization sets out to become a world-class leader in safety, this vision must be clearly articulated to all stakeholders in the organization—internal and external. It must mean the same thing to all stakeholders and the vision must be clearly spelt out in terms of achievable targets that can be broken down into SMART bite-size goals. Essentially, these goals must be

S—**S**pecific
M—**M**easurable
A—**A**chievable
R—**R**ealistic
T—**T**raceable and time bound

World-class safety performance and standards is achieved over time through the dedicated efforts of all workers, including contractors.

The important thing to note is that world-class safety performance and standards is industry specific since some industries are more risky than others. In addition, technological or capital differences may influence the safety performance of the organization and its position as a world-class safety organization. A more capital-intensive or technologically advanced organization may have a lower reliance on manual processes and may be able to engineer out risks to the benefit of its safety performance. On the other hand, a more labor-intensive process, which depends on procedures and work practices to manage risks, may be more prone to human errors and weaker safety performances and by extension, a lower position on the ladder to world-class safety performance or status.

According to Geller (2008), the key to world-class safety performance is leadership. In his view, "the best safety leaders are enthusiastic and passionate, and show respect and appreciation for the people they lead" (Geller, 2008, p. 30). Practical experience has shown that leaders who show empathy and genuine care for workers can retain them longer and motivate them to higher levels of performance. This is true regardless of the industry, and for many of us, we have heard the words *workers do not leave jobs, they leave leaders and bosses.* Poor or weak leadership performance leads to less than optimal safety performance.

Geller identified 6Es in managing safety. 3Es are associated with traditional approaches to safety management and 3Es are associated with people-based safety. Building on Geller's work, we would like to reclassify somewhat differently and add another E to the wonderful list that he has provided as shown in Table 3.1. The last E added to the list—*engagement*—provides workers opportunities to feel that they are a part of the process and are equally responsible for the health and safety of themselves and cowórkers alike. In our view, this is a critical component to any safety management system for which world-class safety performance is aspired.

Emotional intelligence (EI) and situational leadership styles are very important in safety leadership. World class in safety starts with a vision followed by incremental steps designed to move an organization toward this vision. The model designed to move organizations toward world-class safety performance and status is demonstrated in Figure 3.1. As shown in Figure 3.1, an organization must start with improving safety performance at the business unit/area or facility level by having SMART safety goals established. These goals must be designed to support the overall organizational safety perfor-

TABLE 3.1

7Es in Safety Management

E-Word	Relevance	What Do We Mean
Engineering		• Risks and hazards engineered out to protect workers during the design stages
Enforcement	Human factors in safety (traditional)	• Policies and procedures are in place to protect the health and safety of workers
Education		• Training is provided to ensure that workers are competent in performing assigned work
Emotion	Leadership in safety (people)	• Work with passion and demonstrate the passion for safety among workers
Empathy		• Showing workers that you care for them through individual consideration. Knowing each worker and demonstrate genuine care for each individual worker's circumstance
Empowerment		• Providing workers the ability to shut down unsafe work at all times without fear of punitive consequences. They must be able to take charge of their health and safety at all times
Engagement		• Engaging workers to be a part of developing safety solutions for the workplace. Such engagement must also include involvement in implementation of safety processes and systems

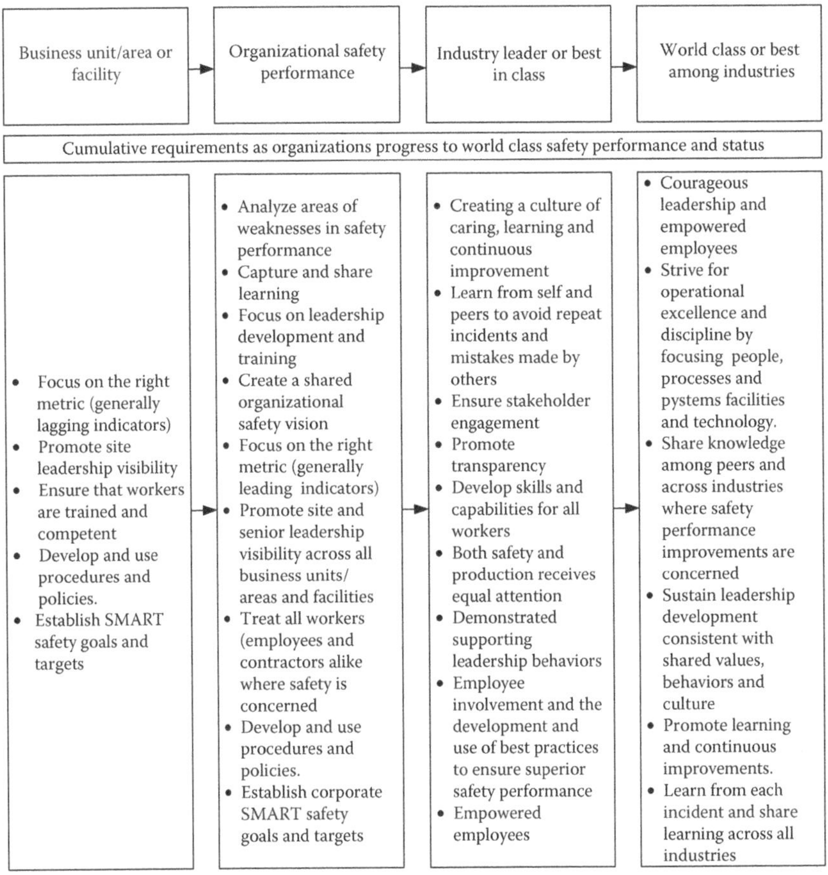

FIGURE 3.1
Model for progressing to world-class safety performance.

mance. Typically, at the business unit/area or facility level, safety performance improvements must be incremental, and as performance improves, targets are set at increasingly higher levels of safety performance that continue to be SMART. Furthermore, recognition of safety efforts, performance management, communication, demonstrated leadership commitment and reinforcement from all levels of leadership are key elements for sustained momentum toward world-class status in safety management systems.

Over time, the collective efforts of the business units and facilities lead to enhanced and ever-improving organizational safety performance. Recognizing safety excellence, empowering employees to make timely decisions, creating a culture of never turning a blind eye to unsafe acts and condition, and creating courageous leadership where leaders can make safety-related decisions without the fear of repercussions are all critical elements for moving an organization along the continuum to world-class performance. Continual learning and improvement ultimately lead to best-in-class and world-class performances. Therefore, the roles of leadership and training managers in EHS and health and safety are of paramount importance in achieving success.

Setting the Safety Vision

A safety vision forms part of a larger corporate vision, which defines the corporation's ultimate aspiration. In some ways, the vision statement is defined as the destination, while mission statements define the route and journey to take the organization to its destination. The vision therefore sets the direction for the organization while the mission establishes the means by which the organization gets there.

It is important to note that the vision of the organization seldom changes over time and is often unattainable in the short or medium term. The mission changes consistently with changes in the business environment and circumstances of the organization. Generally, the senior leaders (board of directors, CEO, and executive leadership team) of the organization establish the corporate vision. On the other hand, strategic leaders, middle managers, frontline supervisors, and workers execute the mission and strategic plans to move the organization toward its vision.

Bell (2007) advises that visions should be designed to energize and inspire the hearts of all workers. A vision statement must give people the courage to jump on board and to challenge the most difficult situations to achieve this vision. According to Bell, "a vision exercises a magnetic pull that irresistibly engages people in its pursuit" (Bell, 2007, p. 18) and "it provides the energy and passion that sustain the morale and maintain the momentum" (Bell, 2007, p. 18). Bell advised further that "an effective vision doesn't predict the future, it shapes it" (Bell, 2007, p. 18). Bell (2007) cites a 1992 Harvard study conducted by Kotter and Heskett that demonstrated that when organizations established vision-led cultures, they performed better than peers without such cultures in terms of annual growth in net income, average returns on invested capital, and appreciation in stock prices. Table 3.2 shows increases in business performance from vision-led cultures versus similar performance from nonvision-led cultures.

Having a health and safety vision encourages similar safety performance improvements among organizations. Visions such as Zero Harm (Petro-Canada), Journey to Zero (Suncor Energy Inc.), and Beyond Zero (Jacobs) have all led to significant improvements in safety performance when instituted by these organizations. Around 2004, Petro-Canada (one of Canada's then largest integrated oil and gas producers) introduced its vision of Zero Harm. Coupled with the right mission steps that were continually improved

TABLE 3.2

Business Performance from Vision-Led versus Nonvision-Led Cultures

Performance Criteria	Managed Culture (%)	Did Not Manage Culture (%)
Revenue increases	682	166
Stock price increase	901	74
Net income increase	756	1
Job growth increase	282	36

Source: Adapted from Reifenstahl, E. 2009. *Fort Worth Business Press*, 25(39), 24.

with each successive year, Petro-Canada was able to continually improve its safety performance for both employees and contractors in each successive year (Figure 3.2). Similar improvement in organizational safety performance has been demonstrated by oil and gas producers (OGP) members.

The effect on safety performance is very profound with a safety vision, much like business performance improvements from a corporate vision as discussed earlier. With a clearly defined and shared safety vision, the leaders and workers are better equipped to develop strategies and mobilize resources to deliver on the vision. Table 3.3 highlights the safety vision statements of several large North American businesses in which the health and safety of workers can be affected. In each case, the goal appears to be zero injuries and

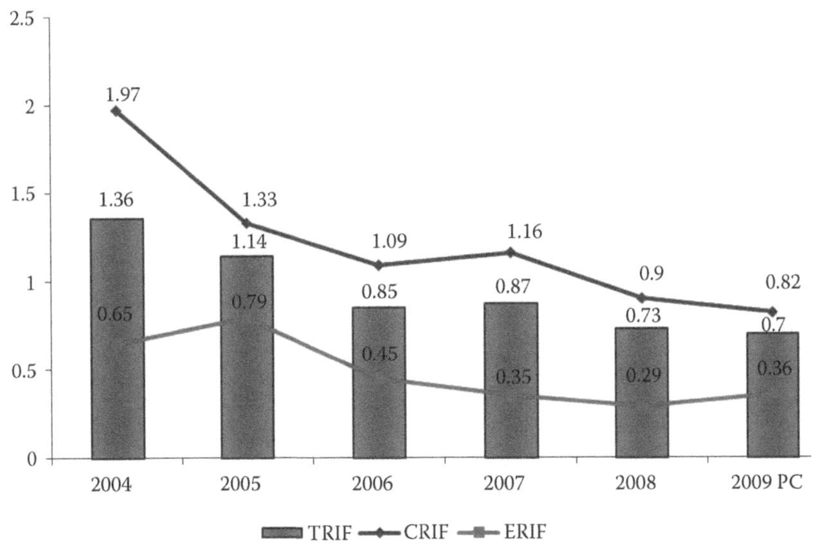

FIGURE 3.2

Petro-Canada safety performance improvements with a Zero-Harm vision. (Derived from data provided at http://www.suncor.com/en/responsible/1199.aspx.)

TABLE 3.3

Stated Safety Visions of Some Prominent Organizations

Company	Industry	Safety Vision
Suncor Energy Inc. (http://www.suncor.com/en/responsible/1983.aspx)	Canada's largest integrated oil and gas producer	"Journey to Zero is intended to embed a safety culture into every aspect of our business. Our aim? To achieve a level of safety excellence that results in an injury-free work site. Suncor wants all work processes and systems to be safe and the company demands employees and contractors take individual responsibility for safety" (Suncor, 2010)
Syncrude Canada (http://www.syncrude.ca/users/folder.asp?FolderID=5717)	Oil sands mining, extraction, and upgrading operations	"Our Safety, Health and Environment Policy sets a clear vision for responsible development of the oil sands resource" (Syncrude Canada, 2010)
Shell Corporation (http://www.shell.ca/home/content/can-en/environment_society/working_safely/)	World's second-largest integrated oil and gas producer	"Safety is always our top priority. We aim to have zero fatalities and no incidents that harm people, or put our neighbors or facilities at risk" (Shell Corporation, 2010)
DuPont (http://www2.dupont.com/Sustainability/en_US/Performance_Reporting/commitment.html)	Multiple industries, including chemicals, manufacturing, energy and utilities, and safety	"We believe that all injuries and occupational illnesses, as well as safety and environmental incidents, are preventable, and our goal for all of them is zero" (DuPont, 2010)
British Petroleum (http://www.bp.com/subsection.do?categoryId=3316&contentId=7066923)	Global integrated oil and gas and energy producer	"Safe and reliable operations are integral to BP's success, and we strive continuously to improve our safety performance" (BP, 2010)

Toyota Corporation (http://www.toyota-global.com/sustainability/report/sr/pdf/sr11_p18_p25.pdf)	Automotive	"Ensuring the safety and health of employees and their family members can be considered the foundation of corporate activities. Based on the philosophy of 'Respect for Humans,' which has been carried out through Toyota's entire history, all of its employees become one to create a 'safe and energetic work environment' and to prevent accidents and occupational illnesses" (Toyota Corporation, 2010)
Emirates National Oil Company Ltd (ENOC) (http://www.enoc.com/ENOCWebsite/)	Largest oil and gas company in Dubai, United Arab Emirates	"The Emirates National Oil Company Ltd (ENOC) is a highly diversified oil and gas holding company. ENOC is committed to organizational excellence by being 'best-in-class' for stakeholder satisfaction, prosperity and sustainable growth. We will design, build and operate to maintain a safe working environment using efficient and cost effective methodologies, products and services that are safe and compliant to local, and where applicable, international market regulations. We will make EHSQ an inherent part of our business planning, strategic and operational objectives" (ENOC, 2010)

Source: Corporate websites of respective companies, Accessed June 17, 2010.

incidents. Reality, however, is that many of these organizations continue to have incidents in which people are injured, the environment continues to be damaged, and assets are destroyed from incidents. Nevertheless, they all continue to aspire for increasingly higher levels of safety performance on an ongoing basis. A compelling vision inspires the hearts of workers, leading to success in the face of adversity and to recovery from setbacks related to each incident through learning and continual improvement.

As mentioned earlier, when an organization achieves its vision, and has excelled in operational excellence derived from focused attention to the interrelationships among people, facilities, and technology, this organization is approaching world-class performance in safety. Such an organization is supported by the appropriate cultural behaviors and values that become a natural part of activities so that no one needs to be reminded of the organizational approach to safety. Transformational leadership behaviors of teamwork, individual considerations, intellectual stimulation, empathy and genuine care for workers, and create and lead through change and vision are all essential ingredients of a world-class safety management system.

Based on the foundation of Suncor's values and beliefs, the Journey to Zero culture is as follows:

- "Building common attitudes, behaviors and outcomes that create and sustain operational excellence. Journey to Zero does this by enabling safety excellence across the organization" (Suncor, 2011).
- "Our beliefs are
    - All incidents are preventable.
    - Safety is a critical part of Suncor's culture—it's how we do business.
    - Each employee is accountable for achieving results the right way, safely. This means protecting the safety and health of our team members, the environment, and conducting our operations reliably and efficiently.
    - No job is so urgent or routine that it can't be done safely. Everyone who works on Suncor's behalf is expected to finish their shift or work day safely" (Suncor, 2011).

4

Have a Safety Management System (SMS)

For most organizations, where people are exposed to health and safety hazards, Safety Managagement Systems (SMSs) are developed to ensure the health and safety of all workers. In this chapter, we explore the requirements of an SMS that provides the requirements for ensuring the health and safety of all workers. An SMS should generally address the health and safety requirements for protecting people, the environment, and assets, in the same order.

According to Broadribb (2010), the findings of the Texas City BP incident suggest that an effective SMS must extend beyond the personal safety and environmental realm to include process safety management for continuous improvements. Broadribb advised that for organizations to have an effective SMS, process safety management must be integrated into "operations, health, occupational safety, and environment (HSE)" (Broadribb, 2010, p. 21).

There are direct and indirect benefits from an effective SMS. Organizations that embrace SMSs see this approach to managing safety as an investment in the profitability of the organization. On the other hand, when organizations see investment in safety as a cost, such actions can be regarded as a compliance requirement for being in business or as a moral or ethical obligation to workers.

Is a Safety Management System Required?

The objective of a SMS is to reduce injuries and to preserve environment and the productive lives of assets. An effective SMS is recognized by many businesses as an essential requirement for remaining in business. Employers recognize that the health and safety of its workers and preservation of the environment and its assets are elements of an effective corporate social responsibility program.

Good Business Sense

In short, an SMS adds to the profitability and sustainability of an organization. It simply is good business to have an effective SMS. Table 4.1 highlights

TABLE 4.1

Quantifiable and Less Quantifiable Benefits of a Safety Management System

Quantifiable Benefits	Less Quantifiable Benefits
• Fewer incidents and injuries	• A more motivated workforce
• Less severe injuries	• Ability to attract and retain the best and brightest
• Fewer work-related fatalities and diseases	• Greater stakeholder commitment
• Reduced absenteeism	• More community support and engagement
• Reduced loss time away from work	• Higher trust and credibility among all stakeholders
• Reduced health care treatment cost	
• Fewer employee turnovers	
• Reduction in insurance cost	
• Lowered operating and production cost	

some of the quantifiable (direct) and less quantifiable (indirect) benefits derived from an effective SMS.

Edic (2009) advised of OSHA estimates between $1.1 and $4.5 in indirect cost for each $1.0 of direct cost incurred from an injury. These estimates, therefore, suggest full cost in the range of 200–550% of direct cost for workplace injuries.

Beyond the direct cost saving impact from fewer workplace injuries, the productivity impacts from a more motivated workforce, the ability to attract and retain great talent, greater stakeholder commitment, and enhanced credibility and trust, although difficult to quantify, often translate into

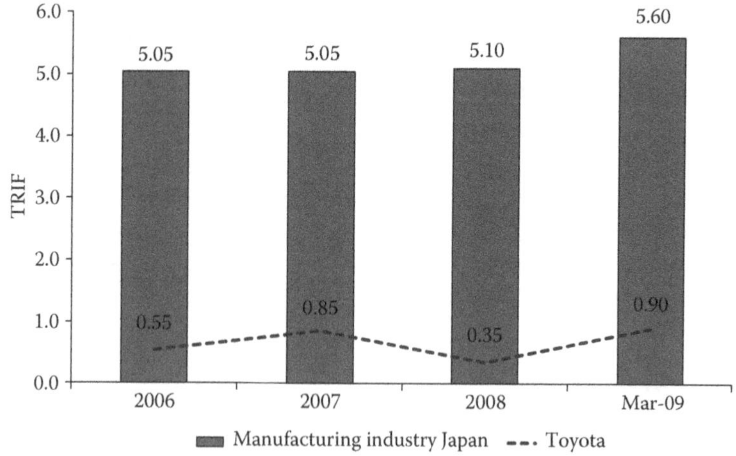

FIGURE 4.1
Toyota safety performance relative to Japanese peers. (Adapted from Toyota 2008 Annual Report.)

excellent business performance in terms of higher productivity, lower cost, and a more profitable organization. Figure 4.1 shows the workplace safety performance of Toyota relative to its peer automotive manufacturing industry producers in Japan. As shown, Toyota's injury frequency falls well below its peers and as we all know, Toyota has been a global leader in automotive manufacturing since 2007.

While Toyota's successes cannot be attributed exclusively to its safety performance, the benefits identified for an effective SMS cannot be overlooked in this instance.

Legal Compliance and Due Diligence

Beyond the profit implications are the legal implications resulting from legislation. The Corporate Manslaughter and Corporate Homicide Act (UK) and associated regulations suggest:

1. Both managers and organizations can be found guilty of an offence if the organizational activities are managed or organized in a way that causes a person's death.
2. Guilty and noncompliant organizations can face unlimited fines—"starting at 5% of annual turnover" (Human Resources, 2009, p. 28).
3. Individuals are required to implement remedial actions and "may be required to publish details of their offences and the penalties implemented" (Human Resources, 2009, p. 28).

In Canada, Bill C-45 places additional demand in demonstrating required due diligence. The new duty, added in Section 217.1 of the Criminal Code, requires that *"everyone who undertakes,* or has the *authority, to direct* how another person does work or performs a task is *under a legal duty to take reasonable steps to prevent bodily harm to that person,* or any other person, arising from that work or task" (Canadian Centre for Occupational Health and Safety, 2011, para. 2).

Furthermore,

- Bill C-45 creates both corporate and individual *criminal liability.*
- Crown proceeds by indictment (the most serious manner of proceeding) and *there is no limit on the amount of the fine* for the corporation or the organization.
- For individuals, the maximum penalty for criminal negligence causing death is life imprisonment, and the maximum penalty for criminal negligence causing bodily harm is 10 years of imprisonment.

Experience has also shown that the top three records sought by regulators and investigation teams when major incidents occur include the following:

1. Standard operating procedures (SOPs)
2. Hazards assessments reflected in
    a. Work permits
    b. Task breakdowns, hazards identification, risk mitigation actions, persons accountable, and when the risk was mitigated
3. Competency assurance process and records that demonstrate
    a. Verification of knowledge
    b. Verification of skills and abilities
    c. Confirmation of work experience

According to the Canadian Centre for Occupational Health and Safety, due diligence is demonstrated when the following conditions are met:

1. The employer must have in place written OH&S policies, practices, and procedures. These policies would demonstrate and document that the employer carried out workplace safety audits, identified hazardous practices and conditions and made necessary changes to correct these conditions, and provided employees with information to enable them to work safely.
2. The employer must provide the appropriate training and education to the employees so that they understand and carry out their work according to the established policies, practices, and procedures.
3. The employer must train the supervisors to ensure that they are competent persons, as defined in legislation.
4. The employer must monitor the workplace and ensure that employees are following the policies, practices, and procedures. Written documentation of progressive disciplining for breaches of safety rules is considered due diligence.
5. There are obviously many requirements for the employer but workers also have responsibilities. They have a duty to take reasonable care to ensure the safety of themselves and their coworkers—this includes following safe work practices and complying with regulations.
6. The employer should have an accident investigation and reporting system in place. Employees should be encouraged to report "near misses" and these should also be investigated. Incorporating information from these investigations into revised, improved policies, practices, and procedures will also establish that the employer is practicing due diligence.

7. The employer should document all the above steps. This will give the employer a history of how the company's occupational health and safety program has progressed over time. It will also provide up-to-date documentation that can be used as a defense to charges in case an accident occurs despite an employer's due diligence efforts (reproduced with the approval of the Canadian Centre for Occupational Health and Safety, 2011).

More importantly, the Canadian Centre for Occupational Health and Safety stresses that "all of the elements of a due diligence program must be in effect before any accident or injury occurs" (Canadian Centre for Occupational Health and Safety, 2011, para. 6).

An effective SMS not only improves the health and safety of workers but also provides a framework for meeting all the due diligence requirements for compliance and stewardship. The organization, as we know, acts through the authority administered by senior officers and authorized leaders of the organization. Authorized leaders can be defined as any employee who has been authorized to make decisions and to supervise the work of other workers and who is responsible for the health and safety of an individual worker or a group of workers. As a consequence, all authorized leaders must take the time necessary to allocate resources to ensure that their actions in executing the organization's work are compliant with the law and do not in any way cause harm to workers. Corporate homicide can be the result of a senior leader's breech of a *duty of care* for workers that may have resulted in the death of a worker or multiple fatalities. An SMS provides the necessary checks and balances to ensure the health and safety of all workers, while providing the necessary due diligence of duty-of-care verification for the business leader.

The Road Map for Improving Safety Performance

An SMS provides the framework and road map by which an organization can systematically improve its safety performance in a sustainable manner. Once the framework is established with the key elements of the SMS and standards are established for each key element, each worker now has a better understanding of how the business is expected to function, consistent with each element or standard.

The road map starts with educating each stakeholder as to why it is necessary to adhere to the conditions, policies, and procedures outlined in the SMS. Human behaviors are such that once answers to "why" is provided, it is easier to gain acceptance and buy-in from all stakeholders. Another key benefit of an SMS is that it defines roles and responsibilities of key stakeholders. Also, roles and responsibilities of senior leaders, middle managers, and frontline supervisors on the tools that workers and contractors need are all clearly defined so that ambiguity among the workforce is reduced.

OHSAS 18000:2007 is an international occupational health and SMS specification developed in response to business demands for a benchmark against which their SMS could be evaluated in providing for the health and safety of workers (BSI, 2007). This British standard follows the universally accepted process of *plan-do-check-act.* OHSAS 18000:2007 standard identifies scope definition, OH&S policies, hazards and risk management, implementation and control, performance management and verification, and leadership oversight as key requirements for an effective SMS.

Elements of a Safety Management System

According to Markiewicz (2009), the key elements of an SMS are as follows:

1. Management leadership and employee involvement
2. Planning
3. Implementation and operation
4. Checking and corrective action
5. Management review

These concepts align well with the continual improvement cycle of *plan-do-check-act.* These elements are not significantly different from those identified by OHSAS 18000:2007 and are intended to provide the organization with the core focus areas required to best ensure the health and safety of all workers.

OH&S Alberta pointed out that a "Health and Safety Management System involves the introduction of processes designed to decrease the incidence of injury and illness in the employer's operation" (OH&S Alberta, 2006, p. 1). As in the case of all successful SMSs, leadership commitment is critical in allocating resources and securing employee participation and involvement. OH&S Alberta listed eight independent components of a health and safety management system as follows:

1. Management leadership and organizational commitment
2. Hazard identification and assessment
3. Hazard control
4. Work site inspections
5. Worker competency and training
6. Incident reporting and investigation

7. Emergency response planning
8. Program administration (OH&S Alberta, 2006, p. 1)

From the list above, management, leadership, and organizational commitment are an essential component critical for success. In the absence of these critical elements, commitment is shallow and an SMS is doomed to failure.

In today's business environment, given the continuing levels of incidents that take place daily, there is increasing emphasis on all employers to develop and maintain an SMS. The importance of an SMS cannot be understated in some industries. For example, all oil and gas producers maintain and sustain SMSs while at the same time requiring all contractors and service providers supporting their business that perform on-the-tools work to maintain similar systems that either meet or exceed the safety requirements of the oil and gas producer.

Once properly categorized, the elements of an effective SMS generally fall into three interrelated categories. These categories are as follows:

1. People management
2. Supporting processes and systems
3. Facilities and technology

Figure 4.2 highlights the elements of an SMS based on the goal of the SMS relative to the safety focus of the organization. As shown in Figure 4.2, this

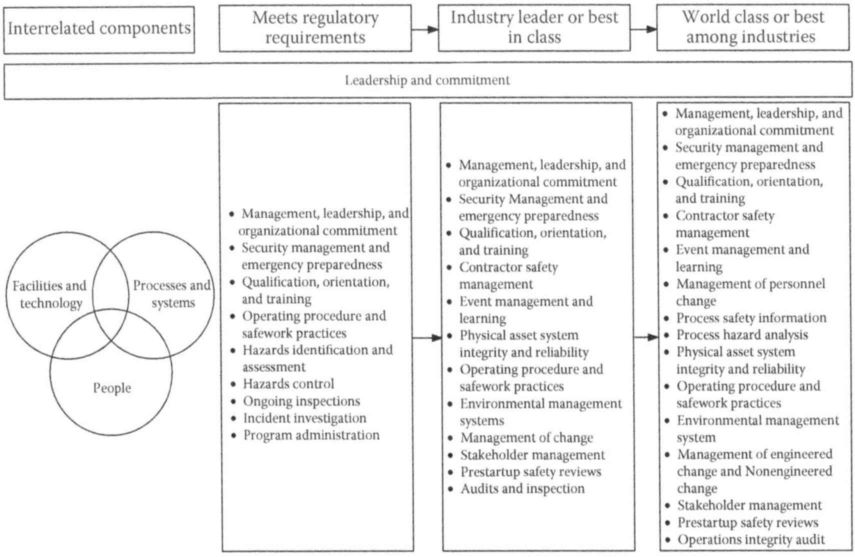

FIGURE 4.2
Typical safety management systems.

may vary from one extreme of meeting the regulatory requirements, within which the business operates, to the other extreme of being a world-class leader in safety. The key to success, however, is recognizing the interrelatedness of all categories and its elements, and stewarding them accordingly.

Implementing a Safety Management System

When implementing an SMS, the Deming business improvement model of *plan-do-check-act* is most commonly used to bring about sustained improvements. In this discussion, there are two approaches for implementing an SMS. The two scenarios to be discussed in this section are as follows:

1. Assuming the organization does not have an organized or structured SMS and is about to implement one.
2. The organization has an SMS and needs to upgrade its processes to promote better safety management and performance.

Each scenario will require different levels of details and attention to ensure sustained long-run performance, and as such, each scenario is discussed separately.

Implementing an SMS in an Organization Where One Does Not Exist

When building an SMS from the ground up, this process is infinitely more complicated than upgrading an existing process. Schaechtel (1997) provides an excellent overview of the requirements for building a new process that points to leadership commitment as a starting point where the safety philosophy of the organization is defined. Standards are developed for each element of the SMS and these standards are then shared across the organization for guiding the behaviors and actions of all workers of the organization.

Implementing an SMS today can be achieved using the same model proposed by Schaechtel. The important thing is to ensure that the SMS is consistent with the overriding safety philosophy of the organization and the key categories of Schaechtel (1997), people management, supporting processes, and systems and facilities and technology, are interrelated and addressed. Figure 4.3 provides a simplified model for building an SMS when there is no existing SMS. The safety philosophy of the organization will determine the level of details required for a successful implementation of an SMS.

The philosophy is also determined by senior leadership of the organization consistent with the safety vision of the business.

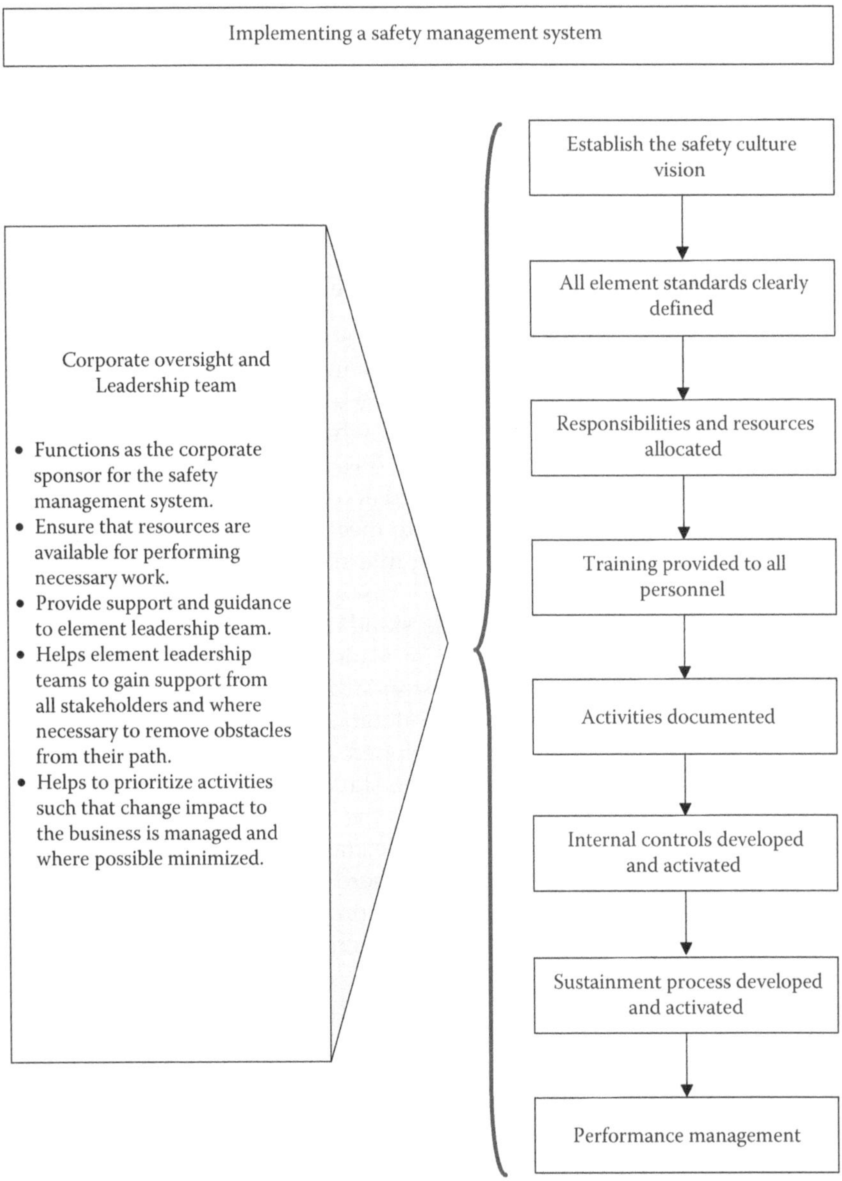

FIGURE 4.3
Simplified model for implementing an SMS where one does not exist.

Establish the Safety Culture Vision

Establishing and communicating the safety culture vision is perhaps the most important step in the process and it begins with leadership. Communication of this vision through continuous messaging and demonstrated leadership behaviors will determine how seriously other members of the organization will support the vision. Leadership commitment, personnel accountability, and stewardship are important components for gaining worker support and commitment to a shared safety culture vision.

All Element Standards Clearly Defined

An effective SMS requires clearly defined standards free of ambiguity for guiding the behaviors of all employees. Standards must be easy to follow, practical, and applicable, and must also be executable across the organization. The standard sets the basis against which the functional units of the business will be audited for compliance. As a consequence, all workers must be fully aware of what must be delivered according to the standards.

Generally, standards are developed to meet or exceed regulatory requirements. Standards are developed using internal (and where necessary external) expertise and knowledge. A multidisciplinary team of subject matter experts are chartered to develop the standard with clearly defined scope. This team will be supported by senior leadership, and adequate resources are provided to ensure that a robust standard is developed. Often, developing the standard will require several iterations of writing, reviewing, and testing before a robust standard is developed. In addition, once implemented, there may be deficiencies or gaps in the standard that may require further revision. The key to success, however, is the need for documentation of the deficiencies and gaps and to upgrade the standard at some predetermined frequency to address the deficiencies. Where immediate hazards exist with the existing standards, a managed change process is required to address the deficiency and to communicate this upgrade immediately to the entire organization.

Responsibilities and Resources Allocated

Execution is enhanced when people who are interested in doing the work, and are passionate about safety, are involved in the execution of the SMS. Organizations must therefore assign responsibilities to people who want to

TABLE 4.2

Deliverables and Accountabilities of a Safety Management System

Deliverables	Accountable
• Safety management standards	• CEO or vice president of HSE
• Assigned stakeholders and subject matter experts for doing the work	• Executive and senior leadership teams
• Personnel training and development	• Business unit leaders
• Effective safety communication tools and methodology	• Corporate communications experts and subject matter experts
• Records management database	• Information technology experts in consultation with business units leaders and HSE
• Budgets and decision-making authority	• Executive leadership team and senior leaders
• Safety responsibilities assigned and performance stewarded	• Business unit leaders in consultation with HSE

be a part of the process and are prepared to do the work necessary to see this type of a project to completion. In addition, the process of assigning responsibilities must ensure that a senior leader of the organization is a member of the execution team. Where a large organization exists with multiple divisions, representation from each division must be sought so that ownership and buy-in are maximized.

Execution of an SMS requires the involvement and support of several teams and stakeholders. Such a large undertaking will necessarily require the sponsorship of a senior executive of the organization, usually a senior vice president or the vice president of HSE. Completion of the work required for full execution should be completed by an element leadership team. Where possible, full stakeholder participation and representation in the element leadership team help to assure all stakeholders that their specific concerns are addressed, during the execution process. Ultimately, success in safety performance lies with line management who are provided support from the HSE organization. Organizations with beat safety performance recognize and embrace this concept and use HSE for support and subject matter expertise.

Deliverables and accountabilities coming out of the SMS may include those provided in Table 4.2.

Training Provided to All Personnel

Where the SMS and standards are concerned, all personnel are required to behave and work in a manner that is consistent with the SMS and standards. They must also be properly trained. Training may vary from general awareness training to training that requires competency assessment and checks. General awareness training may be required for personnel who are not directly involved in the use of standards and for senior leaders who may be required to support the execution of the SMS but are not directly responsible or accountable for any element. Personnel responsible for executing standards for the business must be competent in the use of required standards supporting the SMS. As a consequence, such personnel must be assessed for competency, and follow-up training and competency assessment must be done at some defined frequency.

A training matrix that defines the training requirements for all personnel consistent with each element may be required for ensuring that all personnel are properly trained and assessed. Depending on the scale of operation, the training matrix may vary from a simple Excel spreadsheet (for 10s of workers) to the elaborate individual worker profile, which shows the training needs of each organizational role, in larger organizations (for 1000s of workers). Training needs of roles are recommended since mobility and job changes in larger organizations are more frequent.

Technical safety training is generally provided by subject matter experts resident in the business unit or via corporate HSE-led expert. In some instances, external providers may be sought to provide specialized training where economic advantages support such training. In all instances, however, training must be consistent with requirements of the SMS standards.

Activities Documented

According to Schaechtel (1997), documentation is a key element of any SMS. Well-managed documentation and records fulfill regulatory (OSHA, OH&S) requirements and demonstrate due diligence in the unfortunate circumstance of an event where the organization is subjected to legal action. More importantly, documentation and record provide the framework for continuous improvements and improving the effectiveness of the SMS and safety performance of the organization.

Documentation and records management must be simple and easy to use for best results in an SMS. Documentation and records retention must also be consistent with the regulatory and business requirements within which the business operates. For example, training and inspection records may vary based on the type of operating processes or the state or jurisdiction in which the business operates. Listed below are some typical records and activities that must be properly documented and recorded:

- Training and competency assessments
- Audits and inspections

- Equipment inspection
- Incidents and accidents
- Incidents and accidents investigation reports and remedial actions and closeout history
- Updated process flow and piping and instrumentation drawings (P&ID)
- Standard operating procedures
- Codes of practices and corporate policies
- Management of change records and action items closeout listings

While this list is not all-inclusive, it provides a great start into the types of activities that must be properly documented and recorded for an effective SMS.

Internal Controls Developed and Activated

An effective SMS requires adequate internal controls to ensure compliance to the standards, procedures, policies, and codes of practices. Internal control measures are intended to alert leadership when compliance with the SMS is failing to allow proactive remedial actions to promote compliance. Simple tools must be made available to all users to enable application and compliance with the SMS requirements. Leaders must understand that safety management is one of a series of competing priorities of the business and as such checks and balances must be designed to support, rather than avoid, compliance.

Internal control systems must promote proactive identification of leading indicators so that incidents can be avoided before they occur. For example, the recording and trending of near misses is crucial in the prevention of an accident or incident. Internal controls also include reports that focus on SMART variables to help leadership determine compliance and initiate corrective actions. As a reminder, the SMART principle reflects the following:

S—Specific

M—Measurable

A—Achievable

R—Realistic

T—Traceable and time bounded

Such reports may include both leading and lagging indicators. An example of a SMART report is demonstrated in training records of personnel that may show the amounts of personnel trained over a fixed period. A visible and easily updated training information matrix can be reviewed quickly to assess progress with training, qualification, and competency for all personnel. The use of a training matrix is an easy method for conveying this information. A training matrix provides a snapshot of all groups at the project site, the training required, and the status of each individual's training at a point in time. Clearly identified also are personnel who are designated capable of qualifying trained personnel and also assessing the competency of those trained.

Figure 4.4 shows a typical training, qualification, and competency matrix. Using a vibrant color scheme, it is possible for leaders to determine at a glance—those who are trained from those who are not. By using a training matrix, training can be easily expressed as a percentage completion for assessing compliance while providing an estimate of the magnitude of the training requirements where compliance is required within a fixed period.

The matrix identifies required training for each personnel. It also identifies trainers and can show the date on which the training was undertaken. It can also identify areas where workers have been trained and qualified by competency assessors. Subject matter experts are identified and are required to conduct training and assess the competency of personnel before they are deemed qualified.

A visible training matrix with vibrant color schemes to represent untrained workers provides a visible prompt to all stakeholders of the training needs of the organization as far as the SMS is concerned. Visible and contrasting colors easily attract the attention of leaders and workers alike so that timely intervention can be made to ensure compliance on all elements of the SMS. Changes to the matrix should be limited to a single point of control and ownership; this is required to avoid confusion and delays in updating the matrices.

Training and qualification records must include details regarding the course content, the employee signature, the trainer's signature, and date and time of training. For group training, participation records and group signatures will be required for auditing purposes. Figure 4.5 provides a simplified training record of competency assessment to be retained on a file for due diligence verification and as a means of determining the frequency of refresher training courses.

These examples of internal control measures provide leadership with adequate indicators and measures to determine the status of compliance with the SMS and are essential in guiding remedial and corrective actions. Tools such as action registries can be very useful in stewarding audit compliance and inspections. Trends can be extremely useful in stewarding both leading and lagging safety and environmental indicators.

The use of safety passports is also a practice that various oil and gas companies and oil and gas service companies employ. These are documents that employees retain and that contain an accurate record of their training and

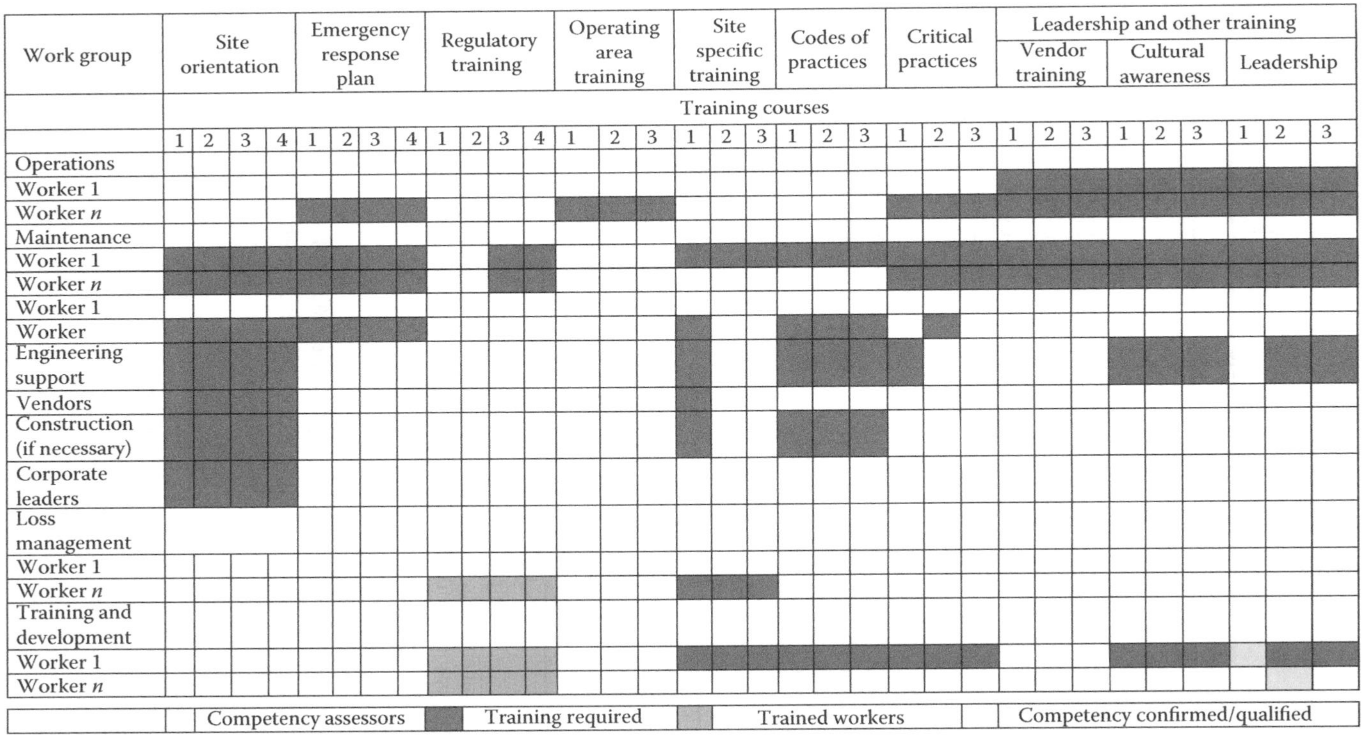

FIGURE 4.4
Training and competency matrix. (Adapted from Lutchman, C. 2010. *Project Execution: A Practical Approach to Industrial and Commercial Project Management*. Boca Raton, FL: CRC Press.)

Employee name:		Operating area/ system	
Job title:			
Assessor name:		Assessment date:	

1. System or area description:		
2. Evidence presented (attach copy where applicable):		
Performed by supervisor	Comment	Feedback and other supporting information
o Observation of employee doing task as per critical practice and SOPs. o Assessment of outcome of employees work. o Task discussions and review of questions. o Discussions on responses to simulated conditions. o Work related assignments.		o Feedback from peers and other workers. o Team outcomes where the individual contribution is evident. o Record of work or training activities.
3. Employee comments:		

Candidate's signature:		Date:
Assessor's signature:		Date:
To be forwarded to training department upon completion.		

FIGURE 4.5
Competency assessment form. (Copyright Suncor Energy Inc., approval from Suncor Energy Inc. is required to reproduce this data. Lutchman, C. 2010. *Project Execution: A Practical Approach to Industrial and Commercial Project Management*. Boca Raton, FL: CRC Press.)

employee assesment and other EHS aspects of their skills and competency at a particular site or company. Experience has shown that safety passports improve:

- Engagement levels of staff resultant from a sense of identity.
- Workers confidence resulting from a reminder that they are trained and qualified to work and operate.
- Verification of training and competency records by regulators and auditors. Safety passports also help employees and frontline leaders to easily access training records without having to revert to databases and archives.

A good business practice is the audit review on some fixed frequency to identify improvement opportunities for the internal control measures of the SMS. Generally, a 1–3-year frequency is recommended. Coming out of such audits will be improvement opportunities identified and suggested measures for gap closure. The business unit leaders, with support from HSE, will be required to evaluate and execute these gap closure measures, leveraging the capabilities of the sustainment process.

Sustainment Process Developed and Activated

Once the SMS is rolled out and functioning, a process for sustainment and continuous improvements is required to move the organization from regulatory compliance through to world-class safety status or that which is consistent with its vision. A sustainment team for each element is required to ensure continued progress. Team composition for each element should include stakeholder representation and should focus on including motivated champions and subject matter experts.

A team charter should be defined for each element team that lists clearly members, expertise, stakeholder group representation, deliverables, authority, accessible resources, and timelines. The team charter should also reflect meeting frequencies and the need for documentation. In most cases, senior leadership sponsorship or endorsement adds credibility to the work of this team and highlights the importance of their work to all stakeholders and the organization as a whole.

This sustainment team should comprise the following groups:

1. Core team—comprised of 3–5 members with a strong leader who has overall responsibility for the functioning of the sustainment team.
2. Network group—comprised of 10–15 stakeholder representatives who function as the worker group of the sustainment team. Capable of performing independent work in subteams as part of the overall team deliverables.
3. Community of practice (functional/frontline subject matter experts)—a resource pool of subject matter experts at the frontline with responsibilities as ambassadors for identifying what works from what does not. This resource pool should also be involved in developing workable solutions for the frontline. This group is also critical in execution of remedial and gap closure actions at the frontline.

The roles and responsibilities of each group within the sustainment team are provided in Table 4.3.

TABLE 4.3

Roles and Responsibilities of Different Groups of the Sustainment Team

	Sustainment Team	
Core Team	**Network Group**	**Community of Practice**
• Defines and prioritizes objectives of team	• Develop workable solutions for gaps and deficiencies identified	• Execute and rollout deficiency and gap closure activities at the business unit or functional level
• Obtains resources required for accomplishing goals	• Identifies and recommends improvement to the element standard	• Steward performance at the business unit or functional level
• Communicates with senior leaders of the organization on progress and develop key messages for stakeholders	• Support functional teams in rollout and execution of deficiency and gap closure activities	• Identify improvement opportunities and mitigation possibilities
• Builds consensus and creates change where necessary	• Share knowledge related to element across organization	• Assist in prioritizing organizational opportunities related to element standard
• Steward the activities of the sustainment team and is ultimately responsible for the performance of the sustainment team	• Develop and refine metrics for stewarding element performance	• Function as subject matter experts to be called upon as required by the sustainment team

Performance Management

A final component of the SMS to be addressed is the issue of performance management. While the focus on leading indicators helps in improving overall safety performance, ultimately, the impact of lagging indicators is the true measure of performance improvements needed. Trends in injury frequencies (TRIF, ERIF, CRIF, and loss time injuries (LTIs)), number of disabling injuries, number of loss time injuries, injury severity, workmen compensation, and insurance are critical measures in gauging the performance of the organization.

Benchmarking against industry peers also helps in motivating the workforce toward enhanced safety focus as improvements become more visible and transparent. Performance management is enhanced when we have the ability to transform data into information. For example, an incident management system that captures annual data on numbers and types of injuries is of little value to a leader, if these data cannot be trended and analyzed, to provide information for proactive and corrective decision making.

Tools such as pie charts and annual trends as shown in Chapter 2 are of great importance to all HSE personnel and leaders to enhance safety performance management. Performance management requires continuous monitoring of data and information and vigilance regarding safety performance at the individual, work group, business unit, and corporate levels. Daily updates on high-risk business activities may be required to ensure that all personnel recognize and respond to the importance placed by leadership on safety.

Where performance falls beneath targets, immediate and prioritized corrective actions are required to demonstrate to all workers that safety is a priority. Procrastination and weak response to the issue of safety infractions and concerns as well as weak funding send the wrong message to stakeholders of the SMS. Focused performance management attention to leading indicators is recommended to avoid incidents and enhance the safety performance and culture of the organization.

Upgrading an SMS in an Organization Where One Already Exists

When an organization has an already-existing SMS, upgrading the SMS is an important process in transitioning the organization from regulatory compliance through to world-class safety performance. In this situation, the work required is significantly greater because workers have become accustomed to doing things a certain way and will now be required to change behaviors. Where behaviors are entrenched, and the workforce is mixed, comprising both Generation X and Generation Y, multiple strategies are required to move the entire workforce forward. As a consequence, intense stakeholder communication and engagement are required to provide a compelling sense of direction to stakeholders. Communication to stakeholders should include the following:

1. What changes are required.
2. Why the changes are necessary.
3. How the changes are executed across the organization or business unit.

A similar model to that provided in Figure 4.3 is required for a comprehensive upgrading of the SMS.

Darby (2010) advised that when change is necessary, the following nine dimensions of change must be addressed to ensure success.

1. Stakeholder relationship management—A clear understanding of groups and individuals affected by the prescribed change is required and the needs of each affected stakeholder properly addressed.

2. Leading the prescribed change—A trusted leader with strong capability of creating a compelling sense of direction is essential.
3. Selecting and managing the change strategy—Clear articulation of the change strategy is essential. For example, the application of a pilot project versus full-scale enterprise-wide application.
4. Communication—Communication and key messages crafted for all levels of the organization.
5. Human resources management—Change will necessarily impact on the skills and behavior of all workers. Careful attention to the personnel needs and development is essential.
6. Training—Training is probably one of the most important aspects of change. Types of training vary across different levels of the organization and the roles provided by stakeholders. Senior leaders may require awareness training whereas frontline leaders and workers may require more detailed training, requiring competency assessment and follow-up.
7. Processes and infrastructural requirements—Depending on the type of changes prescribed, changes in work processes and infrastructural needs may be necessary. There are cascading impacts related to communication, training, and change management in supporting processes and infrastructure.
8. Project management focus—This is necessary to ensure that the concepts of people, processes, and systems readiness are imbedded in the changes process, at each stage of the change continuum.
9. Managing performance—Throughout the change process is the need to manage performance for effectiveness and efficiency. As with all projects, there is a balance between cost and benefits and ensuring that cost effectiveness must be a priority for all changes prescribed.

Consistent with the change dimensions identified by Darby (2010), when upgrading an SMS, these dimensions of change must also be addressed for success. Figure 4.6 identifies worker feelings at each stage of the change process.

Communication and training must address workers' feelings to change behaviors at each level of the ladder in order to create meaningful and lasting changes as workers are moved up the ladder. The ultimate goal is to establish ownership of the SMS.

Strong leadership, creating a shared vision, teamwork, transparency in communication, clearly articulated value proposition, and demonstrated genuine care for improving the health and safety of workers are critical requirements for getting workers onboard and committed to improving the SMS. Upgrading an existing SMS therefore requires careful attention to creating and sharing the safety vision and to the types and frequency of communication provided to stakeholders.

Figure 4.7 provides a model for upgrading an existing SMS. The process is very similar to that outlined in Figure 4.3—model for implementing an SMS where one does not exist. Additional areas of focus (shaded) include the following:

1. Gap analysis completed (new vs. existing standards)
2. Gap closure strategies developed
3. Execution and rollout

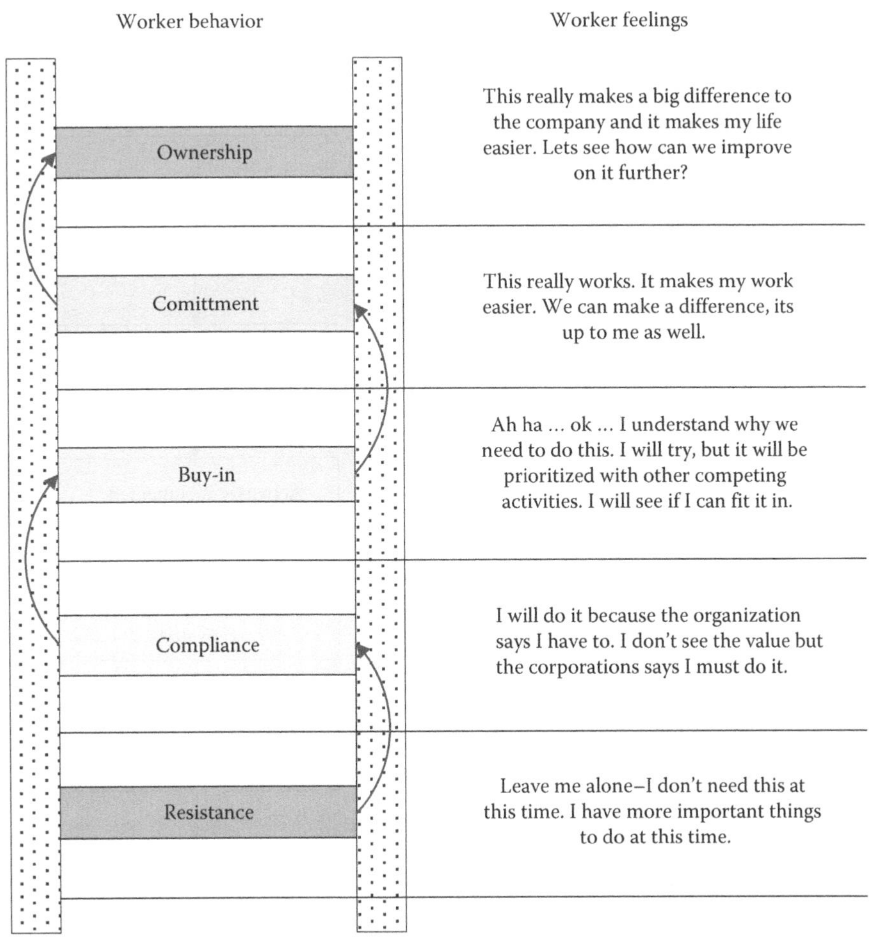

FIGURE 4.6
Behavior changes and worker feelings when creating change.

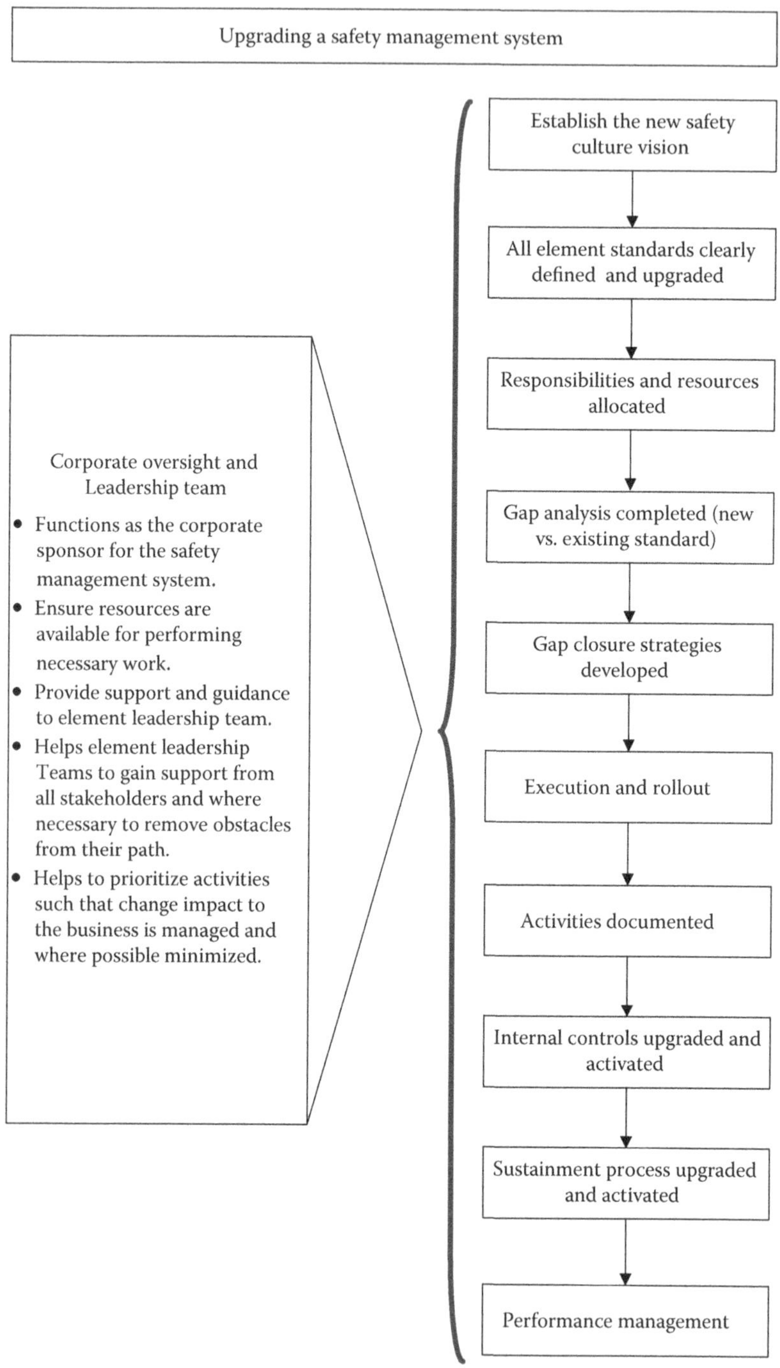

FIGURE 4.7
Model for upgrading a safety management system.

Other activities outlined in this model were discussed earlier in the chapter and the principles for addressing them continue to remain the same. Careful attention to execution and rollout is required to ensure the success of this venture.

Gap Analysis Completed (New vs. Existing Standards)

Performing a gap analysis between the existing element standard and the new standard for the element is an important starting point for effectively upgrading the existing SMS. With the support of senior leadership, a team comprising stakeholder representatives must work together to compare the working of the existing standard relative to the requirements of the new standard. In some cases, the new requirements may be relatively straightforward and easy to execute while in other cases it may be much more complex.

Success comes from establishing a team of users of the existing standard who are capable of interpreting the new requirements of the standard. A simple process for identifying *new* and *upgraded* processes and requirements should be established for guiding the prioritization and execution of the new standard. The number and magnitude of these gaps will determine the amount of resources required and the period necessary for closing these gaps.

Gap Closure Strategies Developed

Once the gaps have been identified and prioritized, gap closure strategies can be developed for execution. In some instances, gap closure strategy and requirements may be quite simple, for example, where the new requirements may call for a documentation system where one did not exist before, this can be easily executed. In other situations, the gap closure strategy may be complex and difficult to execute. For example, for the element—event management and learning, sharing of knowledge and verification that learning takes place across multiple business units may involve a complex process that includes standardizing the format for sharing knowledge, developing a process for sharing this new knowledge, and finally developing a process for verification of competency where necessary.

Strategies developed for closing gaps must be simple and easy to follow, and rollout must involve stakeholders in developing them to ensure ownership. Where necessary, gap closure strategies must be packaged into small projects such that a wide cross section of stakeholder can be engaged and involved. Strategies must be properly evaluated and analyzed for what makes best business sense for the prevailing environment and situation.

Execution and Rollout

The execution and rollout of gap closure measures and strategies must be treated as projects that are properly resourced and managed. A schedule identifying key deliverables and their timelines, resources required, a strong and capable leader, and internal control measures are all essentials for successful execution and rollout. Focused attention to people, processes, and system readiness is required to ensure on-budget and on-schedule delivery.

Lutchman (2010) discussed practical business requirements for execution of industrial and commercial projects that identified the need for breaking large projects into manageable milestones for which people, processes, and system readiness is required to progress from one milestone to another in the execution and rollout of projects. In his work, Lutchman (2010) identified people readiness as personnel trained and competent to allow the project to proceed to the next milestone. Process readiness referred to supporting processes available to sustain the project unto the next milestone. System readiness referred to the status of hardware and equipment required to allow successful transition to the next milestone.

Internal control measures must enable early warning and provide leadership opportunities for course correction when obstacles and constraints are encountered. Care must be taken at all times to assess the amounts of new work required by workers to determine the work–life balance impact to workers. Short-term additional work may be acceptable to many workers. However, if the additional work placed on workers during execution and rollout is heavy, over an extended rollout period, the risk of burnout and demotivation are high. More importantly, we create a situation that is in contrast to the goals of an effective SMS where risk exposures can be increased when workers are overworked and fatigued.

5

An Incident Management System

What is an incident management system (IMS)? What value does an IMS bring to organizational health and safety? The answers to these questions must be clearly articulated to business leaders so that they fully grasp the scope of work required in developing an IMS as well as the benefits derived by businesses from an effective IMS.

An IMS is an organized system or method designed to assist businesses in facilitating the following:

1. Incident reporting—a means for recording all events (accidents and near misses).
2. Assessment of risk and consequence—allows each event to be classified based on risks.
3. Incident investigation and causal analysis—facilitates incident investigation and root cause analyses.
4. Recommendations and corrective actions—facilitates the ability to define recommendations, remedial, and corrective actions by leadership and follow-up on progress of each assigned action.
5. Event historian—a historical database of events that may have occurred in this organization.

In this chapter, we shall discuss how an IMS works and how organizations benefit from an IMS.

Design of an Incident Management System

An IMS can vary from a simple spreadsheet for collecting information related to workplace incidents, risk ranking (or consequence and potential consequence analysis), investigations, and follow-up actions for the small enterprise to an advanced well-developed multifaceted automated system for a large multinational multiple business units organization. At the two extremes, however, the goal is to capture relevant information relating to incidents to proactively reduce incidents in the workplace.

An IMS must be user-friendly and accessible by all employees so that data can be gathered and entered into the IMS as quickly as possible. The types of data to be collected are as follows:

1. Date and time of incident
2. Reported by whom
3. A description of what happened
4. Type of injury or damages incurred
5. Who was injured? Employee or contractor worker?
6. Where did the injury or incident take place?
7. Who was informed of the incident?
8. Estimated cost of incident
9. Entry number for tracking and retrieval purposes
10. SMS element affected

The traditional focus of an IMS has been primarily on lagging indicator metrics, since a typical IMS generally captures information after an incident has occurred. Nevertheless, significant useful information is derived from an IMS when used properly to proactively address safety concerns within an organization. Today, however, IMS has evolved to include inspection records, audit findings, and other leading indicator metrics to improve workplace health and safety.

Figure 5.1 provides a simple spreadsheet for tracking incidents in an IMS that may be usable by small organizations. This simple approach to IMS will require follow-up actions for each incident as noted in column I of the spreadsheet. Regardless of the type of IMS, historical data storage capabilities are essential for data analysis and actions.

A well-organized advanced IMS provides users opportunities to store historical data and generate multiple reports that are useful in managing the safety performance of the organization. These are typically available in larger organizations that seek to address safety concerns beyond regulatory requirements. Figure 5.2 provides a schematic of some of the details typical of an advanced IMS.

The key distinction of an advanced IMS is the ability of the system to maintain an incident database from which useful information can be derived to generate improvements in the safety performance of an organization. Care must be taken to ensure that the IMS does not become or remain primarily a database as opposed to a knowledge-generating machine from which tremendous useful information can be derived. All personnel must be adequately trained in the use of the system for data entry.

Champions or superusers must also be created to maximize value from the IMS. These champions must be trained to extract information that can be used in developing the strategies of the organization over time in combating

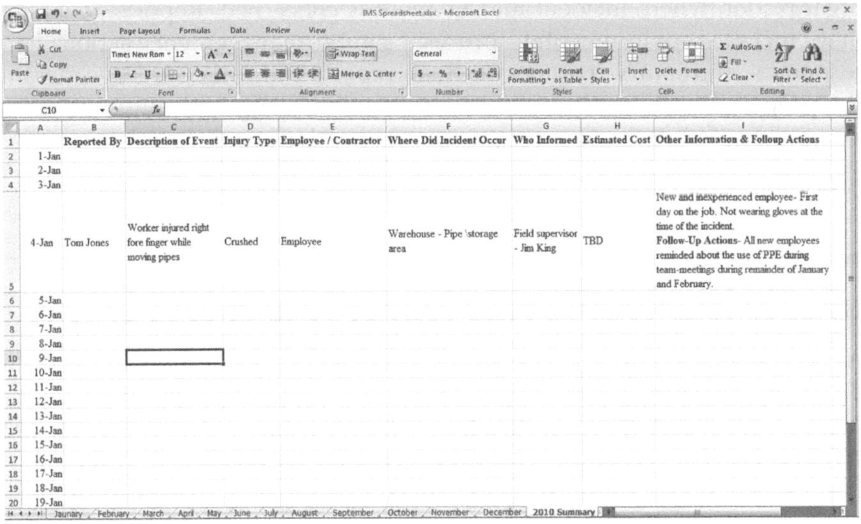

FIGURE 5.1
Simplified Excel spreadsheet IMS for tracking incidents.

incident increases. For example, drill-down capabilities in the incident types can help determine where potential problems lie and the types of immediate and longer-term responses that are required by the organization.

Reports generated by the system are also designed to help business units and corporate leadership better understand the challenges faced by the

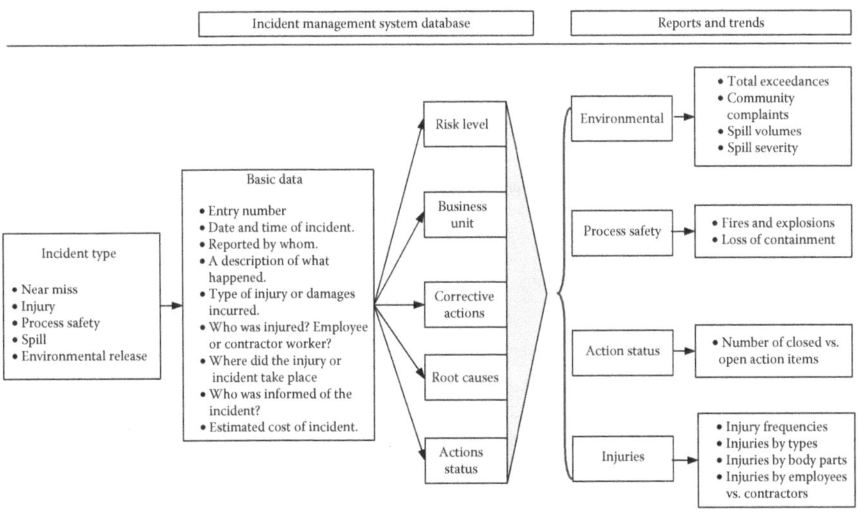

FIGURE 5.2
Details of an advanced IMS typically used by larger organizations.

organization where incident management is concerned. Reports generated may include environmental, process safety management, action item responses, and injury frequencies. Such reports can assist senior leaders in the proper allocation of resources for sustained long-term improvements. *The challenge is that this a rear-view mirror look and response.*

Using an IMS for Short-Term Tactical Safety Responses

A well-organized IMS provides users with opportunities to store historical data as well as current data to better understand and address the safety performance of the organization. Using injuries as an example, we shall explore how an advanced IMS can generate useful information for guiding the responses of safety practitioners in improving the safety performance of an organization.

As shown in Figure 5.3, with an advanced IMS, details regarding incidents can be captured by the IMS such that precise information can be generated for tactical actions by the business. Figure 5.3 demonstrates the drill down potential of the IMS which allows safety practitioners to be able to more precisely identify the root causes of incidents such that appropriate

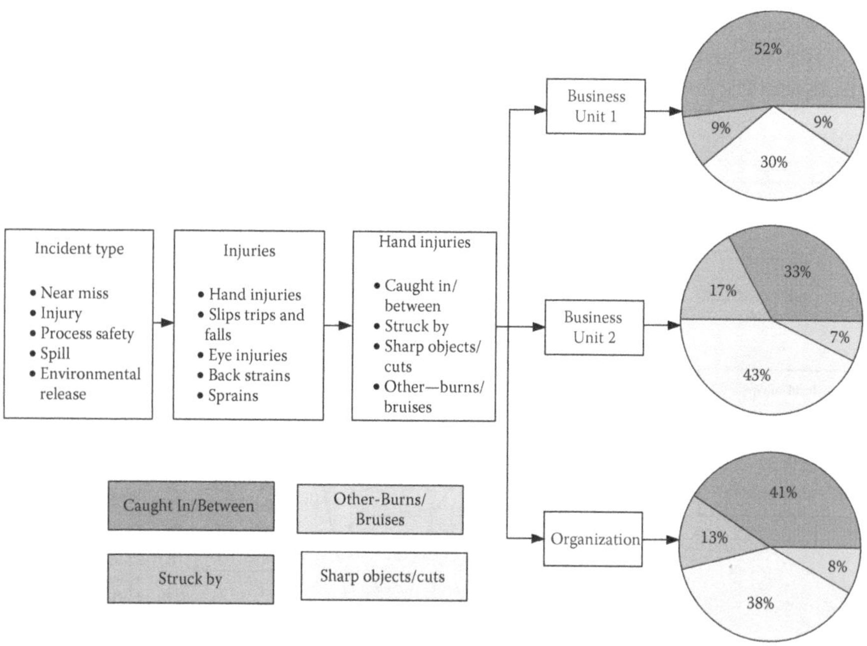

FIGURE 5.3
Information generated from an advanced IMS for tactical action.

corrective actions can be developed and executed. The final outputs of the IMS are patterns that can be used for corrective actions.

In the example demonstrated in Figure 5.3, the focus of the analysis is on hand injuries. In the example provided, the types of hand injuries typically sustained during work include the following:

1. Hands or fingers being caught in equipment or tight spaces or between various equipment and tools.
2. Hands or fingers can be struck by or against various tools and equipment.
3. Hands or fingers can be cut or gouged by sharp objects or edges, tools, or equipment.
4. Hands or fingers can be injured from other sources of energy such as burns, chemicals, or bruises.

As shown in Figure 5.3 and the pie charts generated, both Business Unit 1 and Business Unit 2 are experiencing hand injury challenges as follows:

1. Hands or fingers being caught in equipment or tight spaces or between various equipment and tools.
2. Hands or fingers can be cut or gouged by sharp objects or edges, tools, or equipment.

However, Business Unit 1 appears to be doing better where injuries from sharp objects and cuts are concerned while Business Unit 2 appears to be doing better where injuries resulting from hands being caught in equipment or crushed are concerned. This type of information is very useful in that it provides the organization opportunities to prioritize resources to preferentially address the specific concerns of each business unit, thereby maximizing the value creation from each effort. Furthermore, the organization can leverage learning from both business units to better address the concerns of the organization.

Since the types of hand injuries sustained above originate from different sources, the organization may be able to develop specific response actions for each type of hand injuries sustained by the business. In the examples provided, injuries that resulted from fingers or hands being caught in equipment or tight spaces or between various equipment and tools may be the outcome of improper hazards assessments or workers are rushing to complete work. As a consequence, training programs or leadership messaging can be tailored to address these specific concerns.

On the contrary, hands or fingers being cut or gouged by sharp objects or edges, tools, or equipment may be a reflection of tool selection for the particular job or improper use or type of PPE used in completing the job. In either case, the application of specific and targeted remedial measures has a

TABLE 5.1
Raw Data from IMS Improves Decision Making

	Business Unit 1		Business Unit 2		Organization	
Source of Hand Injuries	**Number of Incidents**	**% of Incidents**	**Number of Incidents**	**% of Incidents**	**Number of Incidents**	**% of Incidents**
Caught in/between	12	52	10	33	22	41
Struck by	2	9	5	17	7	13
Sharp objects/cuts	7	30	13	43	20	38
Other—burns/bruises	2	9	2	7	4	8
Total	23	100	30	100	53	100

greater chance of generating improvements as opposed to a broad-spectrum approach to addressing hand injuries in a nonspecific manner.

Occasionally, it may be necessary to understand more about the data than the charts and trends generated. For example, for the same pie charts above, the IMS should be able to generate the tabulated data as well, as shown in Table 5.1. The raw data provide useful information regarding the magnitude of the problem relative to the information derived from the pie charts and are very useful in tactical decision making.

Using an IMS for Long-Term Strategic Safety Management Decisions

An effective IMS should have the ability to retain historical data and generate trends and reports consistent with these historical data as shown in Figures 5.4 and 5.5. It is also supported by an effective incident investigation process from which root causes can be identified. Addressing root causes derived from the investigation process allows for real and sustained improvements in health and safety. Many tools and processes exist for doing so among which are TapRoot, Reason, 5-Why's, Det Norske Veritas (DNV), and other investigative techniques.

As shown in Figures 5.4 and 5.5, the historical data gathered by the IMS in Suncor reflect the strategic actions taken by the organization consistent with its Journey to Zero strategy for reducing workplace incidents. According to Suncor:

> Journey to Zero is intended to embed a safety culture into every aspect of our business. Our aim? To achieve a level of safety excellence that results in an injury-free work site. Suncor wants all work processes and systems to be safe and the company demands employees and contractors take individual responsibility for safety. (Suncor Energy, 2010)

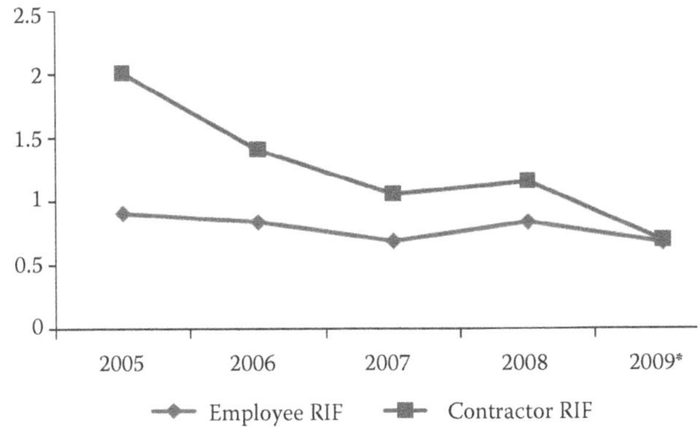

FIGURE 5.4
Recordable injury frequency—Suncor Energy Inc. 2010. *Reflects merged data for Suncor and Petro-Canada from organizational merger. (Adapted from http://sustainability.suncor.com/2010/en/responsible/3574.aspx.)

In the simplified processes and design provided for an IMS in this chapter, we have addressed the needs for incident reporting—a means for recording all events (accidents and near misses); assessment of risk and consequence—thereby allowing each event to be classified based on risks; incident investigation and causal analysis—facilitating incident investigation and root cause analyses; action management and follow-up—facilitating definition of remedial actions by leadership and follow-up on progress of each assigned action; and event historian—a historical database of events that may have occurred in this organization and which is useful in helping define the long-term strategic actions of the organization in its SMS.

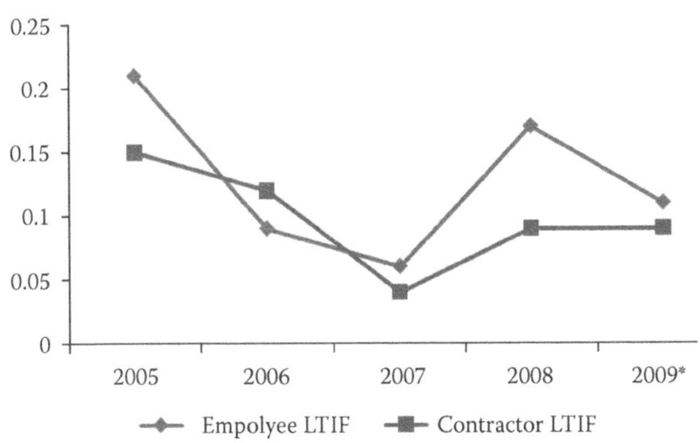

FIGURE 5.5
Loss time injury frequency—Suncor Energy Inc. 2010. *Reflects merged data for Suncor and Petro-Canada from organizational merger. (Adapted from http://sustainability.suncor.com/2010/en/responsible/3574.aspx.)

6

Leadership and Organizational Safety

History has shown worker fatality as a normal part of doing business in the early construction of major projects. For example, during the construction of the Hoover Dam (1931–1936), a total of 96 work-related fatalities were incurred (Hernan, 2009). A similar number for the construction of the Panama Canal, inclusive of all deaths during the period 1880–1914, was ~20,000 fatalities (many of which were mosquito-related illnesses) (American History, 2003). During the actual 7-year construction by the United States, a total of 5609 fatalities from illnesses and injuries were incurred (American History, 2003).

Through the collective efforts of leadership, governments, unions, research scientists, and various regulatory watchdog agencies, workplace safety has improved dramatically such that history of the type described earlier will not be repeated. Stakeholders hold leadership to higher standards of occupational and organizational safety and prevailing laws in many countries make it illegal to compromise safety for profit making.

Today, organizational safety is important to senior leadership because safety makes a significant difference to the bottom line of all organizations. Organizational safety just makes good business sense. Incidents that result in injuries, wastage, damage to the environment, damage to assets, and adverse impact to images all result in both direct and indirect cost to organizations. Many recent incidents, which may have been the result of either personnel or process safety management failures, or both, can be highlighted to emphasize the cost of organizational safety to employers.

Incidents such as those shown in Table 6.1 have been quite costly to both employer and society as a whole and are all due to failures in the SMSs of organizations.

When investigated further, some areas of disastrous incidents as shown in Table 6.2 demonstrate leadership's failure to learn from incidents such that repeats of the very same situation continue to plague businesses.

Regardless of where a company operates in the world today, work-related fatalities and injuries are no longer acceptable in the global community. Moreover, human fatalities in either the construction or operation of any project are no longer acceptable among the global community as a normal component of doing business. Public and media outcry, regulatory enforcement, and stakeholder pressures will force such attempts into immediate closure, and death cries for the company's president, CEO, and heads of leadership teams will follow. As a consequence, leadership must not only appear

TABLE 6.1

Recent Incidents with Significant Personnel and Process Safety Impact

Incident	Date of Occurrence	Personnel Impact
BP/transocean rig explosion in the Gulf of Mexico	April 2010	11 killed
Explosion and fire at the Tesoro refinery in Anacortes, Washington	April 2010	4 killed, 4 injured
Imperial Sugar Company dust explosion and fire, Georgia	February 2008	14 killed, 36 injured
Chemical explosion, T2 Laboratories, Jacksonville, Florida	December 2007	4 killed, 32 injured
Texas City BP refinery explosion	March 2005	15 killed, 180 injured

Source: Adapted from Chemical Safety Board (CSB), 2010a. Completed Investigations. Retrieved May 15, 2010 from http://www.csb.gov/investigations/investigations.aspx?Type=2&F_All=y

TABLE 6.2

Repeat Incidences of Combustible Dust Disasters

Major Combustible Dust Explosions	Date of Incident	Personnel Safety Impact
Sugar dust explosion and fire, Georgia	February 2008	14 killed, 36 injured
Metal dust fire and explosion, Indiana	October 2003	1 killed, 1 injured
Organic dust fire and explosion, Kentucky	February 2003	7 killed, 37 injured
Organic dust fire and explosion, North Carolina	January 2003	6 killed, 38 injured
Organic dust fire and explosion, Massachusetts	February 1999	3 killed, 9 injured
A series of devastating grain dust explosions in grain elevators	1970s	59 killed, 49 injured

Source: Adapted from U.S. Department of Labor, 2010, http://www.osha.gov/dts/shib/shib073105.html

to be but must be actively prioritizing and managing the health and safety of all workers under its employment and supervision.

The Role of Leadership

Leadership plays a very important role in the way safety is managed in the workplace. Listed below are the key responsibilities for leadership where safety is concerned.

- Provide leadership and direction
- Provide prioritization and sufficient resources
- Establish written standards and supporting procedures and define roles and responsibilities

- Set goals, objectives, and expectations for worker performance
- Establish accountability for performance against goals and objectives
- Establish and steward safety performance KPIs
- Audit the EHS management process
- Provide oversight of work and performance

In this section, we shall discuss these roles and responsibilities of leadership that can lead to superior safety performance.

Leadership and Direction

Leading safety from the desk of the CEO and senior leadership is very complex. One of the core requirements for leadership in safety is alignment between what leaders say and their actions for credibility in safety leadership. Often there are disconnects between the messages sent by senior leadership and the messages received by middle management based on the actions of senior leadership. More importantly, messages provided to CEOs by shareholders and other major stakeholders are often complex and require a strong will to make the right decisions regarding EH&S. Figure 6.1 demonstrates some of the challenges faced by leaders and the difficulties associated with making the right decisions regarding workplace safety.

At different organizational levels, middle managers and frontline supervisors look to senior leaders in the organization to set directions and guide the business regarding safety. The actions and behaviors of senior leaders in an organization will determine the level of emphasis and commitment placed on safety. We have all heard the phrase: *what fascinates my boss . . . fascinates me even more.* When leaders focus on production exclusively as opposed to providing a balanced interest in production, facility, people, and safety, what this communicates to middle managers and frontline supervisors is that production takes precedence over all else, including safety. This then becomes the focus of attention, of middle managers and frontline supervisors. What this also means is that if the middle manager or frontline supervisor is disciplined for proactive preventative actions related to repairs to ensure the safety of personnel and equipment, future actions will be halted. Similarly, if rebuked by senior leadership for lost production, future responses of middle managers and frontline supervisors will be to maintain operations until an incident occurs rather than proactively acting to protect assets by shutting down the plant or operations.

Similarly the phrase: *what gets rewarded gets done* is a reflection of the behaviors of middle managers and frontline supervisors in interpreting the behaviors of senior leaders. When middle managers and frontline supervisors recognize that the first thing a plant manager examines at the end of a shift is the production numbers, they soon learn to focus attention primarily on production. If safety performance is not high among the priorities of the

True message	Organizational level	Conflicting messages
Safety is our top priority	CEO and senior leadership	• Market and shareholders reward profits. • CEO loses job when profits are lower than peers but safety record is better. • Not willing to recognize a lag in profits between investments in safety and long-term sustainable profitability
Safety before production	Middle managers	• Production rewarded over safety • Down time and outage frowned upon–correcting safety deficiencies require downtime and outage • Preventative maintenance often deferred based on pricing and profit drivers • Demotions and job loss as an outcome of lower than peer production performance • Weak recognition of a lag between sustained productivity and investment in safety
Safety first	Frontline supervisors and workers	• Frontline supervisors poorly trained • Senior and frontline leaders talk safety but do not demonstrate the behavior ... why should I? • Getting the job done quickly • Nondetected shortcuts rewarded
Safety first	Contractors	• Bid process rewards lowest prices–safety budgets generally the first to be chopped to be competitive • Getting the job done quickly rewarded • Contractors often provided the dirtiest, difficult and dangerous jobs • Punitive consequences for reporting incidents and near misses drives reporting underground leading to lost learning opportunities

FIGURE 6.1
Conflicting messages presented to leadership.

plant manager, safety performance also falls on the priority listing of middle managers and frontline supervisors.

Leadership in safety is no different from leadership in any of the other areas of the business and should be treated as such. In this instance, the goal of senior leadership is to change the behavior of the worker to be consistent with the business needs regarding safety. Targets must be established, resources must be allocated, stewardship is required, and personnel must be held accountable for performance. When treated in this manner, safety performance will improve and ultimately will become a priority in the organization. Success, however, depends on strong leadership in safety. Successful safety leaders must be able to establish aggressive but achievable SMART targets. In addition, leaders must be able to create and share a safety vision among the workforce and create strong teams capable of delivering performance consistent with this vision. Success in achieving safety targets is often reflected in behavioral changes. Leadership is the most important variable in changing behaviors where safety is concerned and leaders must be properly trained to ensure that they are able to fulfill the goal of creating a compelling sense of direction for followers.

A key challenge faced by business leaders today is reflected in their inability to collectively manage cost, safety, and quality (systems and processes). This is so because stakeholders (shareholders in particular) are often overly focused on shorter-term cost and schedule performance. Developing quality process and systems is not only costly but also time consuming and can adversely affect schedules.

Often, shareholders are not fully aware of the impacts that such quality systems and processes have on the long-term sustainability of organizations. Shareholder focus is primarily short term, seeking value maximization in the form of increasing share prices and dividends. Such focus may influence the behaviors of internal organizational stakeholders to underdeliver on quality in their quest for budget and schedule targets. The unfortunate outcome of shareholder short-term focus is that leadership may place greater attention to cost, and schedule and quality may be generally looked after as an afterthought. Noteworthy, when the organization has achieved sufficient maturity to weather the impact of shareholder punishment, or gain shareholder support for investments in improving quality processes and systems, it can do a better job balancing all three dimensions.

Prioritization and Sufficient Resources

Leadership is faced with many challenges when managing worker health and safety in the workplace. In every organization, similar to every business process, safety competes with other business operations for a share of the budget. More importantly, when competing with profit centers in the business operations, the EH&S department of the organization is often perceived to be a cost or expense center and one that does not contribute to profits. In addition, this short-term nonvisionary approach to budget allocation can significantly impact on the organization's long-term profitability and success.

Leadership must, therefore, establish processes that may be used in selecting and prioritizing among the many corporate projects to be funded by the organization and among the many projects that are intended to improve safety performance. Where resources are scarce, as is the case, in many smaller organizations, competing for safety-related spending is often done on a nonlevel playing field. This is so because it is not always possible to reduce the benefits of a safety project to a simple dollar value to be expressed in common project evaluation terms such as a net present value (NPV) or internal rate of return (IRR) for comparison with other more easily quantified projects.

TABLE 6.3

Quantitative and Qualitative Cost Associated with Safety Projects

Quantitative Variables	Qualitative Variables
• Project cost and dollar savings • Health care cost associated with injuries • Worker compensation and insurance cost • Wastage and rework cost • Legal and liability cost	• Productivity impact • Morale impact • Incident investigation time • Worker turnover impact (hiring and training time) • Brand and image impact

Advancing safety-related projects therefore requires relating such projects to a combination of quantitative benefits as well as qualitative benefits. Table 6.3 shows a few of the common quantitative and qualitative variables that can be identified and addressed when evaluating and prioritizing safety projects.

For larger organizations with deep pockets, leaders are easily able to justify spending in safety under the umbrella of simple common sense guidance such as

- It makes good business sense.
- It improves the stakeholder engagement process of the organization.
- It improves the image of the organization.
- It improves safety, and safety is our top priority.

In such situations, a combination of quantitative and qualitative benefits can be easily applied in justifying and prioritizing safety projects.

There is support for these types of senior leadership decision making. According to Liberty Mutual (2001), in a survey conducted among business executives, it was found that 95% indicated that safety improved the bottom line while 61% suggested that for every $1.00 safety dollar invested, a $3.00 return was seen. The same study also indicated that 40% of executives suggested that for each $1.00 direct cost incurred from an injury or incident, there were indirect costs in the range of $3.00–$5.00 (Liberty Mutual, 2001).

Written Standards and Supporting Procedures

For each element of the SMS defined in Figure 4.1 (Chapter 4), leadership has a responsibility for establishing a supporting standard to which the organization must adhere. Leadership is also accountable for developing the supporting procedures for compliance to the standard. Standards and procedures developed must be simple to apply and must address all concerns around the element such that workplace safety can be properly addressed.

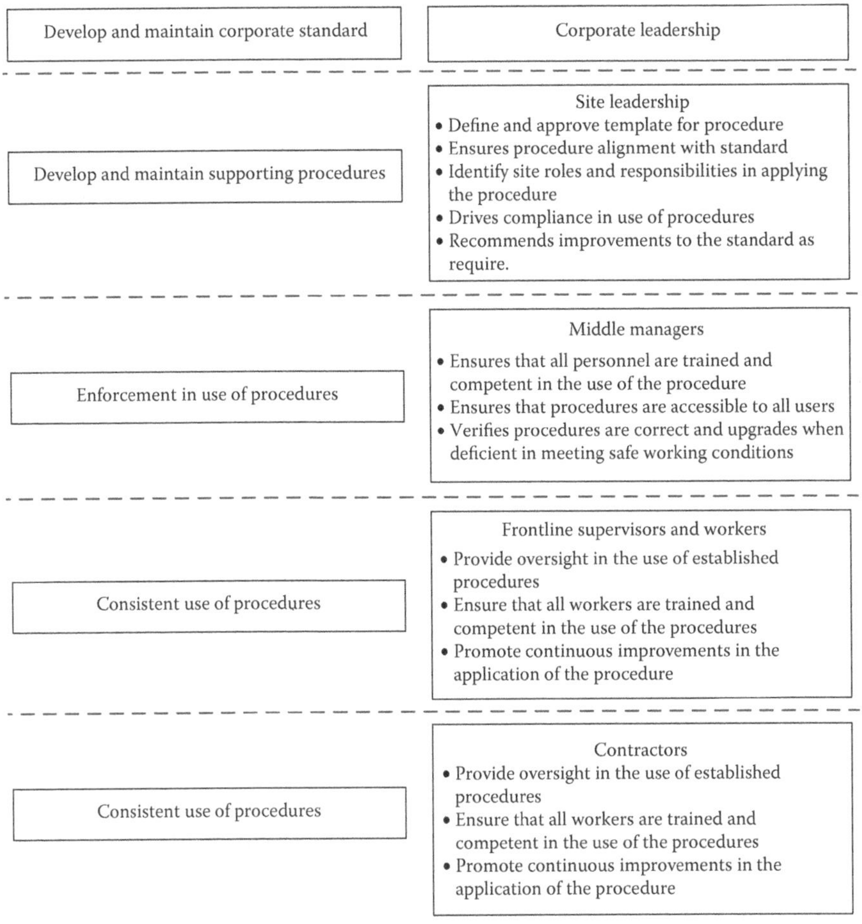

FIGURE 6.2
Roles and responsibilities of various levels of leadership.

Figure 6.2 identifies roles and responsibilities for various levels of leadership in the organization regarding element standards and supporting procedures. According to Figure 6.2, corporate leadership must develop the appropriate standard by which the organization will operate. For example, in the element—*Operating Procedures and Safework Practices*, corporate leadership must develop a standard that meets the organizational needs for ensuring safety and compliance with regulatory and industry standards of performance.

In this example, corporate leadership must develop and approve a standard that guides site leadership toward compliance. In this regard, site leadership may be required to develop supporting procedures consistent with the following attributes:

1. Operating procedures shall be documented.
2. All the required content must be readily accessible by all users.
3. Must be subject to per established review requirements that may be defined in the standard.
4. Clearly articulated contents as defined by the types of work to be undertaken. For example, where hazardous work is to be done, procedures may be required to include the following information:
   i. Environment, health, and safety information such as
      a. Properties of the hazards presented by the materials used in the process.
      b. Worker exposure controls, including engineering controls, administrative controls, and PPE.
      c. Measures to be applied if loss of containment occurs.
      d. Actions to be taken should personnel exposure occur.
      e. Known hazards identified such as reactivity, thermal, or stored energy or pressures.
   ii. Operating procedures and steps for each of the following conditions:
      a. Normal start-up (including facilities readiness reviews).
      b. Start-ups following a facility turnaround or an emergency shutdown (including facilities readiness reviews).
      c. Emergency or abnormal operating situations.
      d. Normal operations.
      e. Planned shutdowns.
      f. Unplanned or emergency shutdowns.
      g. Temporary and abnormal operations as appropriate (e.g., during use of bypass systems or circulation modes).
   iii. Identify those raw materials and other substances critical to the safety of the process.
   iv. Develop quality control procedures to assure material specifications have been met.
   v. Make available applicable operating limits, conditions, consequences of deviation, and corrective steps through the operating procedure.
   vi. Specify inventory limits.

vii. Describe safety systems and their function (e.g., isolation valves, emergency dump valves, scrubbers, flares).
viii. Describe instrument controls, including alarm and interlock set points.
ix. Be readily accessible to personnel who work in or maintain the process.
x. Updated and approved prior to implementing any changes to chemicals, technology, or assets.

Similarly, where safe work practices are concerned, procedures that guide the following may be required:

1. Work permit and authorization.
2. Entrance to and exit from a hazardous process facility by maintenance, contract, laboratory, and other support personnel.
3. Breaking the integrity of process equipment and piping.
4. Lockout/tagout (isolation) of hazardous energy sources.
5. Control of ignition sources (i.e., hot work permit).
6. Confined spaces entry.
7. Movement of heavy equipment relative to equipment containing hazardous materials or high-energy sources.
8. Integrity check (e.g., pressure test) of process equipment at the time of turnover from maintenance and reliability personnel, but prior to acceptance by the operating authority.
9. Bypassing of a safety interlock or alarm or critical systems defeat.
10. Continued operation with an activated safety alarm.
11. Temporary service-to-process flexible hoses for purging or flushing, unplugging, or similar tasks.
12. Ground disturbance.
13. Hazardous waste management.
14. Working alone or in isolated locations.

Set Goals, Objectives, and Expectations for Worker Performance

Consistent with establishing safety priorities, leadership is also required to set goals and expectations for worker performance. Environmental, health, and safety goals must align with SMART principles as outlined earlier in our discussions. Indeed, where multiple business units may exist in an organization, business unit goals may vary but must ultimately be aligned with the corporate goals and objectives and worker performance. Figure 6.3 highlights the alignment of leadership efforts at each level toward the corporate safety goal.

A top-down/bottom-up approach is required to ensure that goals and objectives are properly communicated across the organization and are adequately reflected in the work plans of every employee of the organization. The corporate EH&S goals and objectives must be communicated by senior leaders throughout the organization, including contractors who are required to work for the organization.

Develop and communicate strategic safety goals and objectives	Corporate leadership
Develop and communicate annual or short-term safety goals and objectives	Site leadership • Ensures that annual safety goals are aligned with corporate goals and objectives • Develop tactical strategies for achieving corporate safety goals • Provide oversight and stewardship (e.g. trend analysis) to identify improvement opportunities and resource prioritization
Participate in developing and communicating short-term safety goals and objectives	Middle managers • Develop and steward tools for collecting data and measuring progress relative to short-term goals and objectives • Identify and close gaps that may prevent achieving short-term safety goals
Supervise and manage work safely to ensure that short-term goals and objectives are met	Frontline supervisors and workers • Ensure work is conducted in a way so as to ensure that short-term safety goals and objectives are met • Ensure that work-plan contain tactical measures and strategies to achieve short-term goals • Promote continuous improvements in safety and worker performance
Supervise and manage work safely to ensure that short-term goals and objectives are met	Contractors • Ensure that work is conducted in a way so as to ensure short-term safety goals and objectives are met • Ensure that work-plan contain tactical measures and strategies to achieve short-term goals • Promote continuous improvements in safety and worker performance

FIGURE 6.3
Alignment of goals and objectives to achieve strategic objectives.

Workers respond better to corporate goals and objectives when they know why they are required to work in a particular way, what needs to be done, and, in some instances, how the work is to be done. Once the corporate goals and objectives are established, they must be communicated across the entire organization very quickly. All leaders must seek to establish a line of sight between the works he/she does relative to the corporate goals. In addition, each leader must ensure that his/her followers are knowledgeable about how the work performed by the follower contributes to the corporate goals and objectives.

Establish Accountability for Performance against Goals and Objectives

As discussed earlier, organizational safety should be treated no differently from other business functions. Investments in safety should be justified on a similar basis as other competing projects, and funded accordingly. In doing so, clear goals and deliverables must be established for stewardship. As is the case with production departments and divisions, success is dependent upon leaders being able to deliver relative to preestablished goals.

The difference with safety leadership in large organizations is that often the EH&S department may exist as a support organization with no direct control over field resources for executing safety priorities. As a consequence, performance relative to goals and objectives is controlled by other leaders over which senior safety management leaders may only possess influencing authority.

It follows, therefore, that when safety management is decentralized and imbedded within a business unit, accountability for performance resides within the leadership of the business unit. On the other hand, where safety or EH&S is centralized and supporting several business units, there is joint accountability between centralized EH&S leadership and business unit leadership for performance relative to goals and objectives.

Similarly, there is shared responsibility for accomplishing work defined in the work plans of all workers where safety is concerned. Ultimately, safety is everyone's responsibility and is the outcome of a shared responsibility of every employee to deliver safety performance relative to established goals and objectives.

Establish and Steward Safety Performance KPIs

Almost all organizations establish and steward KPIs for safety. In most instances, there is a combination of lagging or trailing indicators as well as leading indicators (Table 6.4). Success with KPIs depends upon the selection of high-value indicators and the application of focused attention by leadership at all levels of the organization.

In a large organization with multiple business units, there may be several KPIs being stewarded at the corporate level. On the other hand, at the business unit level, leaders may choose to identify 3–5 leading indicators

TABLE 6.4

Sample Lagging and Leading Indicators for Stewardship and Trending

Corporate KPIs	Business Unit 1	Business Unit 2	Business Unit 3
Lagging Indicators			
• Total recordable injury frequency (TRIF)	√	√	√
• Total disabling injury frequency (TDIF)	√	√	√
• Hand injuries frequency		√	
• Slips trips and falls		√	√
• Loss time incidents	√	√	√
Leading Indicators			
• Quality of field-level risk assessments		√	√
• Percentage of employees properly oriented	√	√	√
• Percentage of employees in competency development program	√	√	√
• Number of near misses reported	√		
• Percentage of audit findings and action items closed	√	√	√
• Leadership visibility—percentage of leader's time spent on the frontline			√
Number of self-assessment audits completed		√	

against which resources are applied and focused attention is applied. Using stewardship tools as discussed in the previous chapter, business units may apply trend analyses, pie charts, and other quantitative techniques to determine where greatest risk exposures may exist within the business unit. Generally, at the business unit level, attention is placed on leading indicators because when we address the leading indicators completely, lagging indicators tend to take care of themselves.

It is important to note that the list of leading and lagging indicators provided above is by no means complete and is intended purely to demonstrate the point that business units may focus attention on a few leading and lagging indicators as opposed to all indicators, thereby maximizing value to the organization.

Audit the Environment, Health, and Safety Management System

A goal of leadership is to determine whether its SMS is working optimally. As a consequence, both internal and external audits of the SMS are recommended. Leadership must determine frequencies of such audits. These audits must identify gaps and gap closure measures. Audits must also have a system that ensures that gaps are effectively stewarded such that they are closed in a timely manner. It is imperative that leaders create an environment whereby audits are seen as proactive measures to help the business improve its SMS and performance as opposed to one of punitive consequences. Audits will be discussed in greater detail later in Chapter 16.

Oversight and Supervision of Work and Performance

A requirement of leadership is to ensure that adequate oversight and supervision are provided to all workers at all levels of the organization and in particular where safety is concerned. On an ongoing basis, we provide oversight for production. Is there a compelling reason why similar oversight should not be provided for safety? Adequate worker supervision is required based on the cultural environment and the maturity of the workforce.

We are firm advocates that leadership visibility helps improve safety performance. When leaders are visibly engaging workers in safety and are available to address safety concerns in a timely manner, issues like the following disappear:

1. Weak compliance to following standards, procedures, and practices
2. Poor use of PPE, particularly if the PPE is uncomfortable or cumbersome to use
3. Creation of a work environment characterized by *us* (workers) and *them* (leaders)
4. Shortcuts and chance taking among the workforce

More importantly, workers believe that leaders are more approachable and their welfare is important to leaders.

Leadership Styles and Behaviors: Impact on Safety

In this section, we shall examine some of the common leadership styles and behaviors and attempt to determine those that may have the greatest impact for leading safety. True leadership starts with a good understanding of the impact of your leadership style and behaviors on followers. In this section, we shall discuss some of the common leadership styles that are dominant today across cultures and regions. We shall also attempt to identify leadership styles and behaviors that are better applicable to today's workforce that is characterized by the following trends, some of which were discussed in Chapter 2:

- More educated workforce
- Comprised of both Generation X and Generation Y workers and a growing presence of Generation Y
- Higher expectations from leaders
- Wants to be engaged
- Increasing presence of women in the workplace
- A workforce dominated by multiple cultures
- An ageing workforce in North America
- Continually changing technological environment

Autocratic Leadership Style and Behaviors

Leadership in many developing countries today continues to be built upon the principles of autocratic cultures and styles. In such cultures, leaders tell followers what to do regardless of the maturity of the worker. Power is centrally controlled and all decisions are taken by the leader. Where safety is concerned, autocratic leaders who value safety will produce positive results in managing safety. Table 6.5 shows the strength and weaknesses of autocratic leadership styles and behaviors.

When applied properly, in a stable work environment, safety is well managed and the health and safety of workers are looked after. Workers are trained and follow rules that are designed to protect them. In Western cultures, however, autocratic leadership behaviors are a thing of the past and when applied in workplaces of developed world economies, they will result in worker flight and continuous turnover, leaving the organization vulnerable to extreme exposures from an inexperienced workforce.

TABLE 6.5

Strengths and Weaknesses of Autocratic Leaders

Strengths	Weaknesses
• Looks after the safety of equipment and people • Return structure to a poorly managed area • Make decisions quickly • Credited for running a tight ship	• Relies on threats and punishment to influence employees • Will not trust employees • Employee inputs not allowed • Limits creativity of workers • Creates the Theory X employee—lazy, nonmotivated, here for the paycheck employee • An experienced and competent leader is required

TABLE 6.6
Strengths and Weaknesses of Democratic Leaders

Strengths	Weaknesses
• Can maintain high quality and productivity over long periods of time	• Mistakes when they occur can be quite costly
• Employees like the trust they receive and respond with cooperation, team spirit, and high morale	• The decision-making process can be very lengthy as leaders try to get everyone on board
• Often democratic leaders will recognize and encourage achievements	• The lengthy decision-making process adds to the overall cost of doing business
• Generally democratic leaders encourage employees to grow on the job and be promoted	• Opportunities may be lost while leaders engage other stakeholders to gain support for decisions

Democratic Leadership Styles and Behaviors

Democratic leadership style is one that is characterized by intense team building and team decision making. Leaders who practice democratic leadership style and behaviors involve workers and take their time in making decisions. Where safety is concerned, such leaders generate excellent performance because of the shared involvement in decision making. Table 6.6 shows the strength and weaknesses of democratic leadership styles and behaviors.

Today, many developing and developed nations practice democratic leadership styles and behaviors and the health and safety of all workers under such leadership environments continue to improve and to be well looked after. As these nations continue to seek efficiency, however, democratic leadership style has been evolving toward more effective leadership styles and behaviors.

Servant Leadership Styles and Behaviors

Servant leadership style and behaviors is one that requires a uniformly mature workforce. The goal of the leader is to enable the behaviors of followers such that they may all succeed in the workplace. Table 6.7 shows the strength and weaknesses of servant leadership styles and behaviors.

Servant leadership style and behaviors are ideal for a workforce with homogenous maturity. This style is extremely difficult for improving workplace health and safety given the many characteristics of gender, cultural, education, and age differences among workers in the same work group. Servant leadership styles and behaviors must be adjusted to meet the needs of a very varied workforce and it is extremely difficult for servant leaders to adjust to the needs of an immature or mixed workforce.

TABLE 6.7

Strengths and Weaknesses of Servant Leaders

Strengths	Weaknesses
• Promote empowerment and mutual trust • Possess skills of listening, empathy, healing, awareness, persuasion, conceptualization, foresight, stewardship, commitment to the growth of people • Skilled in consensus making, ethical decision making, and conflict resolution	• Cannot perform in volatile work environments • Needs clear goals for performance • Requires strong relationship between leaders and followers

Situational Leadership Styles and Behaviors

Situational leadership style and behaviors were developed and pioneered by Ken Blanchard and Paul Hershey almost 40 years ago. Today, this is one of the leading models for developing frontline leaders and supervisors. Table 6.8 shows the strength and weaknesses of situational leadership styles and behaviors.

Situational leadership behaviors encourage leaders to adjust leadership behaviors consistent with the stage of development and level of maturity of the worker. Situational leadership as prescribed by Blanchard is a model for improving productivity of the workforce. In the same way, the model can enhance the health and safety of the worker because of its approach to directing, supporting, coaching, and delegating to the workforce based on the maturity level of workers. The model prescribes a lot of directing for immature inexperienced workers at the one extreme and delegating to the experienced and mature worker at the other extreme of the continuum of the worker development. This is a very effective leadership model for improving workplace health and safety.

TABLE 6.8

Strengths and Weaknesses of Situational Leaders

Strengths	Weaknesses
• Develop employees to higher levels of maturity • Create a motivated worker • The complexity of the job and the maturity of the worker influence the leadership behavior when dealing with the worker • Can maintain high productivity	• Can misinterpret the maturity state of the employee leading to unintended consequences • Can lead to situations where employee quits and stay when maturity status and leadership behaviors are poorly aligned • Can potentially discriminate against some cultures in a diverse workforce • Leaders may not effectively operate in all stages of worker development

Transformational Leadership Styles and Behaviors

Transformational leadership style and behaviors are dominant in many Western workplaces today. Transformational leaders seek to invoke the inner good of all workers by inspiring their hearts and minds to strive to higher levels of performance and to work safely. Table 6.9 shows the strength and weaknesses of transformational leadership styles and behaviors.

Transformational leaders focus on learning behaviors that are essential for improving workplace performance. They demonstrate behaviors that suggest genuine care for workers. Transformational leaders generate performance across cultural, gender, generational, and geographical divides. While transformational leadership behaviors generate success in many areas, the worker maturity issue is not addressed with transformational behaviors as completely as it is in Blanchard's Situational Leadership® II.

While there are many more leadership styles and behaviors, these are among the more common styles, used in industry today. The challenge for leaders, however, is to determine what works best, given the industry trends identified earlier and discussed. It is safe to say that no single leadership style or sets of behaviors have the solution for today's complex business environment within which we operate. It is our view, however, that leadership success is derived from the following key traits that must ultimately

TABLE 6.9

Strengths and Weaknesses of Transformational Leaders

Strengths	Weaknesses
• Will create an organizational environment that encourages creativity, innovation, proactivity, responsibility, and excellence • Have moral authority derived from trustworthiness, competence, sense of fairness, sincerity of purpose, and personality • Will create a shared vision; promote involvement, consultation, and participation • Lead through periods of challenges, ambiguity, and intense competition or high growth periods • Promote intellectual stimulation • Usually consider individual capabilities of employees • Are willing to take risks and generate and manage change • Lead across cultures and international borders • Build strong teams while focusing on macromanagement • Are charismatic and motivate workers to strong performance	• Leave a void in the organization if followers are not developed to assume role

become learned behaviors for success in today's business and safety environments.

1. Ethical decision making
2. Provide intellectually stimulating and challenging work to workforce
3. Teamwork
4. Training and competency of workers
5. Demonstrated empathy and care for all stakeholders
6. Involvement of all stakeholders and in particular workers
7. Fair treatment to workers
8. Flexibility and adaptability to a multiplicity of worker maturity levels and working environments

Lutchman (2010) discussed the application of situational leadership and transformational leadership behaviors for the complex project management and execution work environment under the banner of situ-transformational leadership behaviors. Perhaps this model can be used for improving workplace safety also. You be the judge.

Situ-Transformational Leadership Behaviors: Application in Safety Management

Leadership in safety requires leaders to inspire workers to demonstrate safe work behaviors at all times. For example, before work is performed, to ensure that this work can be completed safely, workers automatically perform the following tasks:

1. A field-level RA (FLRA), whereby all workers assess the risks associated with the task and apply risk-mitigating strategies to remove the risk or reduce all risks to as low as reasonably practicable, before doing the work. All workers must do this regardless of the worker maturity or the complexity or risks associated with the job.
2. Conduct prejob meetings to ensure all workers involved in the assigned task are aware of all the associated risks and the criticality of his/her role in ensuring the safety of the job.
3. Be prepared to stop all work if new hazards are identified while the work is taking place.
4. Look after the health and safety of peers and coworker during the execution of any work.

According to Lutchman (2008), "leadership is about creating an organizational environment that encourages worker creativity, innovation, pro-activity, responsibility, and excellence." Lutchman further advised: "Leaders must create a compelling sense of direction for followers and motivate them to performance levels that will not generally occur in the absence of the leader's influence." According to Blanchard (2008), one of today's premier experts on leadership, there is no single best leadership style leading any workforce and this is absolutely true for leading workers in safety.

Lutchman (2010) discussed the application of the principles of the situational leadership and transformational leadership behaviors working together for delivering superior performance in a project execution environment. By leveraging the strengths of both these approaches to leadership reflected in the right level of worker maturity (training, competency, experience, and commitment) combined with a shared vision in safety, the right motivation and inspirational leadership, strong performance in safety management is possible. The attributes of situ-transformational leadership behaviors generate best performance when all leaders starting at the frontline understand the principles of this model and can use these behaviors when it is appropriate to do so.

The Frontline Leader

The first point of contact between workers and the rest of the organization is through frontline leaders and supervisors. The way frontline leaders and supervisors treat workers will influence workers' motivation, commitment, performance, productivity, and loyalty. Commitment to health and safety is critical at this point since the behaviors of the frontline supervisors will be reflected by the workforce. If frontline supervisors fail to wear proper PPE or take shortcuts, followers will emulate these behaviors and safety performance shall suffer.

The onus is on employers, therefore, to properly equip frontline supervisors and leaders with the skills and capabilities to positively influence the behaviors of workers for workplace safety. Frontline leaders must be equipped with leadership skills and behaviors that create a more satisfying work environment for all workers. This in order that they may do their jobs safely at high levels of performance and organizations can retain these workers for longer periods and maximize value-driven performance during their employment tenure.

Worker satisfaction is a leading indicator of employee turnover (Allen et al., 1999; Cooper-Hakim and Viswesvaran, 2005; Price et al., 2007; Slattery and Selvarajan, 2005; Trevor, 2001). Furthermore, research has linked leadership to worker satisfaction and performance (Madlock, 2008; Sharbrough et al., 2006). When workers are provided with a more satisfying work environment, they are less likely to leave, they are more motivated, and they produce more (Bass, 1990; Hoffman, 2007; Owens, 2006). Safe work habits are essential for ensuring that schedules are met and cost is managed. Safety incidents result in downtime, lost productivity, and reduced morale and must be avoided at all times. As a consequence, the leadership behaviors demonstrated by our frontline leaders can significantly influence workplace health and safety.

While there are many variables for creating a safe and healthy work environment, leadership is one variable that can be easily managed through training and development so that leaders can positively influence the behaviors of followers. Leaders who can create a safer work environment for followers can ultimately help in motivating workers to higher levels of productivity, reducing worker turnover, and contributing to the competitive advantage of the organization. Leadership is highly influential in determining the quality of the immediate work environment in which workers perform and is therefore critical in determining workplace safety and the level of job satisfaction derived by workers (Price et al., 2007).

All organizations are generally characterized by a workforce of varying skills and capabilities during the project execution stage. Therefore, leaders must be able to adapt leadership behaviors to meet the requirements of the individual worker that makes up the workforce. Ken Blanchard's Situational Leadership II provides an effective model for leading and developing workers through the various stages of worker maturity. Blanchard (2004) pointed out that when employers invested time and leadership to improve the maturity of workers, such employers experienced strong gains and productivity improvements. Table 6.10 shows the various stages of a worker's development and leadership focus during each stage of the worker's development.

According to Blanchard (2008), Situational Leadership II requires varying levels of directing and supporting behaviors from leaders based on the maturity of the worker. Blanchard (2008) suggests that leaders apply four sets of behaviors relative to worker maturity. Figure 6.4 demonstrates the leadership behavior requirements of each stage of the worker development to maximize satisfaction and worker productivity and workplace health and safety. Strong situational leadership skills and transformational leadership behaviors are required for leading any frontline workforce. Figure 6.4 shows the application of the principles of situational leadership and transformational leadership behaviors, designed to improve the skills, capabilities, and motivation of the worker (situ-transformational leadership behaviors). Figure 6.4 also shows the transformational leadership behaviors required by supervisors at each stage of the worker development. These leadership behaviors help in developing the worker from incompetent and noncommittal through to competent and fully committed. When applied to health and safety, these behaviors serve to transform works from inexperienced, unaware, and uncaring of the health and safety hazards in the workplace to one that is committed to protecting the health and safety of all workers. With the application of the situ-transformational leadership,

TABLE 6.10

Leadership Behaviors with Maturity Status of Worker

Leadership Behaviors	Directing/Supporting Relationship	Worker Maturity
Directing (Stage 1)	High directing/ low supporting	Immature—low competence and commitment—leadership focus on: 1. Telling the worker where, when, and how to do assigned work 2. Key requirements of structure, decision-making control, and supervision 3. Primarily one-way communication
Supporting (Stage 2)	High directing/ high supporting	Immature—growing competence; weak commitment—leadership focus on: 1. Building confidence and willingness to do assigned work 2. Retains decision making 3. Promotes two-way communications and discussions
Coaching (Stage 3)	High supporting/ low directing	Mature—competent; variable commitment—leadership focus on: 1. Building confidence and motivation; promote involvement 2. Allows day-to-day decision making 3. Active listening and two-way communications and discussions
Delegating (Stage 4)	Low supporting/ low directing	Mature—strong competence; strong commitment—leadership focus on: 1. Promote autonomy and decision making and empowerment 2. Collaborates on goal setting 3. Delegates responsibilities

Source: Adapted from Blanchard, K. 2008. *Leadership Excellence,* 25(5), 19.

workers are able to exercise and prioritize leadership behaviors that lead to protection of life, protection of the environment, and protection of assets, in that order, with great commitment to overall welfare of the organization and society as a whole.

Ultimately, the goal of any leader is to create a low-maintenance worker where health and safety, and productivity are concerned. The principles of situational leadership coupled with transformational leadership behaviors help in doing so.

Successful transition of workers from high maintenance to low maintenance requires that frontline leaders be adequately trained to lead the frontline workforce. Such training must include training in both situational leadership principles and transformational leadership behaviors.

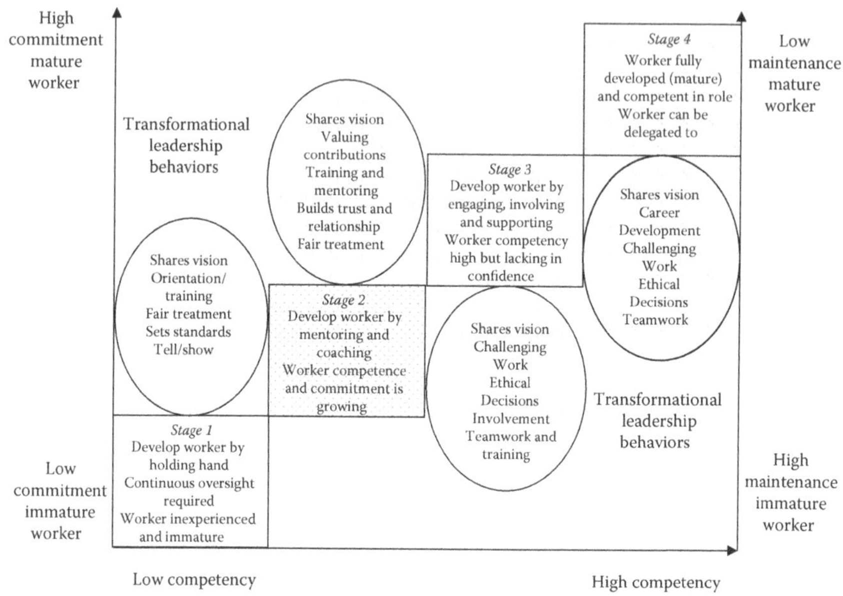

FIGURE 6.4
Situ-transformational leadership. (Adapted from Lutchman, C. 2010. *Project Execution: A Practical Approach to Industrial and Commercial Project Management*. Boca Raton, FL: CRC Press.)

It is noteworthy that the leadership behaviors and traits of the frontline leaders are the outcome of the training provided to them and the culture of the organization. Simply put, the organizational culture is a reflection of the way we do business and frontline leaders are required to develop and encourage behaviors consistent with the values demonstrated by the organization. Training provided by the organization is intended to support these values. Figure 6.5 demonstrates the relationship between the organizational culture, organizational values, and leadership traits.

When managing work, leaders must seek to support the organizational values by demonstrating leadership behaviors consistent with these values. The organizational value of *Safety First* requires demonstrated leadership behaviors of managing risk, setting safety standards, and building talent and capability. For the inexperienced leader, developing these skills and behaviors can be very difficult. Training helps in developing the leadership skills of frontline leaders and must be a high priority for all organization.

Success in best in class safety performance requires frontline leaders to demonstrate immediate behaviors that focus on managing risk, setting standards, and building talent and capabilities. Other essential behaviors for achieving the organizational goals of best in class safety performance include behaviors reflective of the other values of the organization. These may include frontline leaders being proactive, building trust and relationships,

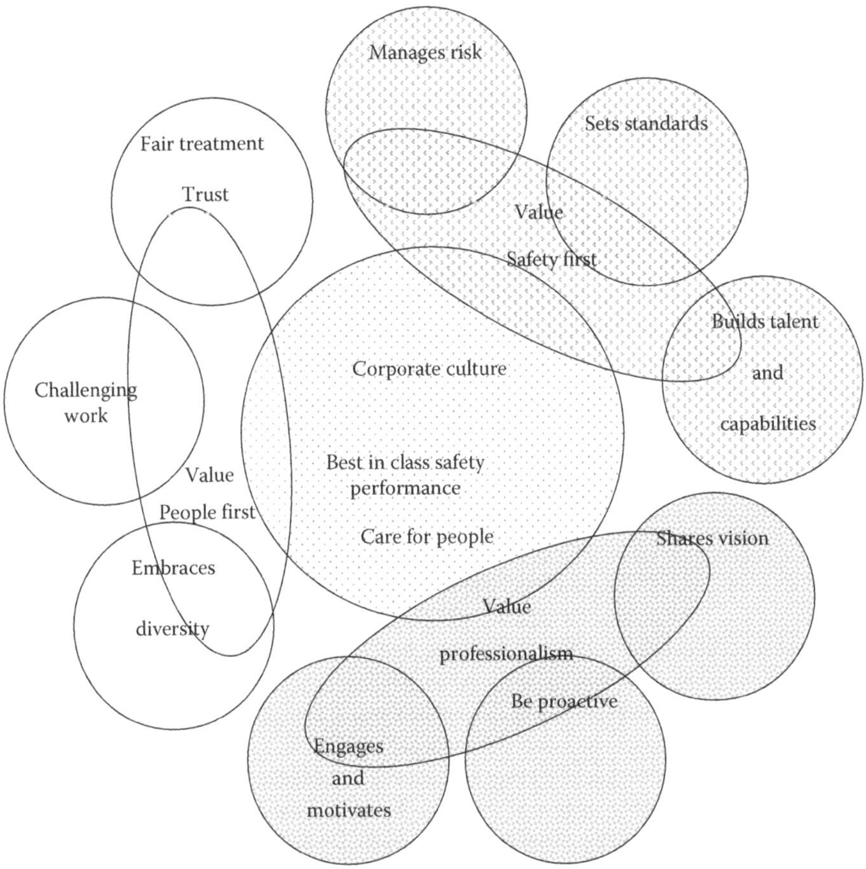

FIGURE 6.5
Corporate culture, values, and leadership traits. (Adapted from Lutchman, C. 2010. *Project Execution: A Practical Approach to Industrial and Commercial Project Management*. Boca Raton, FL: CRC Press.)

engaging and motivating the worker, developing self and workers, and the ability to leverage existing tools of the organization. At the frontline, these behaviors must be reflected in a way that allows frontline workers to understand and respond to the frontline leader.

Frontline supervisors/leaders are the custodians of health and safety in the workplace. They proactively address the needs of the workforce to develop credibility and gain trust from workers. As a consequence, organizations have a responsibility for preparing frontline supervisors/leaders so that they may proactively respond to the health and safety needs of all workers. The principles of situ-transformational leadership allows frontline leaders to recognize the maturity state of the worker and to respond to worker needs with transformational leadership behaviors as shown in Figure 6.4.

Senior Leadership

Senior leadership must demonstrate primarily transformational leadership behaviors. According to Bass transformational leaders encourage followers to "transcend their own self interest for the good of the group, organization or society; to consider their long term needs, to develop themselves rather than their needs of the moment; and to become more aware of what is really important" (Bass, 1990, p. 53). Transformational leaders focus on developing the workforce and creating a work environment where workers feel a sense of belonging, are treated fairly, are motivated, and are provided intellectually stimulating and challenging work. Transformational leaders create strong teams by leveraging the abilities of experienced workers while developing less experienced workers. Lutchman (2008) advised: the "paradigm of the transformational leader is to motivate workers to do more than is expected of them by leveraging their creative excellence."

According to Lutchman (2008), "such leaders will create an organizational environment that encourages creativity, innovation, pro-activity, responsibility, and excellence." They often possess moral authority derived from trustworthiness, competence, sense of fairness, sincerity of purpose, and personality. Transformational leaders are also trustworthy and ethical in decision making. Trust is an earned entity by leaders resulting from consistent demonstrated behaviors of doing the right thing at all times. In addition, transformational leaders have the unique ability to communicate and share the organizational vision. They promote involvement, consultation, and participation. Transformational leaders demonstrate high levels of emotional and cultural intelligence and will successfully lead in volatile, fast-paced, ambiguous work environments that are characteristic of project execution environments.

Transformational leaders have been credited with the ability to promote teamwork and develop strong teams. Research has shown that workers led by transformational leaders exhibit high job satisfaction (Korkmaz, 2007). Hoffman (2007) established that leadership influences employee satisfaction and ultimately employee retention and loyalty. The health and safety standards of a workplace depend almost entirely on teamwork. Failure of any workgroup to maintain high standards of safety ultimately affects the entire organization when incidents result in fatalities, environmental damage, and cause damage to infrastructure or facilities. Leaders must foster and support teamwork among all workers to ensure a safe and healthy workplace.

Motivating Employees

Wren (1994) discusses the early pioneering work of Elton Mayo on how to motivate workers. Mayo pointed out that wages and job characteristics are strong motivators for workers. Fair and competitive wages during project execution motivate workers to remain employed. However, job characteristics are more important in determining the level of productivity of workers during project execution. Challenging and intellectually stimulating work encourages workers to be committed and to produce more. A safe and healthy workplace leads to a motivated workforce. Workers feel safe and this translates into higher motivation and productivity. Ramlall (2004) suggested that motivation affects productivity and employee turnover behaviors.

Gentry (2006) suggested that leadership behaviors can influence undesirable worker behaviors such as absenteeism, tardiness, and turnover. Indeed, this can also be extended to workers attitudes toward the use of PPE, operating procedures, safe work practices, and shortcuts while doing work. According to Gentry (2006), leaders may consider behaviors that reflect a fair treatment of all workers, genuine empathy, teamwork, and involvement in workplace-related decisions to address some of these undesirable worker behaviors. Leaders should, therefore, communicate with followers in a manner that builds trust within the workforce. *Saying what you will do and doing what you said is so very important in building trust.* Workers are motivated to emulate the behaviors of leaders who make ethical and trustworthy decisions aimed at ensuring health and safety in the workplace.

According to the The Ken Blanchard Group of Companies (2010), trust in the workplace is generated from the application of the ABCD model of behaviors as described below.

Able—demonstrate competence:

- Produce results
- Make things happen
- Know the organization/set people up for success

Believable—act with integrity… be credible:

- Be honest in dealing with people/be fair/equitable/consistent/respectful
- Values-driven behavior "reassures employees that they can rely on their leaders" (The Ken Blanchard Group of Companies, 2010, p. 2)

Connected—demonstrate genuine care and empathy for people:

- Understand and act on worker needs/listen/share information/be a real person
- When leaders share a little bit about themselves, it makes them approachable

Dependable—follow through on commitments:

- Say what you will do and do what you say you will
- Be responsive to the needs of others
- Being organized reassures followers

According to the The Ken Blanchard Group of Companies (2010), as leaders, you have a choice in becoming a trustworthy leader as shown in Table 6.11.

The Ken Blanchard Group of Companies (2010) summarized that leaders can successfully develop organizational trust in the following ways:

1. *Demonstrate trust in your people*—"If you want to create a trusting work environment, you have to begin by demonstrating trust" (The Ken Blanchard Group of Companies, 2010, p. 4).
2. *Share information*—"Information is power. One of the best ways to build a sense of trust in people is by sharing information" (Ken Blanchard Group of Companies, 2010, p. 4).
3. *Tell it straight/never lie*—"Study after study show that the number one quality that people want in a leader is integrity" (Ken Blanchard Group of Companies, 2010, p. 4).
4. *Create a win/win environment*—Creating competition among workers leads to loss of trust among all.
5. *Provide feedback*—Hold regular progress meetings with direct reports. Check-in and provide feedback on performance, in particular, in a timely manner to avoid surprises to workers later on.
6. *Resolve concerns head on*—Engage workers in finding solutions.
7. *Admit mistakes*—"An apology can be an effective way to correct a mistake and restore the trust needed for a good relationship" (Ken Blanchard Group of Companies, 2010, p. 5). Some cultures have difficulties admitting mistakes. However, when mistakes are admitted, it makes the leader human and promotes greater team bonding and trust.
8. *Walk the talk*—"Be a walking example of the vision and values of the organization" (Ken Blanchard Group of Companies, 2010, p. 5). If the leader believes in it then so can I and so will I.
9. *Timely recognition of positive behaviors*—Recognizing and rewarding positive behaviors in a timely manner. Such recognition must be specific and relevant. Choosing the right environment for doing so is also very important since some cultures may require pomp and show while others may seek conservatism.

Trust in leadership and the organization is a key outcome from the demonstrated actions of the leader and employer. Communication methods, channels, and behaviors adopted by leaders influence the levels of trust workers

TABLE 6.11

Leadership Behaviors That Build or Erode Trust

Erode Trust	Build Trust
• Lack of communication	• Giving credit
• Being dishonest	• Listening
• Breaking confidentiality	• Setting clear goals
• Taking credit for others' work	• Being honest
	• Following through on commitments
	• Care for your people

will place with employers (Hemdi and Nasurdin, 2006; Hopkins and Weathington, 2006). Trust is high when employers demonstrate genuine concern and care for the health and safety of workers, and empathy for workers (Hemdi and Nasurdin, 2006). Trust in leadership is earned when leaders demonstrate competence, EI, integrity, and ethics in decision making (Davis et al., 2000). Where the safety of personnel depends on the decisions made by leaders, such leaders must earn the trust of all workers. Care for people and cultural intelligence in a multicultural and diverse project execution environment helps in maintaining trust in leadership.

Employee motivation can be high when leaders focus on the health and safety of workers, are trustworthy, and promote teamwork; workers are treated fairly and individually; and leaders act with empathy when communicating with workers. Leaders must say what they intend to do and do what they promised to do. Ramlall (2004) identified the following factors that leaders can address to influence worker motivation in the workplace:

1. The organizational health and safety behavior and performance. Unsafe work environments promote worker flight and reduce morale and motivation.
2. The personal needs of the employee—may include but is not limited to training, PPE, or simple advice on a personal issue. Leaders must be able to recognize the body language of workers who may have unresolved personal issues and respond to them with the empathy required.
3. The work environment characteristics—generally, commercial and industrial project environments are characterized by difficult working conditions. Unpaved roads, extreme temperatures, outdoor work, and supplemental lighting can all influence the productivity and motivation of workers.
4. The responsibilities and duties of the worker—when placed in supervisory roles, worker motivation can be affected either positively or negatively. If unprepared for the role, negative consequences

may be generated. If equipped to do the job and if seen as a promotion, workers may respond positively.

5. The level of supervision provided to the worker—immature workers will necessarily require more supervision and guidance relative to experienced workers who may require less supervision. Telling and micromanagement can lead to situations where workers *quit and stay* when they should be delegated to.
6. The extent of worker effort required to perform assigned tasks—labor intensiveness versus skills intensiveness influences worker motivation.
7. The employee's perception of organizational fairness and equity—perceived unfair and biased treatment is a precursor to worker turnover and level of effort generated from workers. Workers tend to equate work with pay and when faced with possible unfair treatment will find creative means to bring about equity. Shirking on the job, absenteeism, late to work, and poor performance are methods used to level the playing field.
8. Career development and advancement opportunities—most workers have an unspoken personal development plan for the workplace. They will be motivated to perform at high levels if career development opportunities are aligned with their personal plans.

According to Ramlall (2004), employers should seek to respond with empathy to the personal needs and values of the employee. They should also seek to create work environments that are safe, respectful, inclusive, and productive (Ramlall, 2004; Wren 1994).

Worker Training and Development

With a growing population of Generation Y workers in the workplace, training and development are continually being elevated to the highest priority by workers. Workplace training can result in greater "job satisfaction, organizational commitment, and turnover cognition" (Owens, 2006, p. 166). Satisfaction and commitment will ultimately translate into higher organizational performance (Price et al., 2007). Little (2006) found that intent to turnover was lower in organizations that looked after the training needs of workers relative to those that did not. Little (2006) in a research study found 41% intent to turnover within a year in organizations with poor focus on training relative to 12% where training was made a priority by employers. When workers feel they are unqualified for a job and they are unable to learn quickly enough in the role to undertake the task, they will leave. More importantly, when health and safety training needs of workers are ignored, worker turnover is an absolute. When turnover does not occur, increasing numbers of incidents can lead to all sorts of disastrous consequences.

According to Finegold et al. (2002), when training permitted the development of new skills, satisfaction and loyalty were higher among younger and inexperienced technical workers relative to older workers. Finegold et al. (2002) found that across all age groups, satisfaction with continuous skills development showed a strong relationship with organizational commitment and loyalty. Clearly, therefore, emphasis on training and development (inclusive of health and safety training) for all workers provides opportunities for organizations to increase not only the technical competence of workers but also their commitment and loyalty to the employer. Training for all personnel serves to reduce turnover and is an absolute requirement for a safe and healthy workplace. Leaders must recognize trained workers as an asset and not view training as an expense. Training necessarily needs analysis to be conducted for all employees and proactive responses to closing these gaps are required.

In our experience, when workers were trained properly, their confidence levels were increased in the tasks they were assigned to perform. Training not only enabled skills and capabilities but also motivated workers to perform assigned tasks quickly, safely, and efficiently. Training also creates a line of sight between the current state of the worker and future career development roles leading to high motivation to perform in the workplace.

Involvement and Participation in Workplace Decisions

Collaboration and involvement in workplace decisions influences worker loyalty and employee retention (Agrusa and Lema, 2007). Leaders are continually seeking ways and means to improve performance and the way we do work. Involving experienced workers in workplace decisions is critical in the continuous improvement process. According to RoSPA (2006), involving workers in organizational safety can result in improved organizational safety performance. Collaboration and involvement promotes ownership and buy-in, into work activities, and a greater willingness by workers to get the job done right. As a consequence, leaders who involve workers in workplace decisions and carefully consider inputs from their followers will likely see more support for new activities in the workplace.

Involvement does not necessarily mean accepting the input from followers; instead, leaders should give consideration to input with follow-up communication on the justification for exclusion or inclusion of such inputs. Workers are extremely appreciative of feedback and information that explains why their input may not have been accepted and will often lend support for the accepted decisions once properly communicated with. While leaders may find it difficult to allocate time to consider the input of all participants and to provide feedback where input is not considered, the benefits of doing so in building trust, commitment, and motivation are high and should not go untapped.

Research has shown that employee involvement in decision making at appropriate levels influences turnover decisions by workers (Guthrie, 2001; Wilson et al., 1990). According to Riordan et al. (2005), participation in workspace (immediate workplace) decisions led workers to feel more valued by their employers. Riordan et al. (2005) found that worker perception of involvement in decision making influenced organizational effectiveness. According to Lockwood (2007), organizations that involved workers in business decisions benefited from a more diligent, loyal, and motivated workforce. Lockwood (2007) also found that when employers failed to involve workers in business decisions, such employers were more likely to experience higher employee turnover and lower productivity. Involving workers in workplace decisions is a step in the right direction for increasing worker motivation, productivity, loyalty, and safety.

Teamwork

Teamwork (peer and leader support) is a leading contributor to worker loyalty (Pisarski et al., 2006; Thompson, 2002). Teamwork that promotes coworker encouragement, support, advice, approachability, listening skills, and EI led to more satisfying work environments for workers (Pisarski et al. 2006). Teamwork and team building depend on collaboration and sharing among workers (Silén-Lipponen et al., 2004; Rosen and Callaly, 2005). Teamwork and a sense of belonging by workers where worker input is considered with open-mindedness can enhance worker performance and loyalty (Glisson and James, 2002; Pisarski et al., 2006). Transformational leadership behaviors support teamwork and the creation of strong teams.

Harris et al. (2005) found that poor-quality leadership support led to poor job satisfaction and commitment among workers. Leadership support positively influences the workspace of workers and their productivity and should be an important consideration for all leaders when trying to enhance health and safety performance. According to Smither (2003), when workers are treated individually (leaders avoided lumping into a collective group) workers were more motivated and were less likely to leave an employer. Knowing your workers by first names, taking time away from the office to be visible and available to workers, and responding to their needs with genuine interest are strong motivators for all employees regardless of their level in the organization. Smither (2003) recommended responding to the needs of every worker individually with genuine empathy to each worker's personal and career needs. Genuine empathy and honest and open communication free from repercussions serve to motivate workers and contributes to a trusting transformational work culture where workers want to stay and perform (Ramlall, 2004; Smither, 2003). Workers will leave when they feel they cannot communicate with their supervisors or leaders and their opinions are not valued (Thornton, 2001).

Leaders must seek to create a safe and satisfying work environment for all workers (Hopkins and Weathington, 2006; Price et al., 2007). Creating such a work environment depends on the leader's ability to provide empowerment and challenging work, promote involvement and participation among workers, provide equal access to training opportunities, promote teamwork and team-building, and gain the trust of workers (Desselle, 2005; Price et al., 2007).

Critical Thinking Ability

A leader's ability to critically analyze issues and to eliminate personal biases adds to organizational success during project execution. Critical thinking skills are learned behaviors and all senior leaders in a project execution work environment must be able to think critically. The ability to probe deeply and to understand root causes of issues both contribute to overall success during project execution.

Facione (2006) pointed out that the characteristics of a critical thinker include inquisitiveness, a desire to be well informed, self-confidence, open-mindedness, trustful of reason, flexible, fair-minded, honest about personal biases and prejudices, open to opinions of others, and not afraid to change course when reflection suggests its necessary to do so. Harris (1998) suggested that leaders must be competent in recognizing and removing obstacles to critical thought that may impair the conclusion and results of work, if not addressed in the decision-making process. According to Harris (1998), obstacles may include behaviors such as prejudices and stereotyping, defensiveness, language interpretation difficulties, emotions fixation, channel visioning, learned helplessness, and psychological blocks. Critical thought promotes creativity in the workplace and is absolutely desirous during project execution.

Michalko advised: "creativity deviates from past experiences and procedures" (Michalko, 2000, p. 18) thereby creating room *for thinking outside the box*. How can we enhance our capability to think critically? Leaders who develop habits that recognizes and addresses prejudices, stereotyping, and logical fallacies enhances their critical thinking capability and skills development. Leaders should be taught skills to improve critical thinking skills for maximum health and safety performance.

Critical thinking is the *art* of forming the right conclusions from a body of information provided on any particular subject. The specific application of the word *art* suggests that critical thinking is a learned behavior that encourages the critical thinker to clearly navigate through information with cognizance to avoid errors of fallacies, emotions, language influences, and other obstacles to critical thinking in developing conclusions from information. Leaders can inadvertently develop sacred cows and construct empires when critical thinking skills are weak.

It must be noted that developing critical thinking skills starts with an acknowledgment of one's own self-limitations. There must also be a genuine

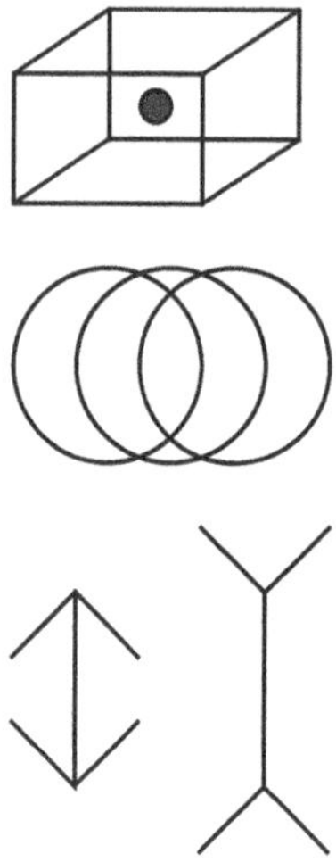

FIGURE 6.6
Critical thinking and perceptions. (Adapted from Results Based Interactions, 2004. *Development Dimensions International, Inc., MCMXCIX.* Pittsburgh, Pennsylvania.)

desire to improve capabilities in critical thought processes. Leaders should be enlightened about the benefits of critical thinking. It is in response to perceived benefits of the critical thinking process that people are motivated to aspire to be better critical thinkers. The learning process starts with basic training in recognizing obstacles that get in the way of the right decision or conclusion. Figure 6.6 provides a simple exercise to demonstrate the impact of perception on decision making. The examples provided demonstrate that some situations may have different answers based on individual perspective as in the case of the dot on the face of the cube and the circular rings. The parallel lines example highlights the importance that things may not always be what they appear to be.

7

The Safety Challenge: Why Is Organizational Safety Important?

Is organizational safety important? Is it all about profitability? How important is corporate social responsibility? In this chapter, we hope to address some of these concerns and provide a better understanding as to why organizations must continue to spend money and allocate resources to continually improve their safety performance. The safety performance of an organization has a huge impact on the image of the organization, and in today's business environment can make the difference between continuing to be in business and bankruptcy.

As we all know, the rate at which information travels today is phenomenal and often without control. The quality and types of information transmitted through the various media, Internet, and information technology channels have a profound impact on organizational image and successes. On March 12, 2009, within minutes of a helicopter ditching event off the East Coast of Canada, several blogs were already established to provide a real-time account of the incident that killed 17 workers and crew members with only one survivor (CBC News, 2009) from the event. Indeed, several of the blogs and information suited conveyed different and often misleading information describing the same event.

Furthermore, the cost of incidents to organizations can be staggering and in many instances can lead to cessation of business activities. Some recent examples that come to mind are the 2005 Texas City BP incident and the 2010 BP Gulf of Mexico spill. Both these incidents cost the organization billions of dollars. CBC News (2010) advised that the latest estimated cost of the Gulf of Mexico spill was approximately US$40 billion and the ultimate cost of the incident was not yet known. For many organizations, recovery from an incident of such magnitude will be impossible.

Table 7.1 highlights the 20 top losses from workplace incidents in U.S. dollar value. The cumulative loss to the global nation was estimated at $13.65 billion dollars. These losses do not include loss associated with injuries and fatalities nor do they include loss of production resulting from accompanying outages and repairs to assets.

TABLE 7.1

Global 20 Largest Losses

Date	Type of Operation	Type of Event	Location	Country	Cost in US$[a]
07/07/1988	Upstream	Fire/explosion	North Sea	UK	1600
23/10/1989	Petrochem	Vapor cloud explosion	Texas	USA	1300
19/03/1989	Upstream	Fire/explosion	Gulf of Mexico	USA	750
12/09/2008	Refinery	Hurricane	Texas	USA	750
04/06/2009	Upstream	Collision	North Sea	Norway	750
23/08/1991	Upstream	Structural failure	North Sea	Norway	720
15/05/2001	Upstream	Explosion/fire/sinking	Campos Basin	Brazil	710
25/09/1998	Gas processing	Vapor cloud explosion	Victoria	Australia	680
15/04/2003	Upstream	Riot	Escravos	Nigeria	650
24/04/1988	Upstream	Fire	Campos Basin	Brazil	640
21/09/2001	Petrochem	Explosion	Toulouse	France	610
25/06/2000	Refinery	Vapor cloud explosion	Mina Al-Ahmadi	Kuwait	600
04/05/1988	Petrochem	Explosion	Nevada	USA	580
19/01/2004	Gas processing	Fire/explosion	Skikda	Algeria	580
05/05/1988	Refinery	Vapor cloud explosion	Louisiana	USA	560
01/11/1992	Upstream	Mechanical damage	North West Shelf	Australia	470
14/11/1987	Petrochem	Vapor cloud explosion	Texas	USA	430
25/12/1997	Gas processing	Fire/explosion	Sarawak	Malaysia	430
27/07/2005	Upstream	Fire/explosion	Mumbai High field	India	430
20/01/1989	Upstream	Blowout	North Sea	Norway	410
					13,650

Source: Marsh Energy Practice 2010.

[a] Cost inflated to December 2009 dollars.

Marsh Energy Practice (2010) also provided cost estimates for the top 100 hydrocarbon-processing incidents during the period 1972–2009, which amounted to US$30 billion inflated to 2009 dollars.

While capital losses continue to be of concern from workplace incidents, the more alarming concern is in the area of workplace injuries and diseases. Applying its workplace safety index (WSI) to the U.S. labor force, Liberty Mutual (2009) suggested that "the most disabling workplace injuries and illnesses in 2007 amounted to more than $52 billion dollars in direct U.S. workers compensation costs, averaging more than a billion dollars per week." According to the same source, 2007 cost estimates represented almost 9% increase from the previous year. Figure 7.1 shows the historical WSI injuries and illnesses cost estimates for the United States.

A correlation analysis between gross domestic product (GDP) growth and cost associated with workplace injuries and illnesses generated a correlation coefficient of –0.77. This suggests a strong inverse relationship between economic growth and injury and illnesses costs. What can we infer from this relationship? In the absence of a detailed scientific analysis, we offer the following plausible reasons for this relationship:

1. When economic growth starts to decline, the organizational emphasis is on producing more with fewer resources—workers are asked

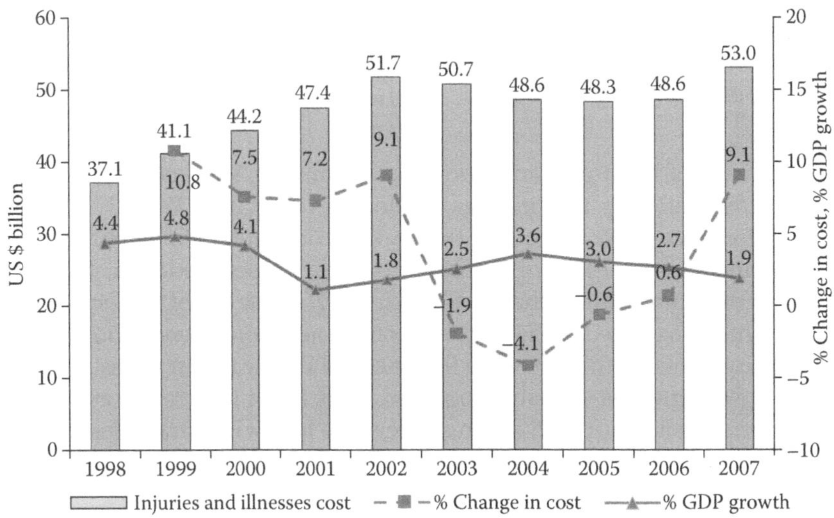

FIGURE 7.1
Direct workplace injuries and illnesses cost and % GDP growth—United States. (Adapted from Liberty Mutual. 2009. *The Most Disabling Workplace Injuries Cost Industry an Estimated $53 Billion*. Liberty Mutual Research Institute for Safety 2009. Retrieved November 13, 2010 from http://www.google.ca/search?hl=en&source=hp&q=Liberty+Mutual+Workplace&btn=Google+Search&meta=&aq=f&aqi=&aql=&oq=&gs_rfai=)

to do more during the same period, which generally results in an escalation of injuries.

2. During periods of economic decline, budget cuts generally result in spending cuts in support services, particularly in areas such as training and development and safety management.
3. Economic downturn generally leads to the suspension of high-cost consultants and experts associated with development work. The remaining expertise is often less experienced and not as competent as the high-priced expertise. This situation contributes further to errors, mistakes, and incidents.
4. Strategic decisions may result in extending the run periods of equipment and machinery before preventative maintenance is done thereby increasing the vulnerability of both the equipment and personnel operating these equipment.
5. During economic downturns, the accompanying job cuts and budget contractions impose undue stress on workers who are unsure about their future. This results in minds being away from the tasks while work is being performed and a higher chance of workers being injured during such periods. In addition, workers tend to do more to ensure that they remain employed with possible risk exposures from overexertion.
6. During economic downturns, contract work tends to be the first type of work that is reduced. This often results in fierce price competition, which ultimately translates into shortcuts and reduced spending on safety management.

While each of the arguments above are defensible, we would like to draw particular attention to the findings of the Liberty Mutual WSI (2009), which highlighted overexertion as one of the leading contributors to the direct cost of injuries and illnesses. Figure 7.2 shows the costs associated with the top 10 causes of disabling injuries that occurred in 2007. At the top of the list is overexertion, which accounted for 24% of the costs. In addition, overexertion has consistently remained at the top of the list since the WSI was first released in 2000.

Let us examine this snapshot together. In its quest for greater efficiencies, organizations have pursued among others the following strategies:

1. Just-in-time technology
2. Driven decision making down to the lowest acceptable level and promoted flatter organization structures
3. Lean organizations
4. Capital-intensive technologies that require expert capabilities and competence to maintain continuous operations

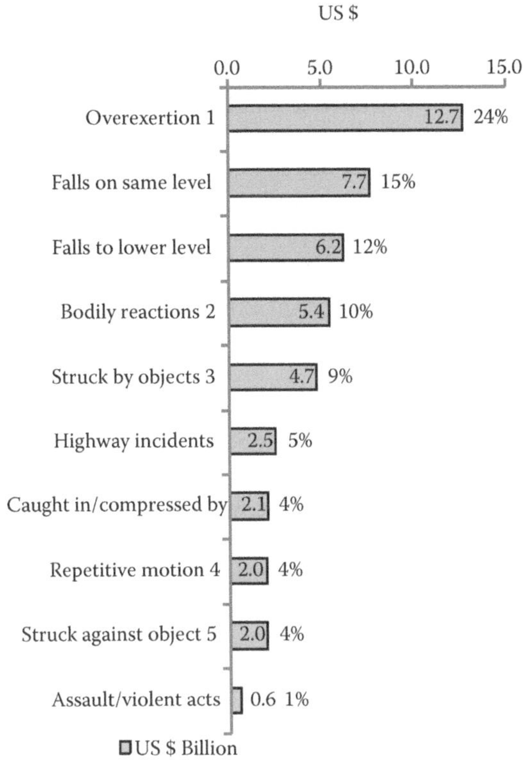

FIGURE 7.2
Top 10 injuries and illnesses cost and % of cost—United States, 2007. 1—Injuries caused from excessive lifting, pushing pulling, holding carrying, or throwing. 2—Injuries caused from slipping or tripping without falling. 3—Such as a tool falling on a worker from above. 4—Injuries due to repeated stress or strain. 5—Such as a worker walking into a door. (Reproduced with permission from the Liberty Mutual Research Institute for Safety. Research to Reality Quarterly Update of the Liberty Mutual Research Institute for Safety, Vol. 12, No. 3, 2009, pp. 10–11.)

5. Continuous 24/7 operations
6. Increasing scales of operation

More importantly, the mantra of businesses today is *cheaper ... better ... faster*. The authors of this book argue that while organizations pursue these illustrious goals, the focus generally gravitates toward workers doing more, which often results in the related overexertion concern. Other incidents can generally be attributed to the speed of doing work and the use of unproven techniques and less rigorous risk analyses to get work completed faster. When compounded with organizational responses to weaker economic growth, workers are exposed to an even greater burden of needing to do even more, which leads to further overexertion and increasing injuries and their related costs. Furthermore, when safety and training budgets are reduced during these periods, the problem is worsened leading to further increasing incidents and injuries and their related costs.

While our discussions so far have been limited to the direct worker compensation cost associated with workplace injuries and illnesses, when health care spending, overtime and lost production, and productivity are factored in, the real cost of workplace injuries and illnesses can be astronomical. Leigh et al. (2000), in an examination of 1992 injuries and illnesses cost estimates, concluded the following:

1. The total direct and indirect costs associated with these injuries and illnesses were estimated to be $155.5 billion.
2. Direct costs included medical expenses for hospitals, physicians, and drugs, as well as health insurance administration costs, and were estimated to be $51.8 billion.
3. The indirect costs included loss of wages, costs of fringe benefits, and loss of home production (e.g., child care provided by parent and home repairs), as well as employer retraining and workplace disruption costs, and were estimated to be $103.7 billion.
4. Injuries were responsible for roughly 85% whereas diseases accounted for 15% of all costs.

Some sources indicate the annual cost of treating workplace injuries and illnesses in the United States alone is estimated at ~US$170 billion. According to Medical Benefits (2002), work done by Liberty Mutual in 1999 on injuries and illnesses cost estimates suggested US$40 billion of direct costs with indirect cost ranging between US$80 billion and $200 billion, which ultimately results in a total injuries and illness cost impact in the range of US$120 billion to $240 billion. Pennachio (2009) pointed to cost estimate of $170 billion based on the information provided by the American Society of Safety Engineers. We can conclude that employers and organizations must be concerned about organizational safety because of the cost impact to the bottom line and the cumulative effect on economies.

More importantly, there is a desperate need to change the mindset of leaders to recognize that during periods of economic downturns, the strategies of choice, which often focus on budget cuts in the areas of safety and training, are precisely those that must be avoided to reduce operating and capital replacement cost. A trained and competent workforce at all times and a well-maintained and active safety management system save lives, reduce incidents, prevent equipment failures and downtime, and ultimately lead to sustained improvements in operating cost, productivity, and profitability.

Great Safety Performance Equals Great Business Performance

The contributions of a great personal and process safety management are often not quantified during normal business operations. For many organizational leaders, money spent in maintaining a safety program is often looked upon as an expense with no real value contributions to the immediate cash inflows of the organization. There is nothing further from the truth than these beliefs. While value contributions from investments in personnel and process safety management are often difficult to identify and quantify, it can be done in a very credible manner that can influence the beliefs and value systems of doubting senior leaders.

Expenditures in personnel and process safety management can be rationalized in a manner similar to which investments in other production processes are rationalized and approved. On the assumption of a limited pie for sharing across multiple stakeholders, seeking investment capital, leaders in personal and process safety management must make similar representation for a share of this pie based on credible, justifiable projects, and safety expenditures. A key measure for determining the income associated with safety-related investments is a realistic estimate of the protected business integrity in the event of a failure resulting from failing to incur the investment in safety. Measures such as NPV, IRR, pay-back periods, and contributions to cash flows supported with well-articulated qualitative measures provide an excellent basis for selling investment in personnel and process safety management.

Following the logic that when there are fewer workplace incidents, there is less damage, fewer injuries, greater uptime, and a less stressed and often a more motivated and productive workforce, ultimately, these conditions transform into stronger business performance. Burtch advised that this is because "resources are more efficiently used, employee turnover is reduced and manufacturing plant/company operations run more efficiently with enhanced profitability" (Burtch, 2008, p. 2). Burtch (president, DuPont Safety Resources Asia Pacific and vice president, Global Energy Practice, DuPont Safety Resources) also indicated that "the key truth that Safety Excellence means Business Excellence. No company can excel until it makes safety a way of life—and a way of doing business" (Burtch, 2008, p. 2).

The relationship between organizational safety and business or operational excellence is best demonstrated in Figure 7.3. As shown in Figure 7.3, the safety vision of the organization impacts the safety management system of the organization, which in turn influences the business management model of the organization, which ultimately results in operations excellence. Operations excellence is reflected in the organization becoming an employer of choice, profitability and financial performance, and how the organization is perceived from the perspective of corporate social responsibility.

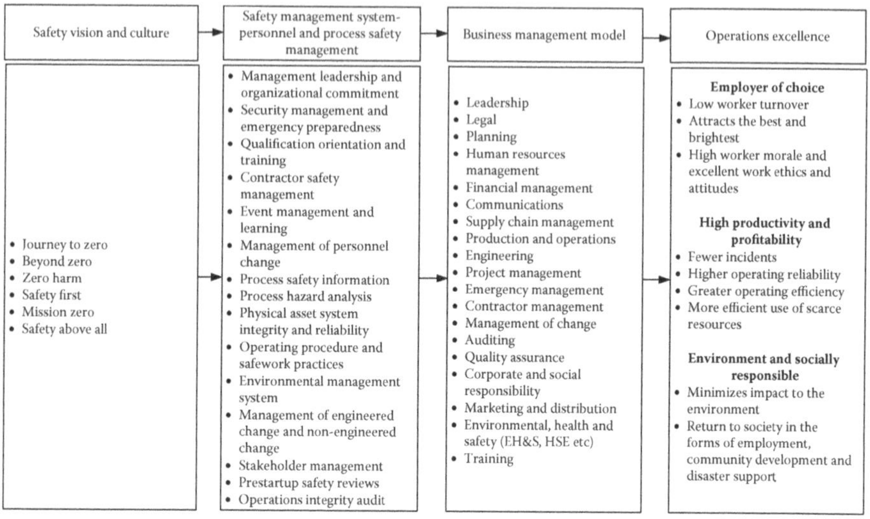

FIGURE 7.3
Safety culture and its influence on operational excellence (OE). (Adapted from Suncor Energy Inc., copyright of Suncor Energy Inc., approval from Suncor is required to reproduce this work.)

In the absence of an effective safety culture and safety management system, an organization generally fails to achieve operational excellence. From the experience of the authors, when safety is not ingrained in the workforce, organizations are generally in a firefighting mode and are unable to have one incident resolved before the other occurs. The impact on business performance is tremendously negative and the toll on workers' morale and productivity is even greater. Such organizations create a revolving door for employees (our most valued assets) who will remain in such roles until the next job becomes available. In such situations, the organization must adopt a comprehensive approach to addressing its risk and safety concerns before focusing on productivity and profitability.

Great Safety Performance Helps to Attract and Retain the Best and Brightest

One of the most valued assets of an organization is its human resources—people. It stands to good reason, therefore, that employers are continually seeking to hire and retain the best and brightest workers that may be available. People (Generation Y in particular) like to associate themselves with organizations that excel in corporate social responsibility. A core component of being socially responsible is the ability to look after the health and safety of all workers and the well-being of the environment.

Table 7.2 highlights the unique composition of the workforce today. According to Lowe et al. (2008), with this complex workforce, the solution for attracting and retaining workers is rather complex. This complexity is driven by the different values of the many generations in the same workplace at one time. From Table 7.2, therefore, the larger an organization becomes, the more complex it becomes to manage the health and safety of all generation groups and by extension, the workforce. Indeed, with each passing year, the traditionalist and baby boomers generations are replaced by an increasing number of Generation Y workers in the workforce. As a consequence, attraction and retention strategies must be designed to primarily attract and retain Generation Y workers while at the same time satisfying the needs of the baby boomers and Generation X workers.

According to Dols et al. (2010), Generation Y knows what it wants from the workplace, unlike baby boomers and Generation X who are intent on pleasing leadership. Generation Y workers are prepared to ensure that leaders are aware of their inherent training and expertise limitations and the risk exposures to all generated from such conditions. Dols et al. (2010) advised when leadership supported safety in the nursing industry, this was a means of "showing respect and recognition, two important factors for development of morale" (Dols et al., 2010, p. 72). McKayn (2010) advised that Generation Y and subsequent generations are all concerned about a greener global culture. This places demands on organizations to not only be safer but also to be socially more responsible in its investments.

The ability of organizations to attract and retain the best and brightest of available workers helps to determine the ultimate competitive position of such organizations. To achieve this goal of the best and brightest, organizations must differentiate themselves from competitors in its safety management systems and safety performance, its impact on the environment, and its social programs. As a consequence, organizations that are safe and socially responsible provide themselves opportunities for hiring the best and brightest among the workforce that ultimately translates into a strong competitive advantage for such employers.

Great Safety Performance Maintains and Elevates Organizational Image

In the industrial world, when we think of a leader in organizational (organizational and occupational health and safety), the first company that comes to mind is DuPont. According to MacLean (2004), at DuPont, safety is in the company's DNA. More importantly, the emphasis laid on safety in DuPont is equivalent to the emphasis laid by other companies on production numbers. This safety focus identifies DuPont as a leader in personnel and process

TABLE 7.2

Values and Behaviors of All Generations in the Workplace Today

Generation	Period	% of Workforce	Values/Behaviors
Traditionalist	Pre-1945	10	Family and patriotism Manages information carefully Satisfied by a job well done Organizational loyalty Experienced and knowledgeable Prefers a power control and hierarchy Seeks control Strives to do what is right
Baby boomer	1945–1964	45	Competitive and enjoys change Ambitious, loyal, and ruthless in business Respects authority; seeks equal treatment Dislikes authoritarianism and laziness
Generation X	1965–1980	30	Self-reliant and seeks work–life balance Fast paced and embraces change Values personal development and working conditions Expects fair and immediate treatment from leaders Believes in work smart principles
Generation Y	Post-1980	15	Family focused, moral, and patriotic Leverages technology to its fullest Well educated and team players Blurs the race and gender gaps Requires engagement and consultation Holds leaders to words Wants to be challenged Wants to be treated as partners and remove bureaucracy Drives and encourages change Wants to be associated with a positive organizational image Will change jobs several time throughout career—loyalty binding required and it must be earned by the employer Plays particular attention to the management of the environment and nonrenewable resources

Source: Created from Lowe, D., Levitt, K. J., and Wilson, T. 2008. *Business Renaissance Quarterly*, 3(3), 43–58. Retrieved November 20, 2010 from EBSCOhost database; Dols, J., Landrum, P., and Wieck, K. L. 2010. *Creative Nursing*, 16(2), 68–74. Retrieved November 17, 2010 from EBSCOhost database.

safety management and has promoted DuPont to world-class safety performance and status. This profile has led to DuPont being highly sought after for workplace safety solutions.

In contrast to DuPont, when we think of the world's least safe organizations, incidents such as the Texas City BP incident, the Gulf of Mexico oil spill, Bhopal, Exxon Valdez spill all come to mind and by association the responsible organization may be deemed unsafe. Capital Confidential (2010) pointed to OH&S data that indicated when compared with its peers, BP was significantly less safe than other oil and gas companies.

In today's fast-paced information and media technology environment, good news travels fast, but bad news travels even faster. Organizations that seek to protect their images must not only seek to be safe and socially responsible, but must also be transparent and capable of withstanding the scrutiny of media and other special interest groups intent on unveiling violations, deviations from regulatory requirements, and kinks in their safety management systems.

The 2010 Gulf of Mexico spill as an example led to a dramatic drop in the value of BP, and shareholders saw the price of BP's stock drop to <50% of its preincident value. Analysts predicted that it would require several years before the share price recover to preincident value. Skoloff and Wardell (2010) suggested a clean-up cost estimate for BP in the range of $40 billion dollars. According to the Associated Press (2010), BP's image "took an ugly beating after the Gulf oil spill." Other stories surrounding the BP Gulf of Mexico spill regarded the CEO of BP as the most hated man in America while others called for his arrest and jailing.

The image impact is dependent on the organizational history and the size of the event. The greater the environmental and safety impacts, the greater the image impact. Lessons learned from the 1989 Exxon Valdez oil spill and the 1984 Union Carbide Bhopal incident tells us that the Gulf of Mexico oil spill will remain with BP for a very long time and shall continue to be a grim reminder to the organizations of failures in its safety management systems. Organizations must, therefore, pay particular attention to personnel and process safety management to avoid such incidents that can potentially hurt its image.

Table 7.3 highlights the insights of successful CEOs regarding health and safety within their respective organizations. National Safety Council (2011) conducted interviews with CEOs with a range of organizations that have demonstrated strong safety performance. These organizations recognized bottom line impact from strong safety performance and provided focused means by which they encourage safe behaviors in their respective organizations.

The use of safety moments or safety discussions before meetings, shared learning in safety, worker engagement and empowerment in safety, and demonstrated commitment to safety by senior leaders were common contributors to a safer organization (Table 7.4). These interviews also highlighted

TABLE 7.3

Health and Safety Perspectives of Organizational CEOs (2011)

Health and Safety Consideration	Bahrain National Gas Company (LPG Producer/462 Employees)[a]
Why is safety a core value at your company	A corporate objective Employees most valued asset Employees trained, fit, alert, and working safely
How to instill safety in your employees on an ongoing basis	Build and reinforce safety culture Demonstrated positive leadership and leadership visibility in safety Executive safety role models Effective EH&S communication internally, between industries, and to worker families Consultation and decision-making involvement of employees and unions on EH&S issues
Safety obstacles identified	Attitude—safety has to be personal
Overcoming obstacles identified	Leadership walking the talk consistent with corporate safety culture vision
How safety pays	Safety as a reward itself Higher productivity and morale; better compliance to regulations, lower direct cost (insurance and worker compensation) Competitive advantage
Measuring safety	Safety performance indicators
Safety improvement opportunities	Not identified
Importance of off-the-job safety and types of off-the-job programs offered	Off-the-job safety can sometimes cost more than workplace safety Involving employees and families in safety competitions

Health and Safety Consideration	Hess Corporation (Integrated Oil and Gas Producer 13,300 Employees)[a]
Why is safety a core value at your company	To be a trusted global energy partner Safety for all the right reasons A belief that organizations are judged on how business results are achieved Safe operations owed to workers, partners, and communities
How to instill safety in your employees on an ongoing basis	Visible support of culture by leaders Safety integrated in values, code of business conduct, ethics, policies, and management systems Personnel ownership and accountability Stewardship against established targets Continuous improvements Priority to celebrate success

TABLE 7.3 (continued)

Health and Safety Perspectives of Organizational CEOs (2011)

Health and Safety Consideration	Hess Corporation (Integrated Oil and Gas Producer/ 13,300 Employees)
Safety obstacles identified	Sustaining momentum Avoiding complacency
Overcoming obstacles identified	Enterprise wide communication and recognition programs President's award for safety excellence Safety appreciation day to celebrate safety success Transparent and high-quality reporting to stakeholders Strong company reputation
How safety pays	Lower cost, improved reliability, and higher worker productivity Community support for continued business Strong company reputation
Measuring safety	Not identified
Safety improvement opportunities	Not identified
Importance of off-the-job safety and types of off-the-job programs offered	Not identified

Health and Safety Consideration	Suncor Energy Inc. (Global Integrated Oil and Gas Producer/12,000 Employees)[b]
Why is safety a core value at your company	• Intrinsic personal motivator (keep self safe, keep colleagues safe)—high purpose • Capture hearts and minds of our employees—engagement • You cannot be excellent at business without being excellent at safety (operational excellence)
How to instill safety in your employees on an ongoing basis	• Consistent leadership focus and behaviors • Personal capability development (skills, knowledge, experience, and commitment) • Visibility (e.g., safety moments, performance metrics, safety observations, and communications) • Goals alignment and translation • Celebrating success and rewarding behaviors
Safety obstacles identified	• Weakest link is the absence of brave and courageous leaders • Nonstandardized approach to risk tolerance leading to shortcuts • Root cause underlying behaviors—rushing, frustration, fatigue, complacency • Weak field-level risk assessment, management of change (assets, people, processes)

continued

TABLE 7.3 (continued)

Health and Safety Perspectives of Organizational CEOs (2011)

Health and Safety Consideration	Suncor Energy Inc. (Global Integrated Oil and Gas Producer/12,000 Employees)
Overcoming obstacles identified	• Risk-aware culture supported with solid systems, processes and people • Focus attention on personal and process safety elements (building into the culture) • Risk prioritized (e.g., hazardous energy, leadership, new workers, and contractors)
How safety pays	• Greater employee engagement • Higher operational reliability • More motivated workforce from demonstrated care and concern for people and assets—risk aware capability
Measuring safety	• Leading and lagging indicators—benchmark • Celebrate success, align behaviors, and rewards
Safety improvement opportunities	• Developing transformational leaders—leading and managing change • Leader and employee engagement—capturing hearts and minds • Strong supporting safety systems and processes • Creating a culture of operational excellence and continuous improvement
Importance of off-the-job safety and types of off-the-job programs offered	• Visibility in company newsletters, safety moments • Periodic home mailouts

Health and Safety Consideration	Parsons Corp. (Global Engineering, Construction Technical and Management Services/12,000 Employees)[a]
Why is safety a core value at your company	To emphasize the importance of stakeholder safety Showcase safety as a component of culture as opposed to a priority Requirement of clients and provide a competitive advantage to Parsons Corp.
How to instill safety in your employees on an ongoing basis	Demonstrated senior leadership commitment Online and instructor-led safety training Safety moments to start meetings Near-miss reporting culture Audits and inspection Make safety a part of the business
Safety obstacles identified	Complacency and taking great performance for granted
Overcoming obstacles identified	Continuously reminding workers of safety, weekly safety bulletin, safety committees, and employee participation in safety processes
How safety pays	Reduced direct cost (insurance) Preferentially selected over peers by owners Increasing stock prices
Measuring safety	Safety performance indicators

TABLE 7.3 (continued)

Health and Safety Perspectives of Organizational CEOs (2011)

Health and Safety Consideration	Parsons Corp. (Global Engineering, Construction Technical and Management Services 12,000 Employees)[a]
Safety improvement opportunities	Not identified
Importance of off-the-job safety and types of off-the-job programs offered	Not identified

Health and Safety Consideration	CH2M Hill (Global Consulting, Design, Construction, Operations, and Management Organization/23,500 Employees)
Why is safety a core value at your company	Collaborative and caring organizational spirit Personalizing discussions around injuries Employees assuming ownership for safety in a genuine way Organization obsessed with safety
How to instill safety in your employees on an ongoing basis	Underscoring importance of safety in a team and employee-owned organization Collaboration and sharing across the entire global corporation Start every meeting with a safety discussion Plan tasks and projects with focus on safety Share learnings in safety across the organization Empowered employees to recognize and halt unsafe work Reward exemplary behaviors
Safety obstacles identified	Distracted employees Complacency
Overcoming obstacles identified	Genuine concerns about whether or not more can be done for safety\leading to heightened continuous awareness More emphasis on reporting of near-miss incidents
How safety pays	Employee and worker welfare Changing the behaviors of organizations in countries where safety is not a priority
Measuring safety	Ten leading indicators (not named) Project audits Trend health and safety communication
Safety improvement opportunities	Not identified
Importance of off-the-job safety and types of off-the-job programs offered	Not identified

[a] Constructed from 2011 CEO's who "Get it" http://www.nsc.org/safetyhealth/Pages/211CEOsWhoGetIt.aspx

[b] Personal interview with a senior executive of Suncor Energy Inc.

common concerns around complacency, taking safety performance for granted and the need to personalize safety as potential obstacles to success in safety performance.

Suggestions for addressing these constraints included continuous communication and engagement. Essentially, organizational leadership must possess the will power to continually seek increasingly higher safety performance both at work and away from work in the quest for a safer workplace. The benefits of a safe workplace are tremendous and while it may be difficult to provide an exact dollar value, the cost of continuous improvements in safety is well justified from every injury prevented and every fatality avoided.

TABLE 7.4

Summary of Health and Safety Perspectives of Organizational CEOs and Senior Leaders (2011)

Health and Safety Consideration	Summary
Why is safety a core value at your company	Employees most valued asset—people first Employees assuming ownership for safety in a genuine way A belief that organizations are judged on how business results are achieved Safe operations owed to workers, partners and communities Intrinsic personal motivator (keep self safe, keep colleagues safe)—high purpose Capture hearts and minds of our employees—engagement You cannot achieve operational excellence in business without being excellent at health and safety
How to instill safety in your employees on an ongoing basis	Demonstrated senior leadership commitment, focus, and behaviors Start every meeting with a safety discussion or safety moment Share learnings in safety across the organization Empowering employees to halt unsafe work Rewarding exemplary behaviors Employee engagement and consultation on safety—create employee ownership Audits and inspections and goals alignment and translation Personal capability development (skills, knowledge, experience, and commitment)
Safety obstacles identified	Worker attitude—the challenge of making safety personal Distracted employees Complacency and shortcuts Weakest link is the absence of brave and courageous leaders Root cause underlying behavior—Rushing, frustration, fatigue, complacency Weak field-level risk assessment, management of change (assets, people, processes)

TABLE 7.4 (continued)

Summary of Health and Safety Perspectives of Organizational CEOs and Senior Leaders (2011)

Health and Safety Consideration	Summary
Overcoming obstacles identified	Genuine concerns about whether or not more can be done for safety leading to heightened continuous awareness Continuously reminding workers of safety, weekly safety bulletin, safety committees, and employee participation in safety processes. Risk-aware culture supported with solid systems, processes, and people. Focus on personal and process safety management (build it in) Risk prioritized (e.g., hazardous energy, leadership, new workers, and contractors). Developing transformational leaders—lead and manage change Leader and employee engagement—capture hearts and minds Strong supporting safety systems and processes Culture of operational excellence and continuous improvement
How safety pays	Employee and worker welfare Changing the behaviors of organizations in countries where safety is not a priority Positive impact on the organization's bottom line, corporate image, and motivated workforce Operational reliability
Measuring safety	Leading indicator focus Project audits Trend health and safety communication
Safety improvement opportunities	Sharing learning in safety in an organized manner Leadership development

8

How Can We Improve Health and Safety Performance?

Managing the health and safety of workers is a never-ending 24/7 job that requires continuous attention and continuous improvements. Improvements in workplace health and safety come about from a concerted effort by leadership and a shared vision by all workers. From an organizational perspective, there must be a ground swell of interest in safe work practices that results in a movement toward a safer workplace where all workers accept the safety of himself/herself and colleagues as a part of his or her responsibility.

Factors that contribute to the improvement of workplace health and safety in an organization include the following:

1. Making everyone responsible for health and safety
2. Maintaining a working and effective SMS
3. Establishing and stewarding its RM philosophy
4. Embracing process safety management (PSM) as a component of its SMS
5. Focused attention on contractor safety management
6. Leadership at the frontline
7. Shared learnings within and across organization, within industry, and across industries
8. Maintaining a trained and competent workforce
9. Ensuring an adequate audit and compliance processes

In this chapter, we shall discuss these elements and their influence on the health and safety improvements within an organization.

Making Everyone Responsible for Health and Safety

When an organization is capable of creating a shared safety vision that spans from the CEO through the lowest level of operation, then meaningful improvement in the organization's health and safety performance is possible. Sharing and supporting this safety vision, however, require the following:

1. Continuous communication from leadership at all levels of the organization such that every level of workers understands and accepts the safety vision of the organization.
2. Demonstrated behaviors from all levels of leadership on safe work practices. By this we mean leaders *walk the talk* where safety is concerned. They use the right PPE during field visits, they support health and safety priority over production, they allocate resources for ensuring that safety concerns are addressed in a timely manner, and they demonstrate commitment to safe work practices at work and at homes.
3. Holding every worker accountable for safety in the workplace. At the very basic level, each worker is responsible for full and consistent use of PPE. Each worker is also a buddy for fellow workers where proper use of PPE is concerned.
4. Ensuring that before any job is started, a field-level RA (FLRA is completed to identify and address all risks and hazards that can result in injuries to workers or harm to the environment or business.
5. Workers must be empowered to stop all unsafe work and must feel free to do so without fear of persecution. By unsafe work, we mean that known or overlooked hazards are apparent and can result in injuries to workers or harm to the environment or business.
6. We establish proactive SMART safety performance indicators for which we are collectively accountable. Our focus should be aimed primarily on leading indicators. With a leading indicator focus, lagging indicator stewardship becomes a measure to tell us the extent of our improvements.
7. We hold ourselves accountable to be properly trained for the jobs we perform and our competency in its execution. We also mentor trained but less competent workers until they are fully competent in their roles.
8. We follow the organizational policies, standard operating procedures, critical practices, and safe work practices at all times. We do not take shortcuts and we never sacrifice quality in the performance of our duties.
9. We report all incidents and near misses. By doing so, we collectively provide the workforce opportunities to learn from the incident or near miss and reduce the probability of its recurrence.

This list can be expanded further; however, the intent of this discussion is to provide readers the need for making health and safety everyone's responsibility and to establish that you are your brother's keeper while at work. Our collective safety depends on our ability to look out for each other while at work for identifying hazards one may miss while performing work that others may identify and have addressed before an incident occurs.

Maintaining a Working and Effective Safety Management System

In our earlier discussions, we outlined the requirements of various SMSs for achieving best in class and world-class safety performance. To improve safety performance, an effective SMS that addresses the safety requirements for people, processes, and systems as well as the facilities and technologies of the organization is essential.

In our earlier discussion, our eventual goal is to achieve the following:

- Strive for operational excellence and discipline by focusing on people, facilities, and technology.
- Share knowledge among peers and across industries where safety performance improvements are concerned.
- Sustain leadership development consistent with shared values, behaviors, and culture.
- Promote learning and continuous improvements.
- Learn from each incident and share learning across all industries.

An effective SMS, as discussed earlier, has the potential to deliver on all of the goals identified above. Maintaining an effective SMS, however, takes a lot of effort, time, resources, commitment, stewardship, and above all leadership commitment. Nevertheless, when the value derived from a healthy and effective SMS is presented, from the health and safety as well as profitability perspectives, leadership commitment, and buy-in are generally high.

Establishing and Stewarding the Risk Management Philosophy

Having an RM philosophy guides the business in terms of the risk tolerance of the organization. The use of a comprehensive risk matrix and accompanying standard provides guidance to all workers in terms of how risk is assessed and mitigated by the organization. A risk matrix is a simplified process that examines the severity of an incident and the probability of its occurrence and focuses on the worst credible scenario. The intent of the matrix is to encourage workers in the prejob RA to identify the types of risks present that can lead to incidents and to develop methods and strategies for reducing these risks to as low as reasonably practicable, before any job or task is undertaken.

Despite the presence and use of this and other RM technologies, major incidents continue to take place. The 2005 Texas City BP incident shall remain with us for decades to follow. Similarly, the 2010 Gulf of Mexico spill shall be a case study for generations and the impacts of both these incidents have changed the way we do business today and will continue to do so in the future. Taleb et al. (2009) identified six common mistakes made by executives in RM:

1. Managing risks by focusing on the worst credible outcome (predicting extreme events) may be misleading us. According to Taleb et al. (2009), by doing so, organizations miss the opportunities for addressing risks that do not fall into this category and as such become vulnerable to them.
2. A conviction that "studying the past will help us manage risk" (Taleb et al., 2009, p. 79). While their reviews focused primarily on the financial and economic failures from this mistake, the concept applies similarly as health and safety too. We have become obsessed with reviewing PSM incidents over time for shared learning to prevent recurrence. However, almost every major incident that takes place in the industrial sectors continues to carry similar root causes of prior incidents. Common among these findings are generally worker training and competency, and management of change. Our challenge is that we fail to take corrective actions within our businesses consistent with findings from past incidents.
3. Organizations "don't listen to advice about what we shouldn't do" (Taleb et al., 2009, p. 80). The idea here is that the impact of telling organizations what not to do is more effective than telling them what to do. When RM is presented as a contributor to profits as opposed to a cost center, appreciation of risk and risk mitigation requirements are greater.
4. The way we measure risk can be confusing. For many frontline supervisors and leaders, the concept behind how risk is measured is beyond their understanding. As a consequence, organizations must come up with a simplified method for capturing and packaging all risk scenarios and possibilities. This can be a very challenging exercise and often injects subjectivity into the risk analysis and interpretation.
5. A gap in appreciation "that what's mathematically equivalent isn't psychologically so" (Taleb et al., 2009, p. 80). Taleb et al. (2009) found

that the way risk is framed either increased or decreased risk aversion behaviors.

6. Organizational "efficiency and maximizing shareholder value don't tolerate redundancy" (Taleb et al., 2009, p. 81). Many organizations have removed redundancy from its operations. As a consequence, there is now greater emphasis on extending the operating time between preventative maintenance of equipment. For example, when a pump or compressor seal starts leaking and there is no operational redundancy, a risk mitigation strategy is put in place to extend the run time of the equipment. This situation is worsened if market opportunities are high. High product and output prices force senior leaders to continue operating in this situation for a longer period before required repairs are undertaken. Redundancy helps to reduce risk exposures and drive the right behaviors.

In spite of these limitations, having a consistent process for assessing and evaluating risk that is well understood by all workers will help in reducing incidents.

Embracing Process Safety Management as a Component of the SMS

Traditionally, PSM has been viewed by many organizations outside the realm of an organization's SMS. Essentially, health and safety falls under the control of the EH&S organization, while PSM falls under the control of the engineering organization. When considered in its entirety, EH&S and other similar acronyms such as HSE, health, safety, environmental and quality (HSEQ), and environment, safety and social responsibility (ESSR) should be associated with addressing personnel safety as well as PSM. The authors believe that this is so because generally, failures in PSM lead to personnel safety incidents. Figure 8.1 provides an overview of how the SMS can be subdivided into the following components to address both personnel and process safety management at the same time:

1. People
2. Processes and systems
3. Facilities and technology

Until organizations recognize the interrelatedness between personnel safety and PSM, and seek to address health and safety in a holistic manner with both activities as components of the SMS, true success in improving health and safety performance in the workplace will be an elusive goal.

Management leadership and organizational commitment			
Safety management system			
Personnel safety	Process safety management		
Personnel safety	*People*	*Processes and systems*	*Facilities and technology*
• Safety vision and mission (e.g. Beyond Zero, Journey to Zero) • Safety culture focus • Personnel protective equipment (PPE) • Worker engagement and involvement in safety-ownership	• Security management and emergency preparedness • Qualification orientation and training • Contractor safety management • Event management and shared learning • Management of personnel change	• Environmental management system • Management of engineered change and non-engineered change • Stakeholder management • Prestartup safety reviews • Operations integrity audit	• Process safety information • Process hazard analysis • Physical asset system integrity and reliability • Operating procedure and safework practices

FIGURE 8.1
Personnel and PSM addressed collectively in the SMS.

Focused Attention on Contractor Safety Management

Historically, organizations have left the safety of contractors to be stewarded by the contracting organizations. Today, while most organizations include contractor injuries into their frequencies, the health and safety of the contractor workforce continue to be stewarded by the contracting organizations with injury numbers being fed into the organization for accounting purposes. Data have shown that contractor injuries and injury frequencies have historically lagged employee injuries and injury frequencies. Meaningful improvements in the health and safety of all workers can only be realized when an organization treats contractor injuries as if these injuries were that of its own employees and addresses the root causes in a similar manner.

The guiding principles for contractors have, in the past, tended to include the following:

1. A strong desire to please the organization for which work is being done.

2. Get the job done as quickly as possible and move on to the next job or assignment.
3. Increase contractor workforce when work is available and downsize during lean periods.
4. Hire from union shops when needed to increase workforce.
5. Retain a few key roles on a long-term basis. Other roles are temporary.
6. Some critical activities such as training can be deferred when work is abundant and the contractor is trying to meet the organization's needs.
7. May seek to conserve training cost if workers are likely to be employed for a short term.
8. Generally, will not verify the competencies of subcontractors.

While this list can be extended further, these guiding principles all contribute toward weaker safety performance.

Organizations must, therefore, seek to shift these principles toward those that support improved safety performance. Achieving this goal can be extremely difficult unless a coordinated joint effort between contractor and organization is applied. With the recognition that in many instances, one cannot survive or sustain its operations without the other, a coordinated approach to the health and safety of the entire workforce is required. Contractor safety management is discussed in greater detail in Chapter 11.

Leadership at the Frontline

Frontline supervisors/leaders are very relevant and critically important to the way safety is managed at the worksite and the safety performance delivered from both the employee and contractor workforces. As a consequence, frontline supervisors/leaders must be adequately trained and competent to function in their roles. Frontline supervisors/leaders require technical, safety, and leadership competencies to generate strong safety performance for the organization. Their abilities to create and sustain relationships, build trust and confidence among the workforce, lead and manage change, and motivate workers to higher levels of safety performance and productivity are essential capabilities for effective leadership at the frontline.

Shared Learning within and across Organizations, within Industry, and across Industries

An area of weakness in safety for many organizations is the way learning from safety incidents is shared across the organization. There are two components to shared learning in safety. These are

1. Having an effective incident investigation process and team
2. Having an effective process and system for sharing learning derived from the investigation process

Incident Investigation

Incident investigations can vary from the simple verification of what happened and the obvious root causes and contributing factors to the elaborate and sophisticated process. In both cases, however, the goal is to get to the root and contributing causes and to identify ways and means to prevent the recurrence of a similar or the same incident in the future. There are many reliable tools for incident investigations. Once done properly, incident investigations and reporting can generate very valuable learnings. Among the key requirements for good incident investigation are the following:

1. Be prepared—have a plan of action.
2. Be timely. Get to the incident site as quickly as possible.
3. Preserve and secure the evidence—cordon off the incident site and control access and egress.
4. Gather the information:
    a. Take lots of pictures of the evidence and site.
    b. Interview witnesses as quickly as possible.
        i. Be polite and prepared.
        ii. Do not draw conclusions before the investigation is completed.
        iii. Demonstrate care and respect for people—those injured and affected by the incident.
        iv. Probe deeply with open-ended questions.
        v. Avoid confrontational behaviors. Create a collaborative working environment.
        vi. Stick to the facts.
5. Develop a timeline and sequence of events.

6. Analyze the event:
    a. Root causes.
    b. Contributing causes.
7. Develop a preventative action plan and recommendations for improvements.

Sharing of Learning

In most instances, when sharing does occur, only limited sharing within business units or among a few key personnel is undertaken. Furthermore, the focus is primarily on the sharing component of the process as opposed to the much more important component of learning. At present, shared learning within organizations is at best in its early stages of development. Shared learning within industries is at best highly disorganized and in most cases absent. Shared learning in safety across industries is almost entirely absent.

Few industries have benefited from the collective learning within the industry in the areas of safety with the exception of regulatory-driven industry improvements. The 2005 Texas City BP incident is one such example where regulatory intervention forced introspection among many organizations that led to collective improvements in PSM among oil and gas producers. Until organizations begin to learn more effectively from the collective industry learning in safety, we are doomed to continue to repeat the errors and overlook hazards experienced by peers that led to severe incidents. *Shared learning in safety is perhaps one of the last remaining low-hanging fruits to be exploited for immediate improvements in safety performance within organizations and industries and across industries.* Shared learning in safety is discussed in greater detail in Chapter 13.

Maintaining a Trained and Competent Workforce

For many major incidents, when root and contributing causes are identified, a common theme that appears is worker training and competency. In many instances, training records are available but an adequate process for verifying competency is generally not available. A well-trained and competent workforce helps in reducing workplace incidents, and more importantly, proactively identifying potential hazards and removing them before incidents occur.

Today's highly mobile and diverse workforce is generally characterized by the following:

1. Varying age and generation composition
2. Different learning methods and habits

3. Cultural differences
4. Women in nontraditional roles
5. Experience and knowledge levels

In response to these characterizations, organizations have had to develop creative means for providing training for all workers. In the quest for cost reduction, organizations have dramatically shifted their focus on training toward e-learning. However, e-learning does not cater well for all groups within the workforce. Generation X, which has a preference for face-to-face learning, is adversely affected. Similarly, Generation Y, whose preference is for e-learning, loses the value of interactive discussions on shared field experience that just cannot be captured fully from e-courses.

Ensuring an Adequate Audit and Compliance Processes

For many frontline leaders, audits are perceived to be *big brother looking over one's shoulders*. Health and safety audits serve a powerful role in proactively identifying gaps and high or unacceptable risk exposures in the SMS of an organization such that they are closed or risks mitigated to low and acceptable levels before incidents do occur. Health and safety audits must be timely and at a reasonable frequency so as to avoid audit fatigue among frontline supervisors/leaders. More importantly, a change in the image of auditors from policing to supporting and enabling is required for organizations to derive the right value from audits and compliance processes.

Audits and compliance management is extremely important among the contracting workforce in order to bring about meaningful improvements in organizational safety performance. Audits are costly but when done properly with adequate follow-up actions for closing gaps and mitigating risks identified, safety performance improves dramatically. Much like incident investigations, audits must be executed as follows:

1. Audits must be properly planned.
2. The right personnel must be engaged in the audit process.
3. Auditors must possess some subject matter expertise and supported by a lead auditor.
4. Business unit leaders must be forewarned well in advance and allowed to perform a self-audit such that they can evaluate themselves relative to their own imposed plans and targets.

5. Audit performance must be compared against prior audit performance to gauge improvements and identify areas of persistent gaps and weaknesses such that resource prioritization can occur.

Audits and compliance shall be discussed in greater detail in Chapter 15.

9

The Challenges of Risk Management

How do we define risk? There are many variations in the definition of risks. Where safety is concerned, the definition proposed by the authors of this book is as follows: Risk is the probability or threat of an incident occurring that may result in damage, injury, liability, loss, or other adverse consequence resulting from either external or internal weaknesses in the organizational SMSs that can be reduced to acceptable levels through coordinated actions. All organizations are faced with numerous business risks. Common among these risks are financial and operating risks. In this chapter, we would like to discuss how risks are introduced into the business and strategies for managing these risks. It must be pointed out that the most destructive types of risks are those that are not known since there is no means of preparing for the unknown.

In this discussion, the authors are not concerned with market, economic, and financial risks. However, financial risk imposes pressure on organizations in the way operating risks are stewarded and managed, and should not be excluded entirely. The authors seek to package risks differently from conventional approaches to risk management. We see risks originating from four main sources. These are as follows:

1. Residual or static risks
2. Introduced risks
3. Operating risks
4. Incremental normalized risks

Some of the terms used in this discussion may be new to the concept of RM and are intended to provide a better understanding to readers about these risks and the RM strategies.

Residual or Static Risks

What do we mean by residual or static risk? Residual risk means exactly what the term suggests. It is that remaining risk associated with any business or operation that organizations are prepared to retain while doing business. Some may suggest that this is the risk that organizations are prepared to live with on an ongoing basis and for which there is no apparent cost benefit advantage for reducing the risks further. In some instances, this may be an economic or financial decision and it varies from business to business.

When a new plant or facility is built, there are inherent risks to this facility that will remain with the business throughout its operating life. These risks are acceptable to the organization and may be mitigated further through various actions by the organization. As we are aware, many industries are considered to be of higher risk than others, and organizations pay premiums to attract and retain personnel to function in these industries. For example, nuclear operations, offshore exploration, refining, power generation, and sour facilities may be considered among higher-risk businesses. In such businesses, designs are such that risks are reduced to an acceptable level that organizations are prepared to live with. This risk is termed as residual risk in our discussions.

Introduced Risks

How do we define introduced risks? Introduced risks are those risks that are brought into the business from various sources as the organization progresses. Introduced risks are brought into the organization primarily from its growth strategies. In a competitive business environment and based on stakeholder interests, organizations are continually seeking to grow. The growth strategy adopted may influence the types and amount of risks introduced into the organization. Growth by acquisition, diversification, and Greenfield activities, adopted by the organization, will necessarily define the amounts of new risks introduced to the organization. Let us take a moment to examine how new risks may result in organizations from the growth strategy adopted or pursued.

Growth by Mergers, Acquisition, Takeovers, and Joint Ventures

As is expected, stakeholder interest encourages organizations to grow. Similarly, market conditions are very important in influencing the rate of growth of the organization. Over the last decade, we have witnessed unprecedented growth rates in the energy and oil and gas sectors. Many organizations have sought to increase market capitalization through mergers and acquisition and joint venture strategies. While mergers and acquisition are a very effective strategy for very quick growth, care must be taken to ensure that all stakeholder interests are satisfied.

Mergers, acquisition, and takeovers introduce a series of difficult concerns that must be addressed when this strategy is pursued. Among them are the following:

1. Age and operating conditions of the acquired facilities and assets—it follows to good logic that organizations will not normally sell new and well-performing assets unless compelled to do so, as is the case, in a takeover situation. Where a takeover target possesses multiple assets, organizations pursue acquisitions and takeovers as a growth strategy and will generally receive a mixed bag of assets when the entire organization is acquired. Where selective acquisition and purchases are made, organizations should expect to receive assets that have expended a great part of its working life and are getting to the stage where upkeep and maintenance cost can be high. When organizations sell selective assets, this decision is no different from the sale of your car that you may have had for 20 years. As the vehicle ages, the maintenance cost increases. For many car owners, we practice meticulous maintenance and upkeep practices. However, this behavior changes as the vehicle ages and new upgraded models are introduced to the markets. At some point in time, we are no longer as meticulous as we were, we change the oil less frequently, and we do major overhauls based on cost benefit analysis. At some point in time, the overall cost of maintenance starts to increase and it is at this stage we normally decide to replace the asset or do something different. The new owner must now either undertake a large investment to return the vehicle to top operating condition or may choose to repair as required when the vehicle breaks down. When the choice is to repair as the vehicle breaks down, a new set of safety challenges become inherent to the process. These circumstances are generally the same for commercial acquisition of assets. Recall BP's 2005 Texas City incident. According to the U.S. Chemical Safety Board, BP acquired this asset when it merged with Amoco in 1999. According to the Chemical Safety Board (2007), "cost-cutting in the 1990s by Amoco and then BP left the Texas City refinery vulnerable to a catastrophe." This facility has been in operation since 1934 and according to Lyall, the facility was "poorly maintained and long starved of capital investment" (Lyall, 2010, p. A1).
2. Prevailing safety culture in acquired facilities—the prevailing safety culture of acquired facilities and organizations determines the types and extent of change required to prevent workplace incidents and injuries. Changing the safety culture of an organization is a lengthy process often spanning several years. Extensive leadership commitment and considerable resources are required to achieve this goal. If the safety culture is weak in an acquired asset, expect incidents to be prevalent. An indication of the safety culture

of an acquired organization is again reflected in the Texas City BP incident. Lyall cited texts from a report prepared 2 months before the accident by the Telos Group on a safety review of the refinery in question, and which said: "We have never seen a site where the notion 'I could die today' was so real" (Lyall, 2010, p. A1).

3. Personnel concerns and disenchanted workers—during every merger, acquisition, and takeover, there will be stakeholders who are affected either positively or negatively. One of the most affected stakeholder groups are workers in both the takeover target and the acquiring organization. Jobs are generally lost as duplication is removed and in the process, many workers become highly disenchanted and demotivated. In many instances, acts of sabotage from those who may feel they were unfairly treated are very much possible. Sabotage may range from the destruction of relevant and safety-critical information to the actual physical acts designed to create damage. Swift action is required to identify and determine potential safety concerns from these sources and to eradicate them as quickly as possible. Failure to achieve this goal predisposes the organization to unnecessary safety vulnerabilities.

4. Cost-cutting strategies—typical of any merger, in acquisition or takeover, there is the need to remove duplication and pursue cost-cutting strategies to ensure that synergies are captured and stakeholder (primarily shareholder) expectations are met. Removal of duplication may result in entire departments being removed. Alternatively, selected individuals may be systematically removed in a right-sizing process. Regardless of the approach to removing duplications, human resources are generally most affected. Moreover, in many instances, more of the middle management group, generally with significant experience, is removed from the process. Middle managers are removed because they are the most vulnerable group. They fall between the frontline group essential for continuing to operate the asset and the senior leaders who are the decision makers. Obviously, the decision makers will not remove themselves. However, middle managers are more closely aware of the hands-on operations and how to operate them safely. In addition, experienced middle managers are very well paid and are generally expensive to maintain. During cost-cutting strategies, older, experienced middle managers are the first group of workers to be released and replaced with less experienced younger managers who are often very inexpensive relative to the experienced middle managers. The challenge for organizations is that a significant amount of critical knowledge is thrown out of the door without consideration for its impact on the safe operations of facilities or assets. Striking the right balance between experienced worker

retention and cost-cutting measures is essential for long-term success and RM resulting from mergers and acquisitions.

5. What are the unknowns—in many instances of mergers and acquisitions, there are many unknowns that are to evolve during the merging process. Mergers and acquisitions are not simply on/off switches. These business activities are very lengthy processes that can take 2–5 years easily to achieve alignment between the target and acquiring organization. During this time, many prior unknown situations, circumstances, and events will arise that shall influence risk associated with the safe operations of merged or acquired assets and facilities.

From the discussion presented, stakeholder interests influence the risks introduced into an organization. The experiences of BP cannot go unnoticed. As shown in Figure 9.1, BP demonstrated phenomenal growth by pursuing growth strategies that focused primarily on shareholder satisfaction as opposed to value maximization. This strategy led to many major acquisitions during a very short period. Indeed, with these acquisitions came introduced new risks to BP that were not fully understood and which ultimately led to several major disasters among which were the now famous 2005 Texas City BP incident.

In the absence of a balanced stakeholder focus on value maximization and the quest to satisfy shareholder, leadership is often likely to take on the burden of increasingly larger risks since with risks come rewards and often higher returns. However, investments and acquisitions in high-risk assets only are not sustainable in the long term and must be properly balanced to generate stability in the organization. When high-risk assets are pursued with cost-cutting strategies, this becomes a potentially lethal concoction that

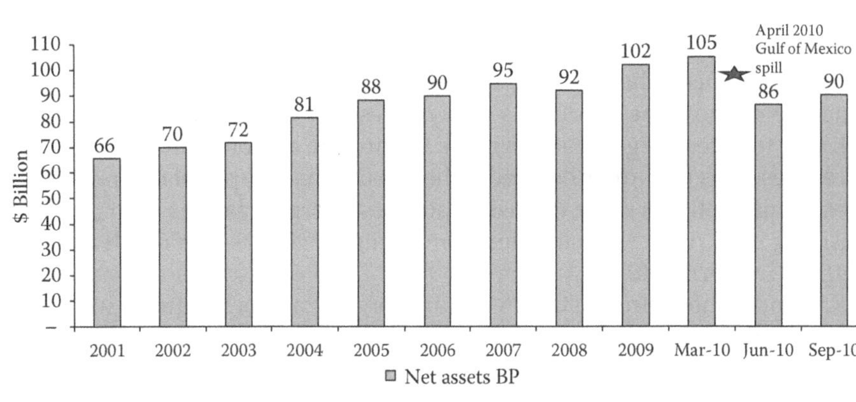

FIGURE 9.1
Net asset growth—BP2001–2010. (Adapted from trend generated from BP annual reports.)

can be worsened by the rate of cost-cutting activities. Ultimately, shareholder satisfaction as a focus strategy is not sustainable.

Organizations must focus on value maximization as its business management strategy. Value maximization focuses on satisfying the needs of all stakeholders' needs and leads to a more stable growth strategy that is inclusive and addresses the health and safety requirements of the organization.

The April 20, 2010 Gulf of Mexico spill was the outcome of an explosion sinking of the Deepwater Horizon, owned by Transocean and leased to BP, which resulted in the deaths of 11 rig workers. The rig was among the most modern and was drilling in 5000 feet of water in the Gulf of Mexico. This single event almost led to the total collapse of BP and was projected to cost the company >$40 billion dollars.

Growth by Venturing into Unchartered Territories

Many organizations pursue growth by venturing into untested technologies and unconventional business areas. The oil and gas industry is one such industry that has forced businesses to explore new technologies and to pursue resources that are no longer easily accessible. Steam-assisted gravity drainage (SAGD), a process dominant in the oil sands of Alberta, is a relatively new technology designed to minimize the environmental footprint and to access reserves too deep to access with conventional mining technologies.

SAGD has had its own share of incident albeit minor relative to many of the other major incidents that have affected the oil and gas industry globally. However, as organizations seek to find cheaper sources of steam and reduce the use of natural gas, the riskiness of the business increases. In addition, as new wells are drilled further and further away from the capital-intensive central processing facilities, the potential for severe incidents to occur increases.

The situation in the oil and gas industry parallels the whale oil industry that saw the shift from on-land processing of whale oil to the creation of high-risk whaling ships, complete with processing facilities, as the supply of whales dwindled from overexploitation. The April 20, 2010 Gulf of Mexico BP oil spill, one of the largest and most costly of its kind, was essentially an outcome of the same model. As easily accessible oil reserves become more difficult to access, organizations move toward the exploitation of difficult-to-access reserves that are miles under the ocean. These types of business activities significantly increase the risk exposures of organizations with potential business bankruptcy possibilities from a single event as was nearly the case with BP in April 2010.

RM under these circumstances requires organizations to adopt a well-coordinated approach to RM and health and safety. Stakeholder engagement is critical at each stage of the process and cost-cutting strategies should arise only after the business venture is operating successfully with an opportunity to capture learnings. Learning must be captured and exploited during repeated similar investments. However, when businesses venture into unchartered territories for the first time, there should be no expense spared. Indeed, such activities may be quite costly. However, where value maximization is pursued, shared capital cost can lead to a less risky proposition for all stakeholders.

Operating Risks

Operating risk for many businesses are addressed using risk matrices and other health and safety tools and practices. Operating risks according to Vasilash are "internal to the business and relates to its ability to achieve its chosen strategy" (Friend and Zehle, 2004, p. 136). Friend and Zehle (2004) identified the following sources of operating risks:

1. Personnel in critical roles resign or are poached by a competitor.
2. "Unforeseen problems occur in the production process" (Friend and Zehle, 2004, p. 137).
3. Equipment maintenance and process problems.
4. Stocks become damaged.
5. Security, fire, theft, and floods.
6. Information technology and information management-induced risks.
7. Sabotage and unproductive actions of rogue employees that ultimately result in large liabilities for the business.

Organizations have made the operating RM process as simple as possible for leaders by developing a simple tool called a risk matrix for guiding the responses of all leaders when faced with operating risk. The risk matrix is a simple tool that attempts to categorize a risk into high, medium, or low risks based on the probability of occurrence and the severity of an incident outcome. Figure 9.2 is a simplistic version of a risk matrix. Once a risk is categorized, the organizational requirement may be to reduce all high and medium risks to low risks through risk mitigation actions.

Organizations generally address operating risks by applying the principles of the Swiss Cheese Model. According to Macleod, the "Swiss Cheese Model is such a useful tool because it can be used not only in an accident investigation but also to help identify potential errors before they occur" (Macleod, 2010, p. 123). As such, the Swiss Cheese Model is a very effective RM tool for addressing operating risks faced by the organization. Many safety professionals have heard about and applied the knowledge generated from the Swiss Cheese Model developed by James Reason in 1990, which was created to provide an explanation of why incidents occur. Figure 9.3 shows an adaptation of the Swiss Cheese Model as it applies to workplace safety and more specifically, field RM in the workplace.

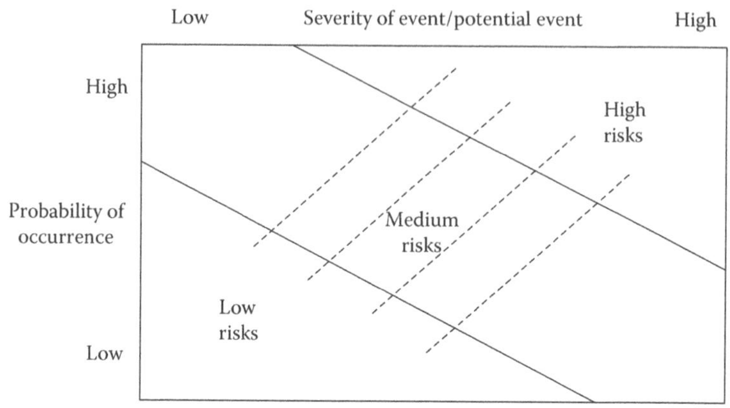

FIGURE 9.2
A simplified risk matrix.

The workplace is filled with hazards and risks and, in many instances, both the organization and employees have agreed that risk mitigation and management is a joint responsibility. When circumstances are right and gaps and weaknesses in all the levels of protection from an organization's SMS are unfortunately aligned, an accident, incident, or loss may occur. As shown in Figure 9.3, the interface between worker, environment, and machinery

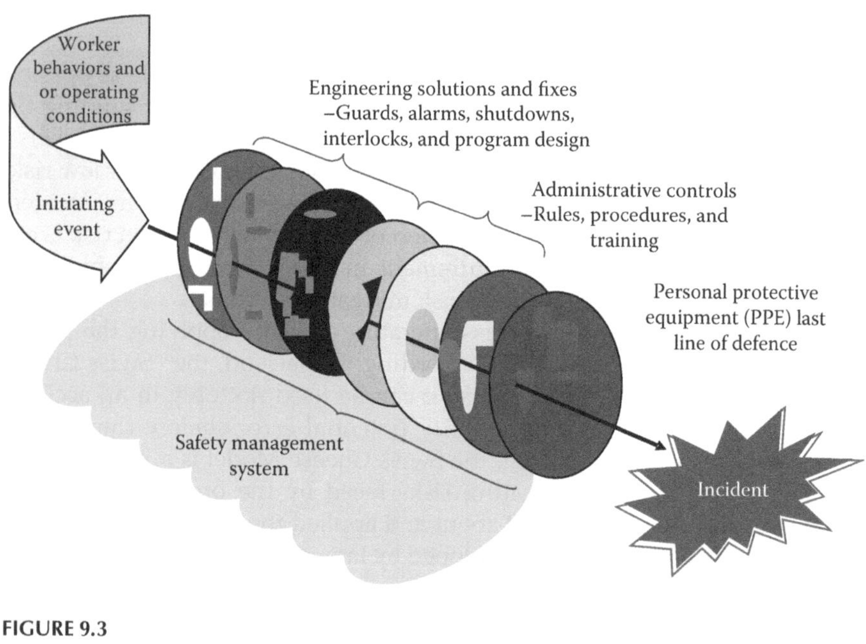

FIGURE 9.3
Adaptation of the Swiss Cheese Model developed by James Reason in 1990.

creates hazards and uncontrolled energy sources that may result in injuries, fatalities, or unacceptable events. Ultimately, the worker behavior or actions and or the operating conditions will create the ideal situation for an initiating event. In many organizations, there are several layers of protection for ensuring the safety of workers. In order of priority, these are as follows:

1. *Engineered solutions and fixes*: This is generally the most favorite and most preferred approach adopted by organizations. Engineering solutions may be extremely costly but the level of protection offered to workers far outweighs the protection offered by other methods of protection.
2. *Administrative solutions*: Administrative solutions refer to rules, procedures, training, and other administrative controls provided by the organization to protect workers during the execution of their duties. Often, this requires an intense effort by the organization to ensure compliance and competency of the workforce. It can be extremely difficult, requiring an intense amount of coordinating, and generally requires repeat training to ensure that the knowledge is provided to all workers on some acceptable frequency.
3. *PPE*: PPE is generally the least preferred option for organizations. Leaders recognize that PPE is the very last layer of protection provided to workers during their performance of work and execution of duties. As such, businesses pay particular attention to ensuring that all workers have access to PPE and that they are trained to use the PPE provided.

The layers of protection in the SMS are designed to protect workers and the organization from risks and are an integral component of the RM system of the organization.

Incremental Risks That Are Normalized over Time

In simple terms, this is called normalizing the deviations. Experience has shown time and again that incidents occur on older facilities during the start-up or shutdown of such facilities. Why is this so? Perhaps a plausible explanation is that over time, new risks introduced with changes are normalized by the organization without a full understanding of the risk exposures faced by the organization. With many older facilities, new risks are introduced each time the facility is shut down for repairs or maintenance and when changes are made. With many older facilities, as technology changes, modifications are made to the existing assets to replace some parts of the facility with newer and more effective technologies that are more robust and reliable than older technologies. With each successive change that is made to the facility over time, the risk of the collective asset increases.

Figure 9.4 demonstrates the increasing risk exposures with each major change that is made to the facility.

As shown in Figure 9.4, when changes are made to an operating facility, new and increasing risk exposures are introduced into the process. Over time, however, the organization continues to focus on the initial residual risk at the start of operations in 1950—R0, without consideration for the other incremental risk that are introduced with each change. Generally, before each change is made, the change is often preceded by a process hazard analysis (PHA) and the risk for that change is mitigated to acceptable levels. The real concern, however, is that in many instances, the impact of the change is not generally done over the entire process to mitigate the new residual risk exposure resultant from the change. The risk impact is generally additive (cumulative), but industry processes have not evolved to assess the overall risk exposure resultant with the new operating position or philosophy. Over time, with each successive change, the introduced risk is normalized as part of the new operation without quantification and mitigation. *The outcome of such situations is generally a catastrophic failure.*

To prevent against such outcomes, organizations must first recognize this normalization of risks as a real possibility in older facilities and adopt proactive measures to reexamine the collective risk increases of all incremental changes introduced into the process. Risk and hazard exposures must be

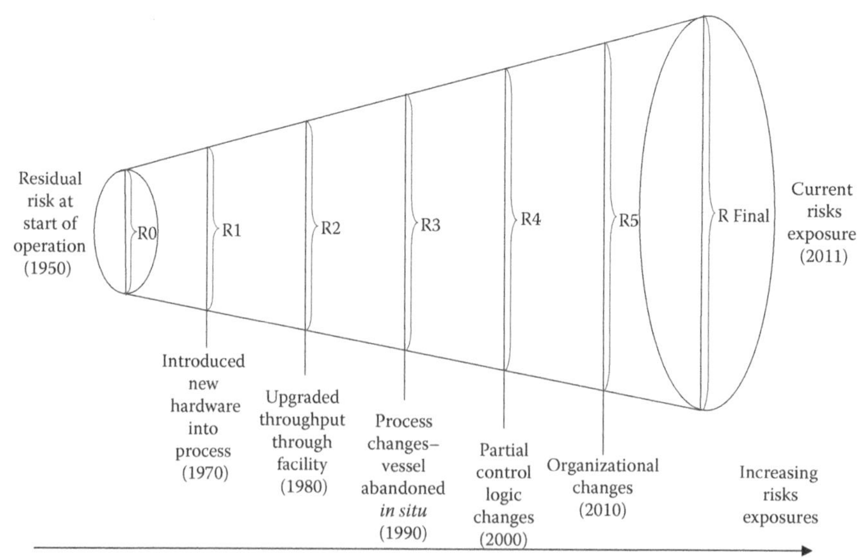

FIGURE 9.4
Increasing risk exposures with successive changes.

prioritized and systematically reduced to acceptable levels. This process is time consuming and requires a significant number of subject matter experts and resources to correct. Nevertheless, when done correctly, risk exposures are reduced to manageable levels and older facilities are less prone to catastrophic failures.

Some of the changes that can result in increased risk exposures are reflected in Figure 9.4. However, key verifications that can mitigate increasing risk exposures are as follows:

1. Ensure that the process safety operating envelope is not exceeded.
2. Promote continuous upgrades to facility such that control logic can be on a single platform that is compatible.
3. Process safety information (PSI) management is critically important to reduce the severity of an incident if it does occur. Therefore, verify and maintain up-to-date P&IDs and process flow drawings (PFDs) and other relevant information.
4. Knowing that protection and fail-safe systems work when needed can only be confirmed through regular testing and preventative maintenance. Test emergency shutdown controls during start-up and shutdown situations for operability.
5. Ensure that a management of change (MOC) process is used in all change situations inclusive of critical roles for people. The MOC process must not be restricted only to the proposed change; however, the impact must be analyzed over the entire process or facility that is affected.

10

Process Safety Management

What is PSM? The authors' view is that PSM is a management process that allows leaders involved in commercial manufacturing and chemical businesses to proactively recognize, understand, and control process hazards, such that process-related injuries and incidents may be prevented and organizations can continue to progress toward its value maximization goals. PSM, when simplified, is the application of a management system and supporting tool to eliminate, prevent, and minimize the impact of process-related incidents and events. PSM forms a critical part of an organization's SMS and is designed to work in conjunction with personnel safety management.

A process safety incident is the inadvertent or unexpected release of energy or process fluids (hazardous chemicals, toxins, flammable, explosive, reactive liquids, or gases) from a system. Process safety incidents have the potential to result in catastrophic failures and are generally associated with major process explosions, fires, and spills. Process safety incidents can be costly in terms of personnel injuries and fatalities as well as facilities damage and impact to production. Furthermore, fines and charges associated with regulatory compliance failures can be astronomical with resultant significant damage to image and goodwill. PSM makes sense, therefore, from both the compliance and value maximization perspectives.

OSHA (2010) advises that the major goal of PSM is to prevent the unintended releases of hazardous chemicals that can cause harm to employees. According to OSHA (2010), a PSM program must be systematic and holistic in its approach to managing process hazards and must consider the following:

1. The process design
2. Process technology
3. Process changes
4. Operational and maintenance activities and procedures
5. Nonroutine activities and procedures
6. Emergency preparedness plans and procedures
7. Training programs
8. Other elements that affect the process

OSHA (2010) further advised that should small organizations address the following in their SMS, there is a greater chance that they will be compliant with regulatory requirements.

1. PSI:
    a. Hazards of the chemicals used in the processes
    b. Technology applied in the process
    c. Equipment involved in the process
    d. Employee involvement
2. PHA
3. Operating procedures
4. Employee training and competency
5. Contractor safety management
6. Prestartup safety review (PSSR)
7. Mechanical integrity (MI) of equipment:
    a. Process defenses
    b. Written procedures
    c. Inspection and testing
    d. Quality assurance
8. Nonroutine work authorizations
9. MOC
10. Incident investigation
11. Emergency preparedness planning and management
12. Compliance audits:
    a. Planning
    b. Staffing
    c. Conducting the audit
    d. Evaluation and corrective actions

While OSHA does not differentiate into the categories defined in this section, it is easy to allocate PSM elements defined by OSHA into people, process, and systems, and facilities and technology. Packaged somewhat differently, compliance to PSM regulations can be achieved as shown in Figure 10.1, which is quite aligned with the authors' SMS proposals for world class in safety performance.

PSM: People

Fulfilling regulatory requirements keep organizations compliant with the law. However, most organizations seek to exceed regulatory requirements in a genuine quest for optimizing the health and safety of workers with its SMS.

Management leadership and organizational commitment			
Safety management system			
OSHA PSM requirements	*People*	*Processes and systems*	*Facilities and technology*

• Employee training and competency.
• Contractor safety management.
• Incident investigation
• Management of change (people).
• Emergency preparedness planning and management.
• Management of change (engineered/Non-engineered).
• Non-routine work authorizations.
• Pre-startup safety review.
• Compliance audits
  • Planning.
  • Staffing.
  • Conducting the audit.
  • Evaluation and corrective actions.
• Process Safety Information
  • Hazards of the chemicals used in the processes.
  • Technology applied in the process.
  • Equipment involved in the process.
  • Employee involvement.
• Process hazard analysis
• Operating procedures.
• Mechanical integrity of equipment.
  • Process defenses.
  • Written procedures.
  • Inspection and testing.
  • Quality assurance.
	• Employee training and competency
	• Contractor safety management
	• Incident investigations
	• Management of change (personnel)
	• Emergency preparedness, planning and management
		• Management of engineered change and non-engineered change
		• Non-routine work authorization
		• Prestartup safety reviews
		• Compliance audits
		  • Planning.
		  • Staffing.
		  • Conducting the audit.
		  • Evaluation and corrective actions.
			• Process safety information
			  • Hazards of the chemicals used in the processes.
			  • Technology applied in the process.
			  • Equipment involved in the process.
			  • Employee involvement.
			• Process hazard analysis
			• Mechanical integrity of equipment.
			  • Process defenses.
			  • Written procedures.
			  • Inspection and testing.
			  • Quality assurance.
			• Operating procedures

FIGURE 10.1
OSHA PSM requirements aligned with proposed SMS for world-class safety performance.

When developing the SMS it is best to ensure that at the very least, regulatory requirements are met. Several elements of PSM are focused on personnel management strategies for ensuring the health and safety of all workers. As shown in Figure 10.1, these include

1. Employee training and competency
2. Contractor safety management
3. Incident investigations
4. Management of change (personnel) (MOCP)
5. Emergency preparedness, planning, and management

Employee Training and Competency

OSHA (2010) requires all workers (employees and contractors) working with hazardous chemicals and processes to understand the health and safety hazards associated with their work. Essentially, workers must be competent to do the work they are expected to perform safely and be able to protect themselves, coworkers, and communities from unintended releases from the process systems. Training and competency must extend beyond immediate hazards of the process fluids to include "operating procedures and safe work practices, emergency evacuation and response, safety procedures, routine, and nonroutine work authorization activities" (OSHA, 2010, para. 35).

Training programs must identify roles that are to be trained, training goals and objectives, and preestablished competency assessment requirements. Competency is reflected in the worker's ability and demonstrated capability of being able to accomplish assigned work consistent with procedural and safe work requirements. Success in training is achieved when workers are engaged and the process is interactive. Compliance in training and competency require organizations to evaluate the training programs on some frequency to determine if the training objectives are being met. Where failure rates are high, the training program must be reassessed and training repeated to ensure adequate knowledge transfer to workers.

The training program must be continuous and must be updated when changes to the process or operating systems have been made. Personnel must also be retrained when changes are made so that they can be competent in their work, inclusive of responses to the new operating requirements. Training and competency is perhaps one of the best investments that organizations will make in its SMS. A trained and competent worker means that there is a reduced likelihood for errors and mistakes and therefore a safer operating system. Furthermore, a trained and competent worker is a more motivated worker who will generally excel in workplace performance. Ultimately, training and competency creates a win–win situation for the organization regarding compliance and safety performance management.

In summary, from an organizational perspective

- The objective of training and competency in PSM is to ensure that all workers responsible for operating and maintaining the operating systems are adequately trained and competent on how to safely do so.
- The training program should include provisions for initial and refresher training, and can provide an auditable process for verifying when training was done, the competency assessment process applied, and when follow-up training is due.
- Trained and competent personnel are an essential requirement for keeping process equipment and machinery operating safely.
- Initial training is intended to provide an overview and the competency required to operate safely with the known process-specific health and safety hazards using operating procedures (including emergency operation and shutdown) and safe work practices.
- Refresher training is intended to maintain the competency of workers over time at an appropriate frequency, generally not exceeding 3 years.

Contractor Safety Management

Where contractors are used in hazardous processes, they must be equally trained and competent as workers to ensure that they can protect the health and safety of themselves and other workers while performing assigned work. An effective contractor prequalification, screening, and verification process is required such that contractors with verified strong safety performance are used to perform work for the organization. Contractors must be similarly trained and competent as employees when performing hazardous work. Their training and competency records must be accessible and verified by organizations. Contractor's work must be authorized and controlled by the organization. In summary, from an organizational perspective

- The objective of contractor safety management in PSM is to ensure that all contractor workers are aware of all known and potential hazards involved in the process and how their actions can affect the health and safety of all personnel working on that particular site as well as those in the surrounding communities.
- Contract language hold contractors accountable for compliance with organizational safety requirements. Organizations are responsible for prequalifying potential contractors before selecting them and require timely corrective actions on contractor safety performance on an ongoing basis once selected.
- Contractors must follow safe work practices to ensure the health and safety of all workers and be compliant with regulatory requirements before work is done for the organization.

- Verification that contractors are trained and competent in the work they perform and have access to all relevant procedures and information for completing work safely.

Chapter 11 has been devoted entirely to contractor safety management.

Incident Investigations

Incident investigation seeks to identify root causes of incidents or near misses such that preventative measures can be established to prevent recurrence. OSHA (2010) requires organizations to conduct investigations into incidents that have resulted in catastrophic events as well as those (near misses) that have the potential to do so. OSHA recommends the development of in-house capabilities for incident investigations by maintaining a multidisciplinary team of trained and competent incident investigators, such that timely incident investigation can be done. Investigations should be consultative and reports generated from investigations should be made available for shared learning. Investigations should also be focused on fact finding as opposed to blame determination. In summary, from an organizational perspective

- The objective of the incident investigations is to provide learning and prevention opportunities such that similar incidents will not be repeated.
- Persistent investigation of all serious and near-miss incidents is essential for continuous improvements in the health and safety and safety performance for the organization.
- Findings must be shared in an organized manner to promote learning and to identify and exploit gap closure opportunities where similar weaknesses in the SMS may exist across the organization.
- Incident investigation must be timely. A multidisciplinary team must be assembled with subject matter expertise comprising relevant involved site personnel and nonsite personnel, including contractors if involved.

Management of Change (Personnel)

MOCP considerations for personnel are generally an area overlooked by organizations that is often left to the human resources department to steward. MOCP is an important component of PSM since the availability of trained and competent personnel is required at all times for safe process operation. MOCP applies to *critical roles* in the following human resources management situations:

- When personnel are promoted or transferred
- When personnel resign or cease to work for the organization

- During situations of worker illnesses or deaths
- During down-sizing and structural reorganization

Ultimately, the goals of MOCP are to ensure the availability of trained and competent personnel at all times to function in critical roles. Short-term solutions such as overtime are not sustainable and effective succession planning is essential to provide a workable solution for critical roles.

In summary, from an organizational perspective

- The safe operations of any process facility require well-trained, competent, and experienced personnel.
- A balanced level of expertise is required at all times for safe operation of process facilities. Workgroups in process facilities must at all times maintain a minimum level of collective experience and knowledge for the safe and continuous operations and maintenance of the facility.
- Loss of minimum levels of experience and knowledge through personnel movements and organizational changes increases the risks and hazards exposures of the process facility.
- Turnover management strategies are critical in MOCP and must be applied to critical roles in the process operations of the organization, such that the required minimum level of skills is maintained.

Emergency Preparedness, Planning, and Management

Emergency preparedness, planning, and management are essential in PSM since unplanned events will at some time occur with the release of energy, process fluids, and hazards from process systems. Organizations must be ready for such events to minimize the impacts. According to OSHA (2010), emergency response is the third line of defense activated when both the second line (control the release of chemical) and the first line (operate and maintain the process and contain the chemicals) have failed.

Emergency management planning requires all workers to know what to do and how to respond in the event of an emergency, and they must be so trained. A warning device is required, and a muster point or several muster points are required generally upwind of the potential hazards releases. Drills and mock emergency responses are required to develop the skills of those who are required to control and manage the response as well as those who are being evacuated. PPE must be available for all responders who will be exposed to the hazards and they too must be adequately trained and competent in the use of the PPE.

Where necessary, community involvement and mutual aid responders should participate in drills. Mutual aid responders are the collective response teams from other organizations who would assist in the event of an emergency in return for reciprocal emergency response support. Mutual aid responders are common in areas where multiple organizations are involved in hazardous businesses in the same region or within close proximity to each other. An incident commander and command post or center that is protected and outside the hazard zone is essential to support the response. Supporting tools such as worker telephone lists, external support resources list, emergency budget, and information technology—computers, radios, and telephones—must also be available in the command post. In summary, from an organizational perspective

- Detailed emergency response planning for potential emergencies is required so that timely and effective response by the organization can protect workers, communities, the environment, and facilities, which may be exposed to hazards during an incident.
- Emergency response planning must consider the many possibilities of emergency situations and must be well prepared to respond to any and all of these possibilities.
- All responders must be properly trained and competent in their response roles. They must also have full and ready access to all PPE required to support the response and must be trained and competent in the use of the PPE.
- Training exercises, simulations, and planned and unplanned emergency response drills are a vital part of emergency planning, which helps to develop the readiness of the response teams.

PSM: Processes and Systems

Success in PSM requires the involvement of many supporting systems and processes. This section introduces some of the key processes and systems that enable PSM. Processes and systems include the following as per Figure 10.1:

1. Management of engineered change and nonengineered change
2. Nonroutine work authorization
3. PSSR
4. Compliance audits

Management of Engineered Changes and Nonengineered Changes

According to OSHA (2010), when we consider PSM as changes, other than *replacement-in-kind*, we refer to the following:

- Modifications to equipment and machinery
- Changes to operating procedures

- Changes in the types of inputs and raw materials to the process
- Changes to operating and processing conditions

MOC (change management) is a systematic procedure that organizations adopt to ensure that all hazards are addressed when a change is made. Generally, when a change is made in process organizations, it may involve changes in process and operating limits—temperatures, flows, or pressure, and change in equipment. When such changes are made, new hazards may be introduced into the process. The MOC process is designed to identify these newly introduced hazards such that they may be adequately mitigated to reasonably practicable levels. Included in the MOC process is the assignment of action items and due dates for the completion of mitigation actions before the facility/process is restarted after the change. In addition, where necessary, training and competency evaluation are required before the change can be activated.

Management of Engineered Change

Engineered change refers to any change that is a specified addition, alteration, or removal of equipment, facilities, infrastructure, or software. Alternatively, it refers to any change to standards or specifications that may result in new components, materials, processes, or procedures being introduced. This includes any alterations to the PSI for a hazardous process. For most organizations, where engineered changes are concerned, the following perquisites are generally addressed before changes are made:

- Purpose of change. Why is the change necessary? Is there a business case for doing so? The business case may include but is not limited to eliminating a process hazard.
- Safety, occupational health, and environmental impact, including whether a PHA is required. When a PHA is required, the completed and approved review must be attached to the MOC document.
- If the change involves a deviation from established standards, the technical deviation must be authorized and included with the MOC, and the PSI amended accordingly.
- Modifications to operating procedures.
- Essential training and communication needed for employees involved in operating and maintaining the process and contract employees whose job tasks are affected by a change in the process shall be informed of and trained in the change prior to the start-up of the affected part of the process.
- Limits for the change (time period and/or quantity of material).

- Approval and authorization requirements.
- Any applicable PSM regulatory coverage.

Trial runs or temporary changes that exceed safe operating limits shall require a management of engineered change. According to OSHA (2010), many temporary changes have become permanent changes and have resulted in catastrophic events. Temporary changes must therefore be carefully managed and proper documentation, review, and evaluation considerations are essential before such changes become permanent, so that the health and safety of workers, the environment, and assets can be preserved. Processes shall be returned to standard operating conditions following an authorized trial period. If permanent changes are recommended as a result of a trial run, a new management of engineered change must be initiated. Areas shall establish and implement a tracking system to provide closure to those changes where there was either a trial or a test period. All engineered changes shall require a PSSR.

- The objective of the MOC process is to ensure that all hazards introduced by the implementation of the change are identified and controlled prior to resuming operation.
- The MOC process is designed to ensure compatibility between process and equipment where changes are made.
- All changes to the equipment, process, and procedures are properly reviewed to ensure that hazards introduced in the process can be and are adequately addressed.
- All personnel required to operate the facility or process after the change is made are properly trained and competent in operating the facility or process.

Nonengineered Changes

Nonengineered changes refer to changes introduced into the process where engineered specifications or process safety requirements are not affected. Nonengineered changes may include, but is not limited to, changes to control systems or operations, graphic configurations, equipment titles, descriptions, equipment, and similar related items. Nonengineered changes may also include small nonimpacting changes that are not *replacement-in-kind*, to assets or software potentially impacting on hazardous processes. While we may consider nonengineered changes as relatively unimportant, they still constitute change, and overtime, with additional changes new hazards and risks may be introduced into the process. As a consequence, all nonengineered changes must be documented in a similar manner as to engineered changes.

Ultimately, the bottom line in using an MOC process is that all modifications are subject to

- Adequate review of the impacts of the change on the process and on process safety considerations
- All changes are appropriately documented
- All changes are authorized by appropriate levels of management
- All personnel are made aware of the changes and where necessary are appropriately trained and competent

As a consequence, at the very least, PSM elements affected must be considered when changes are made to include PSI, PSSR, and employee training and competency.

Nonroutine Work Authorization

What do we mean by nonroutine work in PSM? In process operation, there is routine work and nonroutine work. The daily work of plant and facilities operators is considered routine work and it includes adjustments made by operations personnel to produce the desired final products. Nonroutine work, on the other hand, refers to all work that may be conducted periodically. Examples of nonroutine work may include work related to planned and unplanned maintenance, turnarounds, changes, and modifications to equipment and machinery.

When nonroutine work is conducted, such work must be properly monitored and managed to avoid incidents. Nonroutine work must be properly authorized in a consistent manner. A work permit, which provides a documentation of the scope of the work, hazards identification, and mitigation process are among the required guidance to whoever is performing the work, is essential.

The work permitting process should include verification of the procedure used for, but not limited to, the following:

- Isolation of any energy sources such as electricity, pressure, and temperatures
- Breaking the integrity of a closed system
- Confined space entry
- Hot work processes requirement and authorizations
- Work completion and return-to-service procedure

Nonroutine work authorization is perhaps one of the most important steps in maintaining the health and safety of personnel, environment, and assets during nonroutine work. *We must bear in mind that most incidents occur during instances of nonroutine work, and when incidents occur, these are among the first sets of documents that investigators will acquire to begin an investigation.* Care and attention must be provided to the work permitting process and due diligence requirements are necessary for ensuring that we have done a complete and effective hazards assessment and mitigation exercise and that personnel are provided the right tools, procedures, and oversight when nonroutine work is performed. A well-documented, transparent process is required.

Prestartup Safety Reviews

Experience has shown that process safety incidents occur when plants and facilities are being shut down or when they are being started up. This can happen during Greenfield startups or shutdowns as well. To reduce the possibilities of such incidents, a facilities readiness review (Lutchman, 2010), or PSSR, helps in determining the readiness of the facilities to be started up. A PSSR is a jointly performed exercise between representation of multiple stakeholders and subject matter experts to verify the following prior to the startup of new and modified equipment:

- All new equipment and installation are completed in accordance with design specification.
- All required and affected safety, operating, maintenance, and emergency procedures have been updated and are adequate.
- All workers involved in the operation and maintenance of the new process, equipment, and/or facilities are property trained and their competency is assessed.
- Modified facilities meet all PSM requirements. PSSR provides a final review of any new and/or modified equipment to verify that all appropriate elements of PSM have been addressed satisfactorily and that the facility is safe to operate.
- PSSR are generally conducted by multidisciplinary teams consisting of operations, technical, design, maintenance, and appropriate safety representatives on an as-needed basis.

Compliance Audits

Ultimately, an audit is a process for gathering information that would guide business leaders on the degree of compliance or noncompliance to the PSM element standard and to all PSM standards. Compliance audits are proactive verification of compliance to PSM element standards. OSHA (2010) points to organizational requirements of an impartial, trained, and competent team of compliance auditors (or individual for smaller facilities) for performing comparative assessments between field actions and requirements of the accepted standard by which the organization is guided.

The aim of the audit is to proactively identify hazards that are undetected or ignored by businesses and to take the necessary corrective actions to prioritize and address these hazards and risks exposures of the organization before incidents occur. Audits must be properly planned and done such that facilities being audited do not feel intimidated by the process. Leaders of facilities being audited must perceive audits as a positive process for improving their operations and overall risk and hazards exposures as opposed to a fault-finding process. As a consequence, engagement of personnel is encouraged and organizations must be forewarned and provided adequate opportunities to prepare for the audit. Well-run organizations with strong safety cultures welcome audits and see them as positive steps in improving the health and safety of the workplace.

When audits are conducted, emphasis should be on processes, transparency, and traceability for compliance to the standards. A check sheet for every standard helps in the auditing process so that all components of compliance to the standard can be reviewed. Audits include a series of interviews of personnel in leadership and frontline roles as well as workers at the frontline. Audits also include site inspection and physical inspection of the facilities. Interviews provide stakeholder understanding of the requirements of each standard and their related roles and responsibilities. It also provides the auditor opportunities to identify improvement opportunities and corrective actions.

Audits and compliance provide opportunities for identifying and prioritizing a list of compliance action items for stewardship by the business. Leadership review and debriefing of audit findings and action items are necessary, such that appropriate priorities, timelines, resource allocation, and responsibilities can be established. A historical database of action items is also essential since it provides opportunities to trend responses to compliance requirements and to preferentially allocate resources during stewardship if necessary. Traceability is essential to facilitate the audits and compliance process. In summary, from an organizational perspective

- The objective of the auditing element of PSM is to evaluate the effectiveness of PSM by identifying deficiencies, risks, and hazards exposures of the organization and recommending corrective action. The audit also identifies pockets of excellence that can be leveraged across the entire organization.
- Audits provide a measurement of compliance with the established PSM standard for each element.
- An effective auditing approach requires a cultural shift for many organizations to conduct self-audits supported by audits performed by an independent impartial group of trained and competent auditors.

- Audits of each element of PSM should not exceed a finite period between audits. Typically, this may be 24–36 months. Care must be taken to ensure that audit fatigue does not become a challenge that eventually evolves to a window dressing exercise to demonstrate compliance and fulfill an obligation.

PSM: Facilities and Technology

The safe and efficient operations of facilities and technology are also a core focus of PSM. As a consequence, standards have similarly been developed to ensure the safe and efficient operations of facilities and technology. In this section, we shall discuss these requirements and PSM element standards for facilities and technology. As shown in Figure 10.1, these include the following:

1. PSI
2. PHA
3. MI of equipment
4. Operating procedures

Process Safety Information

PSI refers to the management process for ensuring the availability of complete and accurate written information concerning process chemicals, process technology, and process equipment. PSI is essential to an effective PSM process since it makes available all relevant information to all stakeholders involved in the stewardship, operations, and maintenance of a process facility or technology. PSI management includes the way information is managed regarding, but not limited to, the following:

1. PFDs and piping and instrumentation diagrams (P&IDs): The diagrams are essential in ensuring processes remain within the operating parameters of the process and identifies the controls required for maintaining the process. Accurate and updated P&IDs, in particular, are essential for enabling timely containment and control in the event of an incident. Experience has also shown that P&IDS are generally very poorly updated when changes to the processes and facilities are made and in many instances several versions of P&IDs may be retained leading to major confusion and delays when trying to contain an incident.
2. Policies, codes of practices (COPs), SOPs, critical practices and industry practices: These documents contain and provide information to workers about how work is to be conducted. Policies, COPs, and

SOPs are critical documents for ensuring that work is conducted safely at the facility. These documents must be easily accessible and available to all workers required to do routine and nonroutine work at the facility.

3. Materials safety data sheets (MSDS): MSDS provide information on the hazards of all materials and chemicals used or processed on the worksite or facility. This information is essential to inform all workers of the hazards they are exposed to at the worksite.
4. Technology information: Technology information refers to the information relating to the technology used in the process. Technology information may include but is not limited to preestablished criteria for maximum inventory levels for process chemicals. Exceeding these limits may be considered abnormal operating conditions requiring a risk mitigating strategy.
5. Technical information on equipment, machinery, and interconnecting piping and devices: The focus here is on the codes and standards used to establish good engineering practice.

In summary, from an organizational perspective, PSI deals with

- Compilation of written information that enables workers to identify and understand potential hazards they may be exposed to in the workplace
- PSI includes
  - Documentation of the related chemical hazards exposures and consequences
  - The technology used in the design and operations of the process
  - Documentation of the technical specifications of all equipment, machinery, and interconnecting pipelines
- Importance of PSI:
  - PSI is foundational to PSM
  - Provides essential information for conducting PHA and stewardship of mitigation strategies
  - Critical for generating employee participation, compiling, reviewing, and updating operating procedures
  - Critically important for demonstrating due diligence in audits and compliance

Process Hazards Analysis

OSHA (2010) regards PHA as one of the most important elements of a PSM. PHA is an organized and systematic process for identifying and analyzing process hazards and developing mitigating strategies for reducing them to as low as reasonably practicable levels. PHAs are generally conducted by a multidisciplinary team of subject matter experts, who are led by an experienced facilitator such that all hazards and accompanying protection offered by the system are identified. Where engineering solutions cannot be included in the risk mitigation strategies, successive layers of protection that include administrative controls are developed before the last level of protection—PPE is considered. The PHA team must be capable of putting aside personal goals and focus on finding the best solutions to hazards, designs, controls, and risk mitigation strategies. The team leader must also be skilled in consensus building such that true solutions can be developed for risk mitigation.

In summary, from an organizational perspective

- A PHA team is multidisciplinary and comprises subject matter experts in the process being reviewed.
- The PHA leader and facilitator must be experienced and knowledgeable in PHA methodologies and terminologies.
- The business leadership is ultimately accountable and responsible for determining when PHAs are required and for ensuring quality PHAs are conducted.
- A well-conducted PHA provides assurance that all process hazards are identified and adequately addressed through engineering and administrative controls in that order.
- The importance of PHA cannot be understated:
    - PHAs identify potential hazardous outcomes and events, and require proactive solutions to these hazards.
    - PHAs provide opportunities for the best solutions for hazards—engineering controls.
    - PHAs helps in identifying required actions necessary for eliminating and/or reducing process and operational risks in process facilities.
    - PHAs highlight the interrelationships among PSM elements for ensuring the health and safety of all personnel required to work on the facility.
    - PHAs provide transparency and due diligence to protecting the health and safety of workers, environment, and facilities.

Mechanical Integrity of Equipment

The MI of a process system is the outcome of the following:

1. Process defenses:
    a. Operating principles and guidelines—Essentially, the goal is to ensure that the operating limits of temperatures, pressures, flows,

and other specified operating limits and conditions are not exceeded.

b. Process built in protection systems, such as pressure release systems, vents, overflow systems, and flares. In the event that operating principles and guidelines are exceeded, the built-in protective systems of the facility are activated to protect the facility.

2. Written procedures are intended to support the maintenance requirements of each category of equipment at the facility.
3. Inspection and testing requirements that are determined by manufacturer specifications and applicable codes. Inspection and testing are proactive measures to examine the degree of wear-and-tear experience by equipment during fixed cycles of operation. Erosion and corrosion are common concerns for process equipment and machinery that must be assessed and evaluated to prevent failure. In many instances, inspection and testing frequency may exceed manufacturer recommendations.
4. Quality assurance helps to maintain the integrity of construction by ensuring the proper use of materials and techniques (such as welding). Quality assurance is essential in assuring the defenses of the process.

According to OSHA (2010), key requirements of MI include

1. Naming and categorizing equipment and instrumentation
2. Establishing inspections and testing requirements and acceptance criteria for each piece of equipment and instrumentation and the required inspection and testing frequency
3. Development of required preventative and maintenance procedures
4. A trained and competent maintenance workforce
5. Adherence to manufacturer recommendations
6. Documentation of test and inspection results
7. Adherence to manufacturer recommendations for equipment and instrumentation

In summary, from an organizational perspective

- MI is essential to ensure that the integrity of a process is maintained through adequate preventative maintenance and inspection, and testing procedures, to prevent unexpected equipment failures.

- MI requires demonstrated commitment to using written procedures, and for trained and competent personnel, maintaining the ongoing integrity of the process equipment.
- MI includes a comprehensive approach to equipment inspection and testing, and documentation, of findings such that appropriate frequency can be established. At the very minimum, manufacturer recommendation must be maintained.

Operating Procedures

Operating procedures or SOPs are tools provided by the organization that clearly describe the sequential steps required for tasks to be performed such that all work can be done safely. SOPs also provide technically correct operating limits and normal operating conditions for each process that personnel must abide to. As part of the routine work, SOPs identifies data to be collected, the frequency of data collection, and required samples to be collected. Ultimately, SOPs are provided to all workers to ensure that work is done safely at all times and the safe operation of any facility is maintained.

Through the involvement of those using these SOPs, they are revised periodically and always maintained current, through version controls. If an SOP has been identified to have an obvious error that can jeopardize the health and safety of workers, immediate steps must be taken to correct the deficiency in the SOP and to communicate the change to all users. When changes are required, the established MOC process must be followed. Where necessary, should training be required, competency assessment is also essential. SOPs must provide instructions on all safety requirements for performing each step of the task safely.

In summary, from an organizational perspective

- SOPs are developed to ensure that all nonroutine operating and maintenance work are conducted safely.
- SOPs are very effective training tools for developing the competency of inexperienced workers.
- SOPs must be made available and accessible to all workers (employees and contractors) at all times.
- When nonroutine work is being authorized, all required and relevant procedures for that work must be included in the work package and should be reviewed by the work group before any task is undertaken.
- SOPs must contain guidance around normal operation as well as required guidance around emergency operation.

In this chapter, we discussed the key elements of PSM that deals with

1. Preserving the health and safety of all workers at all times on a process facility
2. The required processes and systems for ensuring all work are conducted safely at all times
3. Maintaining the integrity of the facilities and processes such that the health and safety of all workers are maintained and the integrity facility is maintained

PSM introduces a bold and complete new approach for managing the health and safety of workers and for reducing the impact of capital losses associated with catastrophic failures and disasters in the workplace.

11

Contractor (Service Provider) Safety Management

Should we be concerned about contractor safety management? Contractors are hired to work for organizations (employers) for the following reasons:

1. Contractors provide a pool of resources, skills, and capabilities that are easily accessible to organizations but are not required by organizations on a continuous basis.
2. Contractors may possess specialized competency and capabilities that may not reside within the expertise of the organization.
3. Contractors are readily available and are very competitive and can do work more cheaply than if completed with internal organizational resources.
4. Contractors will often do the *dirty, dangerous,* and *difficult* (*3D*) jobs without complaints.
5. Contractors can do some specialized jobs more safely than organizations themselves.
6. Contractors provide a great opportunity to transfer risk associated with 3D work away from the organization.

Regardless of the circumstances, contractors are here to stay and they form an integral part of the workforce. Their health and safety in the workplace are of equivalent importance as that of employees and they must be treated in similar fashion if real improvements are to be achieved in the workplace.

Historically, organizations have treated contractors as *us* and *them,* and in the past, we were also able to transfer all health and safety responsibilities to contractors through contractual agreements and prime contractor arrangements. Many of these opportunities for risk transference continue to exist today, and are still used as an effective means for transferring risks away from organizations. However, progressive organizations with great safety cultures draw no distinction between the safety of organizations (employees) and contractors. Such organizations demand from the workforce that all work must be completed safely regardless of whether it is being done by employees or contractors.

Contractor safety management is quite possibly the greatest opportunity for improving organizational safety. This is so because contractors incur more injuries than employees on an annual basis. Historical data also show that CRIF is also higher than ERIF for many industries. For the oil and gas industry, records from the International Association of Oil and Gas Producers (OGP) confirm this relationship and also show improvements in both employee and contractor injury frequencies as detailed in Figures 11.1 and 11.2.

In Figure 11.1, contractor frequencies for recordable injuries (fatalities, lost workdays, restricted workdays, and medical treatment cases) were consistently higher than employee frequencies, and varied between 1.1 and 2.6 times that of employee injury frequency for the period 1999–2008. Similarly, Figure 11.2 highlights that contractor fatality frequencies were consistently higher than employee fatality frequency for the period 1999–2008 (except for 2008 where both employee and contractor fatalities were the same) and varied between 1.0 and 2.7 times that of employee fatality frequency.

While the argument may be made that contractors work significantly more hours than employees and exposure to hazards are greater, when normalized into frequencies, this argument is weakened. Experience in the oil and gas industry in Canada suggests excellent improvements in workplace safety, for both employee and contractors. Historical data for one large organization showed that almost 75% of all injuries in raw numbers were incurred by contractors. Similar historical data also confirmed that almost 65% of LTI were incurred by the contracting workforce for this organization.

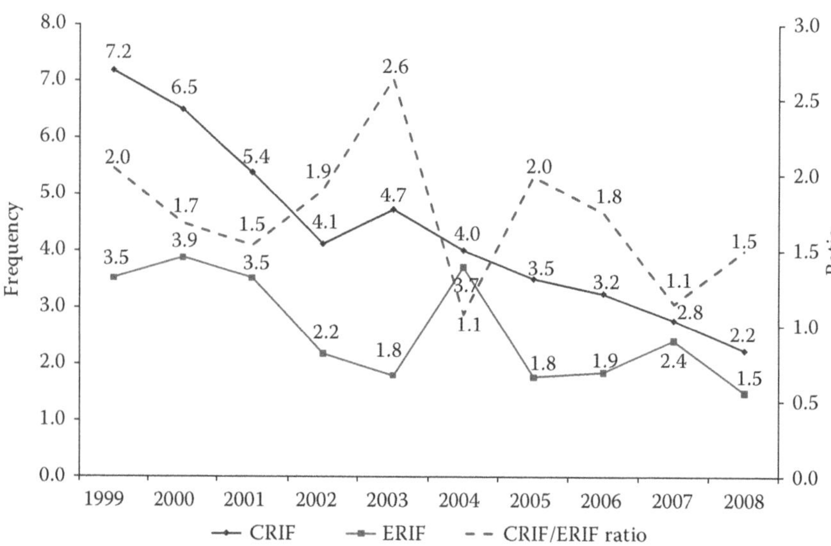

FIGURE 11.1
Injury frequencies for OGP companies (per million man-hours worked). (Developed using data from OGP Safety Performance Indicators 2008.)

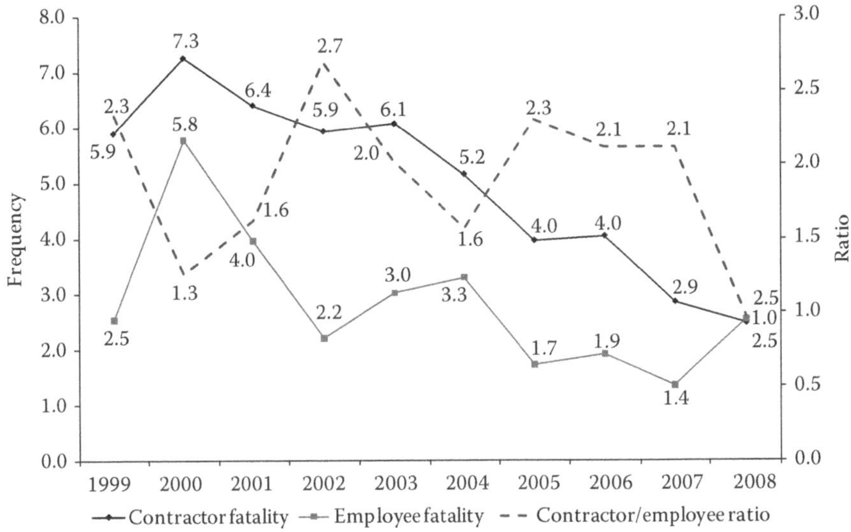

FIGURE 11.2
Fatality frequencies for OGP companies (per million man-hours worked). (Developed using data from OGP Safety Performance Indicators 2008.)

From the perspectives of the authors therefore, opportunities are abundant in the way we manage contractors to generate step changes in organizational safety performance. Contractors are more abundant in the workplace; they work more hours and are more exposed to hazards and 3D work situations. According to ENFORM Canada (2010), the consequences of inefficient, incomplete, or inconsistent approaches to contractor management can include the following:

1. Increased risk of costly delays
2. Costly mistakes
3. Hazards to health, safety, equipment, and the environment
4. In extreme cases, serious injury or fatality of workers can occur
5. Irrevocably damaged corporate reputation may also be possible

ENFORM Canada (2010) argued that investing the time and effort required to standardize the approach to contractor management is foundational to great business sense.

In this section, we shall provide an introduction to contractor safety management. The aim is to enhance the organizational ability to effectively manage the health and safety of a large contracting workforce that is dominant in our global workforce. Effective contractor safety management starts with a strong contractor management process.

Core Requirements of Contractor Management

Effective contractor management requires the combined efforts of multiple stakeholders who must work collaboratively to ensure the health and safety of contractors in the workplace. In large organizations, contractor management involves collaborative work among the following stakeholders:

1. Procurement and supply chain management (SCM)
2. Business units and leaders
3. Frontline supervisors and managers of both contractors and organizations
4. Support service personnel that include EH&S personnel
5. Legal
6. Contractor leadership

Success in contractor management is achieved when the combined stakeholder group interest is met. In addition, supporting processes and systems are simple and easy to follow. The core requirements of contractor management include the following:

1. Ownership for contractor management
2. A corporate standard for contractor management
3. Stakeholder interest map
4. Categorization of contractors for effective treatment
5. A contractor prequalification process
6. Contractor safety management
    a. A RACI chart
    b. Performance management processes and tools
    c. Stewardship of leading and lagging indicators

Ownership for Contractor Management

Who owns contractor management in an organization? Some may argue that it is the business unit utilizing the services of the contractor, while others may argue it is the SCM. The authors suggest that contractor management resides within the SCM organization. SCM is ultimately responsible for preparing the contractual document with input from all stakeholders, verifying that the contractor has the ability to complete the contracted work, and providing the necessary support to the business unit in managing the contract/contractor.

SCM enables the processes for contractors to be selected and contracted to perform work for the organization, and through a coordinated effort, accountability for the contract management process. Traditionally, SCM is responsible for developing the procurement strategy and getting contractors to work for business unit stakeholders. They are responsible for relationship management with contractors and for negotiating pricing structures with contractors.

Corporate Standard for Contractor Management

A corporate contractor management standard that defines roles and responsibilities of each stakeholder group, resource allocation strategies, and workflow is an essential requirement for ensuring standardization of the contractor management process across the organization. This standard also provides the road map to all stakeholders on how contractor management is done in the organization. A corporate standard must be simple to follow and use, while at the same time must be all-inclusive.

Stakeholder Interest Map

A stakeholder interest map provides an analysis of all stakeholders and their specific interests and goals regarding contractor management. Indeed, the stakeholder map can be reflected in a matrix that represents each stage of the contracting life cycle relative to each stakeholder group. Once a stakeholder map is developed, supporting strategies must also be developed to ensure that all stakeholder interests are met or satisfied. In the absence of stakeholder value maximization at each life cycle stage, the entire contractor management process is compromised. Table 11.1 provides a sample stakeholder interest map in which the interest of each stakeholder group is reflected in key words for the life cycle activity of contractor prequalification.

Categorization of Contractors

Not all contractors or service providers to an organization are exposed to operational risks. As a consequence, organizations must develop mechanisms for treating contractors differently, based on the risk exposure of the contractor or service provider who is working for the organization. Categorization of contractors is an essential requirement to ensure effective safety management of contractors. Contractors can be categorized primarily by type of work performed, risk, and spend impacts to the organization. This categorization provides opportunities to more effectively manage contractors such that value maximization can be derived within each category of contractors. Figure 11.3 demonstrates a method for categorizing contractors.

TABLE 11.1

Template of Stakeholder Interest Map for Each Stage of the Contracting Life Cycle

Stakeholder Groups / Contract Life Cycle Activities	SCM	Business Unit Leaders	Frontline Supervisors and Managers Contractors/ Organization	Support Services, for Example, EH&S/ Legal/Finance	Contractor Leadership
Contracting strategy	Satisfaction	Scope of work	Engaged	Engaged	Considered
Contractor prequalification[a]	Quality Technical Finance EH&S Scalable	Simple Reliable User friendly Credible Training	User friendly Simple Training Support	Flexible Scalable Credible	Simple User friendly Inexpensive
Contractor selection	Cost/safety/ quality/financially stable	Reliable/ cooperative/ easy to work with	Follows procedures safe	Collaborative	Considered
Contractor mobilization	Engaged	Lead	Involved	Verified	Satisfied
Performance management	Advised	Leading indicators	TRIF/CRIF/CDIF	Consulted	TRIF/CRIF/ CDIF
Contractor closeout	Share knowledge	Feedback	Consulted	Liability	Repeat business

CDIF-Contractor disabling injury frequency.

[a] Sample key words to reflect stakeholder interest. The entire matrix must be filled out.

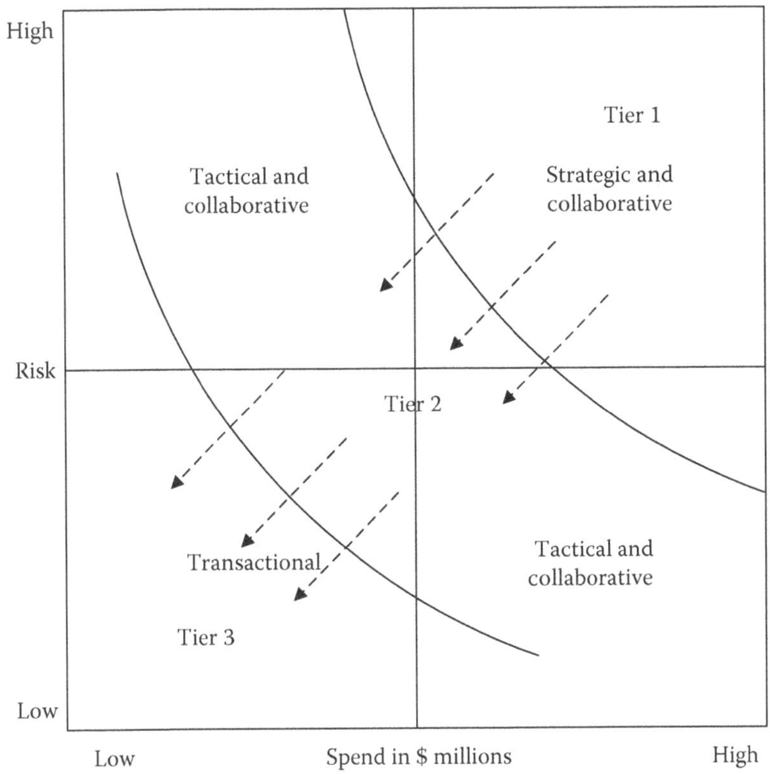

FIGURE 11.3
Categorization and behavior requirements for contractor management.

Once contractors are categorized based on risk and spend, organizations can develop the right behavioral strategies for managing them within defined groups.

The goal of categorizing contractors is ultimately to develop a structured approach to RM among contractors by placing them into categories for more effective group management. While the goal is to move contractors from high-risk to lower-risk categories, streamlining of strategies are now possible. Strategies must also be developed to move contractors from Tier 1 through Tier 3.

Tier 1 contractors are high-spend/high-risk contractors and or service providers to the organization. This is the most important group of contractors to the organization since this group of contractors and/or service providers is more continuous in its presence and involvement in the organization. In most instances, these contractors are almost the equivalent of employees, in their presence in the workplace, except that they are supervised and managed by the contractor leadership. The goal of the organization is to shift this group of contractors into Tier 2 category. Tier 1 contractors and service providers are required to operate with SMSs very similar to that of the organization or which may exceed that of the organization.

Collaborative relationships are required with Tier 1 contractors since they may be difficult to replace where long-term synergistic relationships are pursued. Collaboration is sought in risk mitigation and efficiency strategies where there are effective means for protecting the health and safety of all contractor workers. Strategic planning and collaborative relationships are necessary since business interruptions and adverse financial impacts are possible outcomes from failures in relationship management or safety performance with the contractor.

Tier 2 contractors are more continuous in their presence and involvement in the organization. Risk and spend requirement may vary from high to low. This group represents the second largest group of contractors or service providers an organization may engage with. There may be continuous goods and/or services provided, or multiple repeats of the goods and/or services provided. Risk varies because the contracting workforce may vary from time to time and the hazards may change in the work environment between periods of service from the contracting groups.

Tier 3 contractors are the most common to organizations and require primarily transactional behaviors for management. A good or service is provided and the contractor is paid. Goods and services provided may be one-off in nature. There is little need for relationship management. This group of contractors may account for the largest number of contractors in the organization.

A Contractor Prequalification Process

Contractor prequalification is an effective means for identifying which contractors meet the organization's requirements, to be able to perform work for the organization. Much of the research work on contractor safety management suggests that when organizations are successful in excluding poor-performing contractors from their business, overall performance improves. These performance improvements are reflected in the bottom line performance. Where EH&S prequalification is concerned, when unsafe contractors are excluded from the organization, there is a reduced effort required by the organization's EH&S resources in managing the safety performance of contractors. As such, organizational EH&S personnel are now freed up to support the organization's SMS in a more tangible manner. The EH&S performance of the organization improves dramatically when unsafe contractors are excluded from doing work for the organization.

Organizations may prequalify contractors internally or externally to the organization. Once prequalified, whether internally or externally, the contracting organization can then be invited to bid on work, to be performed in the organization that is being contracted out.

Internal Prequalification

Internal prequalification refers to the development of the capabilities and committing resources by the organization to verify that contractors have met certain threshold requirements to be able to perform work for the organization. Before a contractor is invited to perform work for the organization, the organization requests and assesses various relevant information from the contractor such as SMS, historical safety performance records, insurance records, and training and competency records to name a few. This information is evaluated against preestablished criteria developed by the organization to determine whether the contractor is qualified to perform work for the organization.

External Prequalification

External prequalification refers to a situation where organizations have transferred the prequalification process to specialized service providers who perform this work on behalf of the organization. External prequalification has mushroomed as a new and specialized business area for organizations that contract out a large percentage of its work. Such prequalification service providers do this work on behalf of the organization. Prequalification service providers specialize in gathering relevant information from contractors on behalf of organizations and verify that the information gathered meets the preestablished requirements for the contractor to perform work for the organization.

For many organizations, contractor prequalification is completed primarily for EH&S only. Recent trends, however, have seen a shift in prequalification toward a single platform for contractor prequalification on the bases of quality assurance, finance, technical, and EH&S requirement of the organization. A single platform for prequalification on the basis of the above four criteria provides organizations the opportunity to optimize its contractor management processes by focusing only on contractors that are prequalified in all four categories. Table 11.2 provides a matrix for prequalification requirements based on the above categories for prequalification and the need for basic, intermediate, or advanced prequalification relative to the tiered categorization of contractors.

Prequalification is generally performed on two levels. As discussed earlier, categorization of contractors can generate groups of contractors that may

TABLE 11.2

Contractor Prequalification Matrix Based on Categorization of Contractors

Prequalification Criteria — **Prequalification Required** / **Categories of Contractors**	**Quality**			**Finance**			**Technical**			**EH&S**		
	Basic	**Intermediate**	**Advanced**	**Basic**	**Intermediate**	**Advanced**	**Basic**	**Intermediate**	**Advanced**	**Basic**	**Intermediate**	**Advanced**
Tier 1 contractors			√			√			√	√	√	√
Tier 2 contractors		√			√			√		√	√	
Tier 3 contractors	NA			NA			NA			√		

Note: The advanced questionnaire includes the basic and intermediate questionnaires.

wrequire different treatments. The prequalification required for the various categories of contractors may vary as follows:

Tier 1 category of contractors (high risk/high spend): detailed and advanced prequalification process. An extensive and detailed evaluation process is conducted consistent with the questionnaire provided in Appendix 1.

Tier 2 category of contractors (intermediate risk and spend): intermediate prequalification process. Generally, contractors required to undergo an intermediate prequalification process will be asked to respond to the same questionnaire provided in Appendix 1. However, the level of verification required may be reduced.

Tier 3 category of contractors (low risk/low spend): basic prequalification process. The basic prequalification process is intended to acquire general demographic-type data. Very little workplace risk is presented by this group of contractors.

Indeed, the extent of external prequalification can also vary from partial external prequalification to full external prequalification.

Partial External Prequalification

Partial external prequalification refers to the organizational need to retain certain critical prequalification requirements for final determination, internal to the organization. For example, the organization may reserve the right to determine prequalification based on finance.

Full External Prequalification

Full external prequalification refers to full transference of all prequalification requirements to an external service provider. This is perhaps the most cost-effective approach to prequalification and generates the best results, since external prequalification service providers are specialized organizations that are continually improving the prequalification process. Many organizations have sprung up within the past decade offering full prequalification services to owners. Among the industry leaders in this area of expertise is Pacific Industrial Contractor Screening (PICS, http://www.picsauditing.com/). PICS offers a wide range of prequalification services that extend to the auditing and verification of contractor documents, information, and processes, on behalf of owners. PICS processes are designed to improve the health and safety of contractors by helping contractors achieve organizational compliance and meet regulatory compliance.

Such organizations have removed the challenges of cumbersome paper prequalification systems to a computerized, interactive, and supported process for verifying and auditing the information of contractors consistent with owner requirements to ensure that contractors are fit to perform work for the owner organization. A sample prequalification questionnaire is provided in Appendix 1. In most instances, both Tier 1 and Tier 2 contractors are subjected to an advanced prequalification questionnaire. Tier 3 contractors that generally provide low-risk services are subjected to a basic questionnaire. Often, organizations may choose to exclude Tier 3 contractors entirely from the prequalification process.

Contractor Safety Management

What does contractor safety management mean? The authors propose that contractor safety management refers to the management principles applied for ensuring the health and safety of all contract workers who do work for the organization, and are exposed to hazards, unique to the work environment in which they are asked to work. As discussed earlier, contractor safety management provides opportunities for significantly changing organizational safety performance.

Contractor safety management begins with the recognition that when contractors are hurt in the workplace, the organization (employer) is also affected very similarly or to a greater extent than the contracting organization. Worker morale, lost productivity, and direct cost associated with the incident all impact on the organization. Employing organizations must, therefore, work collaboratively with the contractor and its employees (contractors) to ensure that their health and safety are maintained during the conduct of their work.

Refer to the Texas City BP incident of March 23, 2005 in which 15 workers were killed and over 170 injured. Many of those killed as well as those injured were contractors. For this single event, BP claimed to have spent more than $1.6 billion in compensation to those affected by the incident. In addition, the OSHA-related fine for this incident was the largest in OSHA history. As one can well imagine, while contractors were most affected in this incident from a safety perspective, the fallout from the incident impacted on the organization (BP) to a much greater extent. Indeed, the image of BP was dramatically affected by this incident and it shall continue to impact on BP well into the future.

As indicated earlier, contractors outnumber employees in the workplace and are generally required to perform the *3D* jobs for organizations. More importantly, organizations are not immune to adverse impact and public scrutiny when contractors are injured in the workplace. Furthermore, as organizations continue to develop their safety cultures, contractors play an increasingly larger role in workplace safety. It follows to a good reason therefore that contractor safety management assumes greater prominence in the SMSs of organizations. In this section, we shall discuss the requirements for an effective contractor SMS.

Contractor safety management is a subset of contractor management. Figure 11.4 identifies all the supporting components (pillars) of an effective

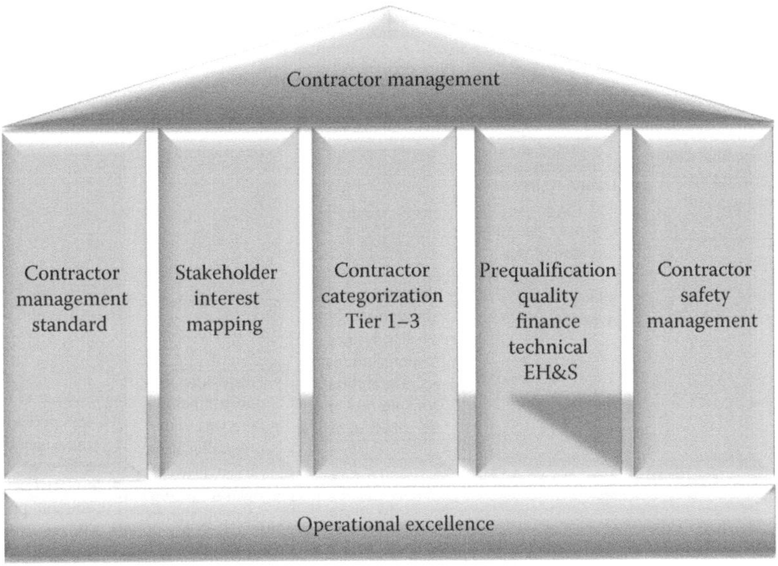

FIGURE 11.4
Supporting pillars of contractor management.

contractor management system, designed to generate operational excellence in contractor management.

Critical to success in contractor safety management is an effective EH&S prequalification process, which is rigorous and dependable and which meets the organization's need for sustained contractor safety performance. Figure 11.5 highlights the key requirements for an effective contractor SMS or process.

Contractor Safety Standard

As with many other business practices, a standard is required to ensure consistent approaches to contractor safety management within an organization. The standard generally highlights what is required, who is responsible for achieving these objectives, and may also highlight how these goals are to be achieved. Once approved by senior leadership in the organization, a contractor safety standard is designed to drive consistent behaviors and processes for improving contractor safety in the workplace for which the business is eventually held accountable against. Appendix 2 is a sample contractor safety standard courtesy of Suncor Energy Inc. 2010.

A RACI Chart

What is a RACI chart and why is it important? A RACI chart is a simple chart that defines who is responsible for the assigned work, who is accountable for

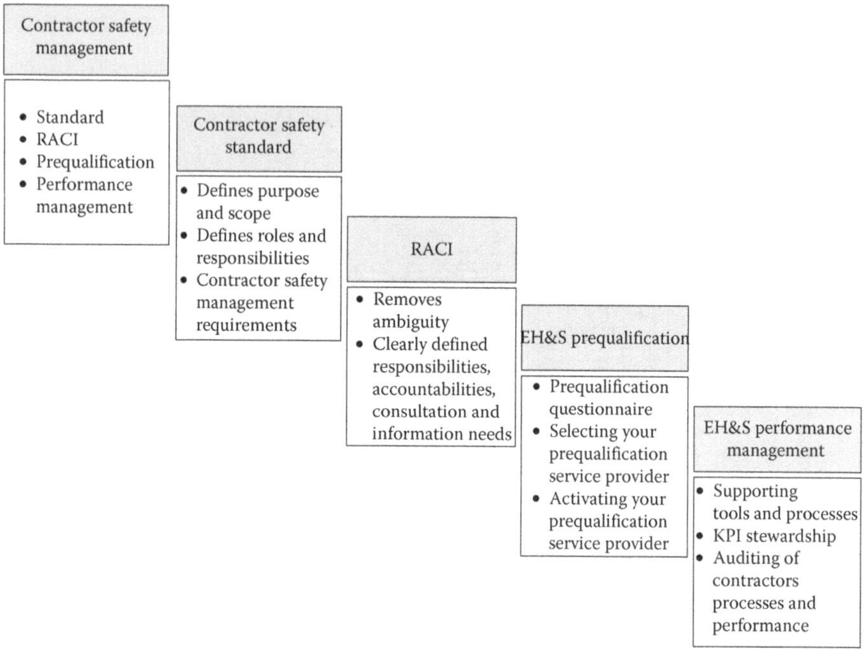

FIGURE 11.5
Requirements for contractor safety management.

the assigned work, who should be consulted, and who should be informed about the status of the assigned work. A RACI chart removes ambiguity about each stakeholder role in contractor safety management. When all stakeholders are aware of what is required of them, there is a reduced likelihood that key activities may slip by unaddressed. A RACI chart is a simple prompt and visible mechanism to remind key stakeholders in contractor management of their roles and responsibilities. This is very useful in avoiding undesirable contractor safety incidents. The acronym RACI denotes the following:

R—responsible

A—accountable

C—consulted

I—informed

A RACI chart is useful in eliminating confusion among personnel in terms of roles and responsibilities and generally serves as a powerful reminder to responsible and accountable participants on their deliverables and is an informal contract among individuals within a workgroup or across workgroups. A RACI chart is extremely important in preventing duplication of work and avoiding work being inadvertently forgotten or incomplete. During the project execution stage of a project, many RACI charts may be employed for different work activities. This simple tool works well in supporting continuity in work and ensuring that work continues at a managed pace.

R: Responsible

Stakeholder-assigned "R" is responsible for a particular activity and will be required to ensure that the activity is fully resolved and managed to completion, during the project execution stage, in the time allocated for doing so. Responsible individuals ensure that resources are obtained for completing the assigned work.

A: Accountable

Stakeholder-assigned "A" is ultimately accountable for each assigned activity and will be required to ensure that adequate attention, priority, and resources are allocated to the activity, to allow the responsible person, to properly execute the task.

C: Consulted

All stakeholder-assigned "C" may have information useful for resolving or addressing those activities for which they must be consulted. The onus is on the responsible person in seeking the input of the groups to be consulted, before the activity can be deemed complete.

I: Informed

Stakeholder-designated "I" is included in the RACI chart, for they are made aware of the status of activities and for decision making within their respective groups. Informed personnel have the opportunity to mitigate within their work groups against any unsafe conditions that may arise during the particular activity (Table 11.3).

Prequalification Questionnaire

A prequalification questionnaire for data collection should be dynamic and targeted. Questions presented to contractors must be relevant to the contractor's type of work and must drill down sufficiently such that adequate information is derived from the contractor to make meaningful decisions regarding the contractor's ability to perform work for the organization.

The prequalification questionnaire should not be a one-size-fits–all approach. It should be scalable to meet the needs of categorization of contractors, into risk and spend as discussed earlier. The questionnaire must also be industry specific and include types of work appropriate for best result as

TABLE 11.3

RACI Chart—Contractor Safety Management

Stakeholders / Key activities	Business unit and functional leaders	Contract coordinators	Supply chain leaders	Contractors	EH&S
Provide sufficient, qualified resources to implement standard	A,R		R		
Audit contractors for compliance to all environment, health and safety requirements in the contract	A,R		R		I
Identify a safety resource qualified to assess contractor safety	A,R				
Designate contract coordinators to oversee contracting process	A,R		R		
Establish a process to evaluate contractor safety performance	A,R		R		
Periodically evaluate the contractor EH&S management process to determine if changes are needed	A,R		R		C
Set objectives and expectations for contractor performance and establish accountability for objectives and expectations	A,R		R		
Set prequalification criteria for contractor safety performance	C		I		A,R
Confirm that contractor's proposal meets the bid package mandatory requirements			A,R		
Confirm that applicable environment, health and safety hazard information is communicated to contractor	R		A		
Document when all safety orientations or communications are provided to contractors	A,R		R		
Complete and document any work-site-specific training for contractors has been completed before work begins	A,R				
Assure that requests to employ contractors includes all safety requirements and limitations, descriptions of work location in reference to hazardous processes, boundaries for any areas where control is given over to contractors or constructors and assure compliance to the prequalification process	R		A,R		

Source: Adapted from Suncor Energy Inc., © copyright of Suncor Energy Inc., approval from Suncor is required to reproduce this work.

Note: R—responsible for the work; A—accountable for work outcome; C—consulted when necessary during the conduct of the work; I—informed after the work is completed

shown in Appendix 1. With the introduction of prequalification on the basis of quality assurance, there is a greater need to categorize contractors based on the types of work performed.

Selecting a Prequalification Service Provider (EH&S, Finance, Quality Assurance, and Technical)

When selecting an external prequalification service provider for EH&S, care must be taken to ensure that competing service providers are evaluated based on the following criteria:

1. The service provider is flexible and adaptable to the organization's needs—In this way, both the service provider and organization can

continuously improve processes to ensure that new knowledge is incorporated in the process for value creation in contractor prequalification.

2. Verification of contractor information—This is quite possibly the most important criterion for selection of a prequalification service provider. In the absence of the service provider verification process, the prequalification service provider becomes a database and registry in which contractors are required to post information that may be accessible by organizations. Information stored by the prequalification service provider must then be verified by the organization before contractors can be prequalified.
3. Trained, experienced, and competent personnel are used in verifying contractor information—Technological advances make it possible to buy off-the-shelf SMSs for satisfying prequalification requirements. It is inadequate for contractors to be prequalified to perform hazardous work purely on the basis of having an SMS reflected in documentation, manuals, procedures, and standards. When available, an SMS must be relevant, applicable to the type of work being undertaken, and easy to use. Trained, experienced, and competent safety personnel are capable of reviewing the contractor's safety management information for compliance to the types of work and possible hazards likely to be encountered during the conduct of the work.
4. User-friendly processes, systems, and tools—A prequalification service provider must provide a user-friendly system such that both contractor and organization interfaces are easy to use and simple. A complex system drives both groups away from the prequalification process leading to unwilling compliance by contractors and frustration in use by organizational personnel.
5. Training support—Regardless of the simplicity and user-friendly capabilities of a prequalification system, training is essential for all users. From the contractor's perspective, training is required to ensure that users are able to input data and information as required. Similarly, from the organizational user perspective, training is required to ensure that personnel are able to access information as required.
6. Technical support—When working with information technology, something will go wrong from time to time. A prequalification service provider must retain the expertise to rapidly respond to problems encountered and resolve concerns of both contractors and organizations.
7. Cost—Cost of service provided by the prequalification service provider is of importance to both owners and contractors. In a competitive environment, there are several prequalification service providers operating within the same market. At the same time, organizations may choose among these service providers the one that best

meets their business needs. The cost of the prequalification service to contractors is important to organizations because a large part of it is eventually passed on to the organization in the form of higher billable charges from the contractor. In most industries, many contractors may work for several organizations within a region. When organizations select different prequalification service providers, for contractors to continue to work for all of these organizations, they are required to become prequalified with several service providers. This creates additional burden and administrative challenges to contractors since the cost for prequalification with different service providers varies dramatically and they may all prequalify contractors differently.

8. Cultural fit and organizational alignment—When selecting a prequalification service provider, care must be taken to look beyond the sales pitch and examine the desires by the service providers to genuinely improve contractor health and safety in the workplace. This is a subjective measure resulting from the service provider's demonstration of empathy for contractor safety and genuine knowledge of safety challenges in the workplace where contractors are concerned.

A weighted scoring process is required to ensure that the best selection is made. Once a prequalification service provider is selected, the next difficult stage is to execute this process across the organization.

Activating Your Prequalification Service Provider

A significant amount of work is required in the activation of a prequalification service provider. As a consequence, when activating a prequalification service provider arrangement, the arrangement should be a relatively long-term relationship of not less than 3–5 years in duration. Care must be taken, therefore, to get it right during the activation and rollout of a prequalification service provider.

Activation of a prequalification service provider is primarily about change management within the organization. Attention must be provided to the following key activation steps:

1. Clearly define the scope of work for the prequalification service provider and execute contractual agreement between service provider and organization.
2. Configure the technological interface between the prequalification service provider and the organization so as to allow full use of the prequalification platform by the organization.

3. Where EH&S is concerned, the prequalification service provider is also required to configure the prequalification questionnaire consistent with the needs of the organization for relevant data and information collection.
4. Develop and execute internal communication plans. Internal communication is required to ensure that all internal stakeholders are fully aware about the reasons for selecting the prequalification service provider. Communication must highlight gaps in existing contractor safety management processes related to prequalification and how the prequalification service provider will assist in closing these gaps. Internal communication must be supported with an execution plan, inclusive of a schedule and resourcing requirements for success.
5. Develop and execute external communication plans. External communication is essential to ensure that all contractors are aware of the new requirements for prequalification. Joint town hall communication sessions with involvement from both the prequalification service provider and the organization, which address contractor safety management gaps and gap closure strategies with upgraded prequalification, are absolutely essential for success.
6. Training for both internal and external stakeholders is required to ensure success in activation. Poor training will usually result in poor recognition and acceptance of the prequalification process. Ultimate success is dependent upon the quality and extent of user training and competency.
7. Identify the best strategy for rollout. Some theories suggest piloting of rollout across the organization, thereby allowing for a learning period. The disadvantage of piloting the activation and rollout of a prequalification service provider is that buy-in and subsequent support from contractors can be weak, resulting in a less sustainable process. Piloting should be avoided where possible, in preference to full-scale rollout across a business or business area. Rollout based on geographical location may also be considered where the impact on the business can be high.

Contractor Performance Management and Control

Regardless of how effective your prequalification process may be, incidents involving contractors will occur. Organizations must, therefore, seek to ensure that adequate measures and controls are in place to prevent incidents. Control measures can be standardized or flexible. For large organizations, where consistent and standardized practices are required for driving consistency in contractor safety management, standardized control measures and tools are recommended.

Control Measures and Tools

A number of tools are available or can be created by organizations for developing consistency in the way contractor safety management occurs across the organization. This is particularly important when the organization may have multiple business units or may be geographically dispersed. The use of consistent and standardized tools and control measures within an organization helps contractors to better understand the requirements for doing work for that particular organization. Moreover, controls provide a further set of checks and balances, for ensuring that risk exposures are proactively addressed before contractors are allowed to work, and during the execution of the work.

Standardized Control Measures and Tools

Some common standardized tools used for contractor safety management and controls of contracted work include the following:

1. Contractor prequalification waiver request—For one reason or the other, a business unit or the organization may be required to hire a contractor to do work while the contractor is not prequalified to do so. A prequalification waiver request is a simple process that allows the organization to hire the contractor under certain conditions. Preconditions for use of a waiver include situations where the contractor is required for emergency-type work or the contractor or service provider may be a single source provider and has the governing power. Waivers should be of a limited duration and should not be repeatable for the same contractor. Finally, waivers should not be easy to access and should require senior leadership (vice president (VP) and above) approval as a means of discouraging its use. In all cases, the use of a waiver must be accompanied by a risk mitigation plan for meeting the EH&S requirements of the work.
2. Prebid meeting check sheet—Before prequalified contractors are invited to bid on contracted work, a prebid meeting check sheet, which highlights EH&S requirements and associated hazards, should be available for review and requirement guidance to all potential contractors. In this way, contractors can provide bid estimates reflective of the cost associated with addressing workplace hazards identified during the prebid engagement process.
3. Prejob mobilization check sheet—When a contractor is selected for the work, before the contractor is mobilized, a prejob mobilization check sheet is used to reiterate some of the requirements discussed in the prebid meeting. Reiteration is necessary since the contracting workforce doing the job may be very different from those at the prebid meeting. At this point also, the premobilization check sheet

provides stakeholders an opportunity to demonstrate prejob readiness for the required work.

4. Prime contractor designation—Often when contracting work occurs in an organization, the workspace can be carved out by the organization such that work can be conducted under the contractor's leadership and SMS. Transfer of ownership of the worksite to the contractor does not remove the organization from any related undesirable outcomes from an incident. Organizational oversight and engagement are essential for long-term success.
5. Contractor responsibilities check sheet—This tool is intended to remind contractors of its responsibilities for providing trained and competent workers, its need for oversight and supervision, and its need for record management and transparency in its activities. Moreover, it is intended as a reminder of the EH&S responsibilities related to the work being performed.
6. Physical condition inspection sheet—This is a simple tool for verifying among other things, the contractor's attitude and compliance toward housekeeping, equipment management, emergency systems, and commitment to the use of PPE.
7. Contractor activation—Activating contractors across the organization should be a simple and standardized process. An effective onboarding process is required such that the contractors understand all requirements of the workplace. This may include site orientation, emergency response, and muster point. Activation also includes ensuring that the contractor knows where to access information relating to the worksite, how to report incidents and near misses, and the general operating requirements of the worksite regarding EH&S.
8. Role description—A critical requirement for contractor safety management is the need for role descriptions and consistency in role definition across the organization. For example, the role of a contract coordinator must be clearly defined and above all must be clear to the contract coordinator himself/herself.
9. Training and competency matrix—A training and competency matrix is a very powerful tool in driving worker behavior. Lutchman (2010) discussed the push/pull effects of a visible training and competency matrix. A training and competency matrix with highly contrasting colors (red—untrained and incompetent; green—trained and competent) located in a high-traffic area of the workplace forces workers to consult with supervision to become trained and competent from peer pressure (pull effect), or, alternatively, supervisors literally force workers into training and competency programs (push effect). In both instances, the outcome is the same,

a visible training and competency matrix leading to a more competent workforce.

10. Contractor audit compliance procedure—A consistent approach to contractor work management system auditing is essential for ensuring the health and safety of all workers. It is not enough for a contractor to say that the contracting workforce is trained and competent to do the assigned work. Due diligence verification requirements are essential. It is not uncommon for organizations to demand that the contractor provide a verifiable process and records of contractor training and competency. As a consequence, the organization should establish a common robust process for auditing the work processes of the contractor. For training and competency, as an example, the audit should establish the presence of training programs and competency evaluation process and records. More importantly, records of this nature should be retained by the contractor for easy verification of compliance with the organizational requirements for a trained and competent contractor workforce. Where gaps in the contractor SMS are identified, gap closure strategies must be developed, resources committed, and the gaps closed in a controlled manner based on the magnitude of the risk exposure.
11. Contractor performance evaluation—Upon completion of contracted work, the performance of the contractor must be evaluated on a consistent basis. EH&S performance evaluation helps to determine whether or not this contractor will be used in future work. Providing feedback to the contractor is essential to ensure that the contractor is aware of any safety management performance concerns that can be proactively addressed before future work becomes available to the contractor.

Flexible Control Tools

Flexible control tools and measures may be business unit or job/task specific. These tools are intended to assist the business unit in stewardship of contractor safety management based on unique hazards associated with the contractor's work. They can be modified to meet the specific needs of the business unit or business area.

1. Field change notice—This is a simple process that engages contractors in addressing changing environmental hazards as the work progresses. It is also a simple tool for communicating the scope of work changes that may result in the introduction of new workplace hazards, which should be proactively addressed.
2. Field observation form—A flexible process for recording field observations on a day-to-day basis. The field observation form allows organizational leaders to capture both good and bad observations in a dated process such that appropriate referencing in performance

feedback and remedial actions can be made. This is an effective process for generating praises for specific work and similarly for addressing specific performance challenges.

3. Workplace observation form—This is a simple and flexible process for identifying, recording, and addressing workplace hazards. With daily-changing workplace hazards, this tool is an effective means for frontline supervisors and leaders to proactively address identified workplace hazards.

Contractor Performance Management

Once a contractor has been prequalified and selected to work, organizations must proactively manage the safety performance of the contractor. Changing environmental and process safety and personnel-introduced hazards ultimately lead to incidents in the workplace. A qualitative review of industry practices, and cross-industry practices highlighted the following key opportunities for improving contractor safety management.

Leadership Visibility Is Essential for Contractor Safety Management

Perhaps the single most important means for improving contractor safety in the workplace is through leadership visibility and oversight. Commitment to safety has to start from the top. That is why senior leadership visibility is essential for creating an environment in which people can perform at their best and reduce trends associated with incidents and accidents. Leadership visibility can be classified into the following:

1. Frontline leadership visibility
2. Senior leadership visibility

Industry knowledge shows that when leaders are visible in the field, safety performance improves. Leadership visibility demonstrates care for workers and is critical in earning trust through demonstrated behaviors of empathy and care for workers from both employees and contractor. Trust is developed when words and actions of leaders are aligned and being face to face with workers is the first step in establishing this relationship.

Frontline Leadership Visibility

Regular leadership presence at the frontline is necessary to build and reinforce the safety vision and ultimately culture of the organization. Being present at the frontline also provides immediate and direct feedback to leaders of the work standards applied, workplace hazards, and EH&S concerns and deficiencies. To adequately manage workloads and cost, a risk-based approach for determining the extent and frequency of leadership visibility and which leaders must be visible at the frontline may be necessary. Consequently, leadership visibility may vary from one team to another. For example, operation line leaders may be required to be visible more frequently than other more senior leaders. Key messages from all leaders must be aligned with the line leader. Alignment on these messages should be in place before a site visit is made and during the preparation stage of the visit.

Senior Leadership Visibility

Visits to the frontline by a VP or CEO generally create a buzz among frontline workers. Almost everyone at the frontline wants to be noticed or to speak with the senior leaders; to be able to do so is almost a badge of honor among frontline workers. This feeling of honor does not discriminate between whether the VP of finance or the VP of operations interacts with the frontline worker. It is imperative that the senior leader takes the time to inquire about the worker's well-being in the field and to seek to learn more about the challenges faced by workers during such visits. Preparation for site visits by senior leaders is essential to maximize value derived from the visit. Maximum impact is derived from joint organization and contractor senior leadership visibility at the frontline. Whenever possible, senior organizational leadership must identify and invite a peer contractor (risk-based key contractors) representative to visit at the same time, to show a joint support for safe work behaviors and practices and to legitimize in no uncertain terms worker behaviors toward safety. The following structured approach to leadership visibility is recommended to maximize benefits:

1. Leadership visits should be fit for the purpose—Visit with a reason; visits must be timely and relevant to the business.
2. Frequency and duration of leadership visits should be risk based as determined by the organization.
3. Site visits must be properly planned with defined guidelines and objectives.
4. Where possible, site visits by organizational senior leadership should be coordinated to be conducted jointly with senior contractor leadership.
5. During site visits, organization/contractor safety messages should be aligned. Preplanning should be leveraged to eliminate or avoid conflicting priorities and focus areas.
6. Leadership skills should be used to embrace contractor involvement and integration.

7. Organizational leadership visits to contractor shops and facilities seek to ensure that safe work behaviors are practiced, both at the contractors' home base as well as at the organization's worksite. The aim is to minimize the changes in worker behaviors, when working at the organization's sites.
8. Senior leaders must seek to communicate more with workers where safety is concerned. The goal is to ensure that frontline workers are engaged. Leaders must listen to and act on the concerns of frontline workers.

Both contractor and organizational frontline leaders are critically important in promoting safe work behaviors and supporting the health and safety of contractors and employees (workers) alike. Their presence and oversight in the workplace provide them with opportunities to proactively take charge of unsafe conditions, behaviors, actions, or work practices before incidents may occur.

The challenge for most frontline supervisors and leaders is to prioritize work more effectively so that they can be more visible and accessible at the frontline. Preparing reports, schedules, and other administrative duties all get in the way of allowing frontline supervisors and leaders to be visible at the frontline. In addition, the skills and capabilities of the frontline leader may prevent him/her from doing an effective job, even if time is available for frontline visibility. We shall discuss this in more detail in Chapter 12.

Demonstrate and Promote a No-Blame Culture for Reporting Incidents

Contractors are sensitive to punitive consequences for reporting near misses and incidents. When the consequences for reporting incidents are punitive, contractors will avoid reporting near misses and incidents. A valuable opportunity for improving safety is missed when incidents are not reported. This is particularly important for near misses since knowledge generated from near misses can be used to prevent actual incidents.

Active Listening Sessions/Lunches: Listening Moment

Within the workplace, contractors at the frontline are acutely aware of new and evolving workplace hazards. Regardless of how these hazards originated, organizational leadership must be made aware of them before they result in incidents. An effective means for proactively obtaining this information from contractors is to promote listening moments. Listening moments are informal sessions (no documentation, with lunch or snacks provided) where frontline contractor supervisors are encouraged to talk about prevailing and undisclosed potential hazards in the workplace. Once hazards have been identified, site leadership is obligated to act on these concerns since failure to do so will result in a loss of credibility.

All Workers Must Be Treated Equally (Where Safety Is Concerned)

The authors acknowledge that organizations cannot treat all workers (contractors and employees) the same in the workplace for liability and other employer/employee reasons. Nevertheless, *where safety is concerned*, we are fully convinced that all workers should be treated equally. Areas where contractors and employees must both be treated the same include, but are not limited to, the following:

1. All workers must be aware of all workplace hazards.
2. All workers must have access to procedures and relevant information required to perform assigned tasks and work safely.
3. All workers must be properly trained and competent to do the work they are assigned to perform.
4. All workers must have access to PPE and must be properly trained and competent in the use of the PPE.
5. All workers must be properly supervised where hazardous work is concerned.

Perhaps the single most important area where improvements can be made on equal worker treatment is in the area of access to information and in particular work procedures that may have been developed by the organization. With the advent of information technology, information is now stored and accessed electronically. However, access to organizational information database is not equal for contractors, thereby placing them at a disadvantage where access to critical work-related information is concerned. This can be easily remedied by maintaining key roles in the organization for contractor/organization interface management whereby all work-relevant information can be accessed.

Initiate Regular Contractor Safety Forums

Contractor safety forms are an important means by which safety knowledge and information can be rapidly disseminated across the contracting workforce. The question faced by many organizations is who will pick up the cost and deal with the administrative nightmares associated with organizing such an event. One practical solution for such events is to have multiple organizational input and shared cost where a common or regional contracting group is used for supporting the organization's business.

Contractor Relationship Management

Relationship management with contractors is a critical component for improving contractor safety. Some of the more common organization/contractor relationships include transactional, autocratic, and collaborative. Where workplace safety is concerned, the most rewarding relationships between organizations and contractors are collaborative relationships.

When both organization and contractor work collaboratively to identify and address hazards in the workplace, identify and resolve training and competency gaps, and perform on a joint platform of value maximization, contractor safety performance will improve. To get to this state, however, organizations must recognize that when contractors are hurt or injured in the workplace, the organization is impacted on similarly to or more than the contractor.

Theron et al. (2008) suggested that strong contractor relationships are built on trust. What this means is that when work is contracted out, the parties involved are required to enter into a partnership built on trust. As we all know, trust is built on fair treatment, credibility, and transparency. More importantly, trust is earned based on behaviors over time. Relationship management therefore requires effort and commitment on both the organization and contractor paths.

Link Contractor Safety Management to Pay for Performance and Bonuses

We have often heard the phrase *what gets measured gets done* and *what gets rewarded gets done*. When the organization's internal performance measures include contractor safety management KPIs, there is greater attention to this business requirement. For maximum success, the organization should focus on a few key contractor safety management KPIs until they have improved to the level of performance required, before expanding the range to include others that may be more peripheral and lower in value-added contributions. Furthermore, the focus should also be on leading indicators, which allow for proactive responses as opposed to lagging indicators that provide a rearview mirror perspective after the incidents have occurred.

Shared Learning

Shared learning in contractor safety management across the organization is a very effective means for improving the safety performance of contractors. There are many ways by which learning can be shared. However, learning identified is dependent upon the quality of investigation undertaken and the effectiveness of the medium adopted for sharing this learning as discussed in Chapter 13.

Stewardship of Leading and Lagging Indicators

Leading and lagging indicator stewardship provide opportunities for significantly upgrading contractor safety performance. When identifying performance indicators, it is important that these indicators are relevant, measurable, value added, and relatively easy to collect and steward. More importantly, it is best to start with 3–5 high-value indicators, perfect performance in these areas, and then expand the stewardship to widen the scope of control.

For greatest success in establishing and stewarding performance indicators, the following should be considered:

1. Method of calculation.
2. Control range of performance indicator (i.e., upper and lower limits of performance for the particular indicator).
3. Where the data were collected and who in the organization is responsible for collecting the data.
4. The frequency of collecting the data. Data collection frequency should be such that it allows for proactive corrective actions by the business.
5. Reported by which stakeholder group.

Lagging Indicators

Lagging indicators are performance measures that report on the indicator after an incident has occurred. Lagging indicators provide a rearview mirror assessment of performance on what is being measured. Long-term trends developed from lagging indicators are very instrumental in communicating progress made by the organization over time. For the contractor workforce, lagging indicators commonly measured and stewarded include those provided in Table 11.4.

Leading Indicators

Leading indicators are performance measures that report on the indicator before incidents occur and are, therefore, proactive measures for stewardship. Leading indicators in contractor safety management have not been clearly defined in industries but variants are very prevalent among many businesses and contractors. Similar to lagging indicators, leading indicators can be trended. Trends generated can be correlated with injuries and incidents and can be used as a very powerful tool for communicating the benefits of proactive focus on the indicator to improve contractor safety in the workplace. For the contracting workforce, leading indicators recommended for stewardship include those provided in Table 11.5.

TABLE 11.4

Commonly Stewarded Contractor Safety Lagging Indicators

Performance Indicator	Contractor Total Injury Frequency (TIF)
Method of calculation	• (Number of contractor injuries × 200,000) ÷ contractor exposure hours • Number of injuries to include first aids, medical treatment, restricted work, LTI, and fatalities
Control range/target	• Trending toward zero—to be determined by the organization based on risk ranking of the contractor category
Frequency of reporting	• Monthly
Data collected by	• Business units collect data and perform calculations
Report by	• Business unit or EH&S. Where multiple business units are available, EH&S will generally collate to present the organizational performance

Performance Indicator	CRIF
Method of calculation	• (Number of contractor injuries × 200,000) ÷ contractor exposure hours • Number of injuries to include medical treatment, restricted work, LTI, and fatalities
Control range/target	• Trending towards zero—to be determined by the organization based on risk ranking of the contractor category
Period of measure	• Monthly
Data collected by	• Business units collect data and perform calculations
Report by	• Where multiple business units are available, EH&S will generally collate to present the organizational performance

Performance Indicator	Contractor Lost Time Injuries (LTI)
Method of calculation	• (Number of contractor injuries × 200,000) ÷ contractor exposure hours • Number of injuries to include LTI and fatalities
Control range/target	• Trending toward zero—to be determined by the organization based on risk ranking of the contractor category
Period of measure	• Monthly
Data collected by	• Business units collect data and perform calculations
Report by	• Where multiple business units are available, EH&S will generally collate to present the organizational performance

Performance Indicator	Workmen Compensation (Insurance) Ratio
Method of calculation	• Generated by insurance organization
Control range/target	• <1.0
Period of measure	• Per insurance frequency but a minimum quarterly
Data collected by	• Contractor receives data and shares with business unit leaders
Report by	• Business unit leaders

TABLE 11.5

Recommended Contractor Safety Leading Indicators for Stewardship

Performance Indicator	Number of Quality Toolbox Talks and Meetings Performed
Method of calculation	• The number and frequency of toolbox talks and meetings to be defined by onsite leadership and will vary based on risk, exposure hours, and requirement to generate improvement. Toolbox talks and meetings are intended to achieve the following: • Definition of scope of work • Confirm resourcing requirements • Verify contractor competency and capabilities • Review of immediate and known risks • Review mitigation plans for risks identified • Reporting can be done by tracking the number of job permits signed, and spot check audits on the content by the business unit
Control range/target	• 100% compliance
Period of measure	• Biweekly
Data collected by	• Contractor collects data
Report by	• Contractor to supply and business units leadership to verify compliance

Performance Indicator	Percentage (%) of Time of Frontline Leadership Spent in Field versus in Office
Method of calculation	• Self-reporting by frontline contractor supervisors will be the primary source of reporting. Audits to verify can be done through outlook calendar, and consultation with business unit leadership
Control range/target	• Recommend a minimum of 50% of time spent in field supervision and oversight
Period of measure	• Biweekly
Data collected by	• Contractor collects data
Report by	• Contractor to supply and business unit leadership to confirm

Performance Indicator	Percentage (%) of Audit Observations versus Corrective Actions Resolved
Method of calculation	• The field audits would be conducted by business unit EH&S personnel. Where risk exposures are identified, corrective actions will be defined and a timeframe for implementing the corrective actions implementation. A percentage of audit findings versus corrective actions implemented on schedule should be maintained • Require business unit to track and contractor responsibility for executing corrective actions assigned
Control range/target	• High-risk corrective actions—100% • 95% of all corrective actions implemented on time
Period of measure	• Monthly
Data collected by	• Contractor collects data
Report by	• Contractor to supply and business unit leadership to confirm

TABLE 11.5 (continued)

Recommended Contractor Safety Leading Indicators for Stewardship

Performance Indicator	Percentage (%) of Employees in Competency Assurance Plan
Method of calculation	• Competency of contractor workforce is to be verified during the premobilization phase. Ongoing audits by organization/business unit and self-reporting by contractors must support competency assurance exceeds threshold 75% competency in workforce • Important to point out that competency extends beyond training and is reflected in competency assessment and signoff by a competent assessor
Control range/target	• Minimum 75% competency in contracting workforce
Period of measure	• Monthly
Data collected by	• Contractor collects data
Report by	• Contractor to provide data and business unit leadership to verify

Performance Indicator	Number of Near Misses Reported with Mitigation Action Completed
Method of calculation	• Daily reviews of *near misses* reports should be followed up and corrective actions should be taken to eliminate the hazard and prevent it from reoccurring. The measure would take the percentage of corrective actions taken as a result of the number of near misses reported • Contractors maintain *near misses* action log and report to organization
Control range/target	• Measure
Period of measure	• Weekly
Data collected by	• Contractor collects data
Report by	• Contractor to provide data and business unit leadership to verify

Performance Indicator	Percentage (%) of Quality Job Safety Analysis (JSA) and Field-Level Risk Assessment (FLRA) Performed Prior to Beginning Work
Method of calculation	• Number and quality of the JSA and FLRAs should be audited periodically to confirm the analyses are done prior to beginning the work and these are effective and contributing to reducing the possibilities of injury. The analyses should have specific activities and expectations defined so they can be reviewed for quality
Control range/target	• 100% compliance. Checking the box does not represent a quality exercise
Period of measure	• Weekly
Data collected by	• Contractor collects data
Report by	• Contractor to provide data and business unit leadership to verify
Method of calculation	• This measurement assesses the experience of an assigned work crew. The purpose of this indicator is to ensure that contractor work crews are not entirely comprised of new workers to the worksite or to the work being performed. This leading indicator may comprise of two parts

continued

TABLE 11.5 (continued)

Recommended Contractor Safety Leading Indicators for Stewardship

Performance Indicator	Percentage (%) of Quality Job Safety Analysis (JSA) and Field-Level Risk Assessment (FLRA) Performed Prior to Beginning Work
	• Amount of time working for the organization at that particular worksite • The collective experience of the workers in performing the specific work • In both cases, it is undesirable to have an entire crew limited site knowledge or work experience with the particular task or work assignment
Control range/target	• Average years of experience on particular work crew. Actual number to be determined by business
Period of measure	• During prejob assessment and crew changeover
Data collected by	• Contractor collects data
Report by	• Contractor to provide data and business unit leadership to verify

Performance Indicator	Number of Joint Field Visits Performed by Senior Organization and Contractor Leadership
Method of calculation	• This is a measure of joint senior leadership visibility. Visits may include a review of the work being undertaken the field or may be a specific site safety audit. In all cases of leadership visibility, worker engagement is recommended
Control range/target	• Target defined by each contract; 100% compliance
Period of measure	• Bimonthly
Data collected by	• Either business unit or contractor to collect data
Report by	• Business unit leadership to verify

Performance Indicator	Number of Safety Alerts Developed (Reviewed) and Shared across the Organization
Method of calculation	• This is a proactive measure of the organization's ability to share learning across the business and it measures the incident reporting effectiveness across both the contractor and organization's businesses. When incidents requiring sharing occur, knowledge generated from investigations should be shared in a way that corrective actions or modified procedures/practices can be put in place to prevent the same incident from occurring elsewhere. Alerts help to facilitate this goal
Control range/target	• All high-potential incidents/serious injuries reported at 100%
Period of measure	• Monthly
Data collected by	• Both business unit and contractor to collect data
Report by	• Contractor and organization

TABLE 11.5 (continued)

Recommended Contractor Safety Leading Indicators for Stewardship

Performance Indicator	Number of Listening Moments and Proactive Safety Improvement Suggestions Submitted by the Contractor
Method of calculation	• This measure contributes toward the safety culture development of the organization and the level of trust and openness being developed between both contractor and organization. The measure best reflects in the ratio of suggestions received versus action on suggestions implemented
Control range/target	• Discretionary
Period of measure	• Monthly
Data collected by	• Both business unit and contractor to collect data
Report by	• Contractor and organization

Performance Indicator	Percentage (%) of New Contractor Employees Properly Onboarded and Oriented
Method of calculation	• Proper onboarding and orientation ensures that all workers are aware of the workplace hazards and emergency response plans. Stewardship of this indicator reflects the % of contractor workers that received a site safety orientation as required prior to beginning work on the site
Control range/target	• 100%
Period of measure	• Monthly
Data collected by	• Contractor to collect data
Report by	• Contractor to provide data, business unit leadership to verify

Source: Adapted from Suncor Energy Inc., © copyright of Suncor Energy Inc., approval from Suncor is required to reproduce this work.

Contractor Audits and Follow-Up

When collaborative relationships are established with contractors, audits are viewed as proactive means for identifying gaps or weaknesses in the contractor SMSs that may result in losses or significant events. When audits are presented in this manner and conducted to generate value-added opportunities for both contractor and organization, significant improvements in contractor safety performance are realistically possible. Figure 11.6 demonstrates the improvement impact of field auditing of contractors. A 50% improvement in contractor injury frequency was recorded over a 2-year period of auditing and collaborative improvement actions.

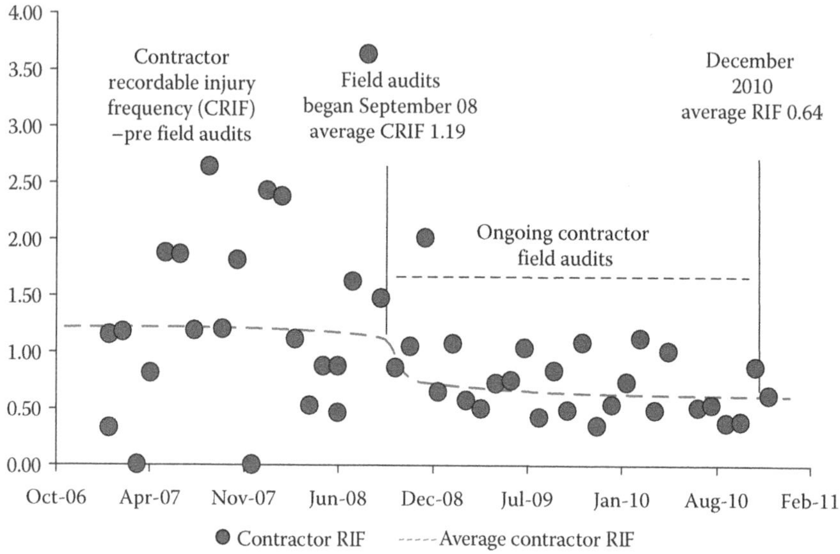

FIGURE 11.6
Impact of field audits on contractor injury frequency. (Adapted from Suncor Energy Inc., © copyright of Suncor Energy Inc., approval from Suncor is required to reproduce this work.)

Contractor audits can be of three types:

1. Self-audits—audits performed by the contractor representative as a proactive means for self-improvement and compliance.
2. Second-party audits—contractor audits conducted by an organization having an interest in the contractor's safety performance such as SCM or a business unit.
3. Third-party independent audits—audits performed by a specialized auditor that is hired to perform an audit of the contractor's SMSs and processes. An unbiased opinion of findings and gap closure recommendations are provided in an audit report upon completion of the audit.

Once treated as an improvement process and tool, audits provide a powerful process for improving contractor safety. Audit findings must be classified according to the risk exposure the findings may present to the organization. Guidance around classification of risks, types of corrective actions, and guidance for classification are provided in Table 11.6.

Contractor safety management is a problem faced by many industries. It requires consistent joint approaches to make significant improvements. Prequalification of contractors on similar platforms have started the shift

TABLE 11.6

Contractor Audit Matrix and Risk Exposures Guidance

Risk Exposure	Actions Required	Exposure Guidance
Unacceptable	• Immediate corrective actions required	• Significant weaknesses are determined in the SMS of the contractor. If left unattended, serious incidents including fatalities can occur
Unsatisfactory	• Corrective actions required within a reasonable period with acceptable controls	• Contractor SMS lacks essential control elements or may be deficient. There are no immediate risks of fatalities or significant incidents but these may arise if the situation is left unattended. Action must be taken within a defined period to prevent losses to the contractor or to the organization
Medium	• Situation can be improved, but controls are adequate at present	• Contractor safety management is adequate with some room for improvements. Controls in place are generally adequate to avoid unplanned incidents. Some control weaknesses, errors, or compliance issues may be present that require timely corrective actions, or management acceptance of risk. Weaknesses, either individually or collectively, do not significantly pose a great risk to either the contractor or the organization
Low or acceptable	• Sustainment	• Contractor SMS is adequate and available controls in place, sound and effective. No immediate risk exposures are identified.

toward industry standardization. However, much work is yet to be accomplished before the victory bell can be rung. Recognizing that contractors are mobile, moving from job to job and organization to organization, until there is a united industry approach to standards, work practices, control measures and tools, training and competency assessments, consistency in language, and performance management approaches, we shall continue to make marginal improvements to contractor safety management. The first step in the right direction is an acknowledgment and recognition of a problem in contractor safety management.

Today, many industries have acknowledged this. The opportunity is ripe for organizations to make a step change in contractor safety management by evolving toward consistent practices. We sincerely hope this chapter provides an opportunity to move us all toward this much-needed standardization for the improved health and safety welfare of all workers in the workplace, employees and contractors alike.

12

Leadership at the Frontline

One of the most overlooked opportunities for improving organizational safety is by upgrading the leadership skills of frontline supervisors. Often, when leaders are selected at the frontline, selection criteria often include, but are not limited to, the following:

1. Length of service with the organization
2. Amount of training the worker has been exposed to
3. Reliability of the worker in completing work
4. Attendance of the worker
5. Technical skills and competence of the worker
6. The age of the worker
7. Relationship with middle managers
8. Personality
9. Availability to assume the role

Some of the most overlooked criteria include the leadership skills and potential leadership skills of the worker and the worker's communication skills. In very active markets, we have also seen availability being one of the key selection criteria. In this chapter, the authors would like to explore the leadership requirement for frontline supervisors and the essential skills required by them to better serve and lead a workforce whose safety ultimately depends upon the direction and guidance provided by frontline leaders.

Role of the Frontline Supervisor/Leader

The authors are of the view that where safety is concerned, frontline supervisors are required to achieve the following:

1. Set the standards for safety.
2. Be proactive in managing risks and the health and safety of all workers.
3. Build trust and relationships among workforce.

4. Build talent and capability.
5. Engage and motivate workers to do the right thing.
6. Leverage existing tools and workforce to ensure that work is conducted safely at all times.

Set the Standards for Safety

Setting the standards for safety is perhaps one of the most important roles of the frontline supervisor. Workers look to the supervisor—their first contact with the organization—for guidance while they complete work. If the frontline supervisor sets low safety standards, then one should expect poor safety performance. Conversely, high safety standards help to preserve the health and safety of the workforce. Setting standards are demonstrated by the following frontline leadership behaviors according to Suncor Energy Inc.*

- Frontline leaders ensure that all workers (employees and contractors) have all the required safety training before being assigned work.
- They ensure that all workers have access to relevant Policies, Standards, SOPs, Codes of Practices (COPs), Critical Practices and Regulatory Training.
- They verify worker competency by observing workers at work.
- They ensure that adequate supervision is provided to all employees based on the maturity of the worker.
- They promote involvement, collaboration, and feedback to workers, on work-related performance.
- They ensure close monitoring and control of safety-sensitive work situations in the workplace.
- They seek help and input when work-related situations are difficult and larger than expected.
- They ensure that all hazards are identified and all workers are aware of them.

The authors add to the above list the following inclusions:

- Adoption of a zero-tolerance approach to wanton disregard of safety processes and tools established to protect workers at the site.
- Shortcuts and failure to follow procedures will not be rewarded if work is completed safely. Rather, such behaviors shall be regarded as a wonton disregard of safety processes and tools unless it is clearly established that the procedure must be updated.

* Copyright Suncor Energy Inc., approval from Suncor Energy Inc. is required to reproduce this data.

- All procedures and work practices must be kept current and must meet the organization's standards for review.

Setting the safety standards is a cultural requirement borne out of alignment with the core values of the organization. If the organization and senior leadership support health and safety of the workforce as a core value as demonstrated through their actions and behaviors, then frontline supervisors are automatically empowered to establish high safety standards and steward toward them. On the other hand, if as we discussed earlier, senior leadership talks safety but frowns upon outage and downtime to improve safety, or rewards production at the expense of safety, frontline supervisors shall be undermined in their efforts to strive for high safety standards.

Be Proactive in Managing Risks and the Health and Safety of All Workers

RM skills are essential components of the behaviors and training of all frontline leaders and supervisors that must be transferred to on the tools workers. Such skills are demonstrated in the following ways at the frontline*:

- All workers unconsciously perform field-level risk assessments before any work is started.
- Frontline supervisors/leaders will take immediate actions to stop unsafe work. They will also empower workers to shut down/intercede in all unsafe work.
- No work will be undertaken until all risks are reduced to as low as reasonably practicable.
- A quantitative measure or tool (risk matrix) is used to determine the level of risk.
- Even when high-risk work is contracted out, frontline leaders do not absolve themselves from responsibilities for the safe completion of the assigned tasks and work.
- Frontline supervisors/leaders will ensure proper permitting process and worksite walk around (before during and after) work.
- They will ensure that all work is preceded by toolbox and prejob meetings that promote worker engagement and where all workers feel free to voice concerns about assigned work.
- They will ensure that proper scoping and risk mitigation plans are in place and reviewed with workers before a job begins.
- Frontline supervisors/leaders will enable all workers to document incidents, near misses, hazards, and mitigations in a formalized process.

A proactive approach to RM requires frontline supervisors/leaders to be empowered to take appropriate actions prior to incidents occurring. Frontline supervisors/leaders must also be competent in FLRAs and risk mitigation strategies. A core requirement of frontline supervisors/leaders therefore is a strong ability to utilize an RM system (risk matrix) for assessing risks consistent with the organization's RM and tolerance levels.

Build Trust and Relationships among the Workforce

Two key elements for trust and relationship among the workforce include leaders being respectful and trustworthy. Being respectful of people requires that frontline supervisors/leaders are aware of the things workers may hold dear to them and in many instances this is influenced by the cultural values of the worker. Cultural awareness and cultural sensitivity are therefore critical skills for frontline supervisors/leaders in a culturally diverse workplace. Indeed, we saw earlier that trends indicate a shift to an increasingly diverse and mobile workforce. This places increasing demands on the frontline supervisor/leader to be culturally sensitive and aware.

Language differences, religious beliefs, taste in foods, and choice of music are all areas of cultural sensitivity in which frontline supervisors may require awareness training. Once the feeling of disrespect has set in among workers, it becomes increasingly difficult for frontline supervisors/leaders to change this perception. An example of the language differences is demonstrated in this simple anecdote. A young and unaware Trinidadian worker promoted to a frontline leadership role in a large Canadian oil and gas company said the following to someone he considered a friend after a training session in the presence of a group of coworkers: Hey *fat-man*, what did you think of the training we just received? The response was horrifying. The friend turned beet red and was apparently very embarrassed by the term *fat-man*. The supervisor was unaware that he had just gravely insulted his friend in the presence of a group of coworkers. In the supervisor's mind he had just embraced his friend because in Trinidad, fat-man is a common term used to embrace a friend, totally independent of the size or weight of the friend. Nevertheless, the damage was already done and it took many months of hard work to repair this apparent disrespectful gesture.

According to some writers, the most common definition of trust is the "willingness of a party to be vulnerable to the actions of another party based on the expectation that the other will perform a particular action important to the trustor, irrespective of the ability to monitor or control that other party" (Mayer et al., 1995, p. 712). At the very basis of trust is the simple ability to tell the truth. Trust is earned over time from displaying fair treatment and consistent behaviors, and promoting transparency in actions. Trust comes from doing the right thing at all times. Trust in frontline supervisors/leaders is critical for success in safety at the frontline.

Inconsistency in the behaviors of frontline supervisors/leaders results in confusion and poor credibility among workers regarding how to respond in uncertain situations. According to McCroskey and Teven (1999), credibility is the outcome of competence, caring, and character. In a recent study on trust, Dunleavy et al. found perceptions of "coworker as more competent, of high character, more powerful, and more trustworthy when the coworker told the truth versus deceived" (Dunleavy et al., 2010, p. 239). They also found that workers "considered coworker to be higher in competence, character, expert power, and referent power when the coworker deceived through withholding versus distorting information" (Dunleavy et al., 2010, p. 239).

Frontline supervisors must, therefore, be able to demonstrate competence in their technical abilities, they must be able to demonstrate genuine empathy and care for workers and they must be able to demonstrate character reflected in their honesty and integrity. According to Suncor Energy Inc.,* behaviors that may indicate trustworthiness and teambuilding capabilities of a frontline leader include

- Frontline leaders promote a collaborative and consultative work environment with all workers for the safe completion of all work.
- Frontline leaders recognize the value of honesty and demonstrate care for all workers in a credible way. They are known for walking the talk regardless of how difficult the decision may be.
- They ensure the safety of all workers are placed above all else while work is being performed.
- They demonstrate genuine care and empathy for workers and embrace diversity in the workforce.
- They promote transparency and are not afraid to acknowledge having made an error or mistake.

Build Talent and Capabilities

Building talent and capabilities is all about developing the competency of the workforce where safety is concerned. This is best achieved through training and development of all workers. Following the worker's stage of development, as discussed in the situ-transformational leadership model, leaders will recognize and respond to the appropriate training needs of workers. Building talent and capabilities is achieved in the following ways*:

- Frontline leaders maintain continuous vigilance on the capabilities of the workforce under his/her supervision. Addressing gaps in skills and training of workers are priorities for the frontline leader.

---

* Copyright Suncor Energy Inc., approval from Suncor Energy Inc. is required to reproduce this data.

- Frontline leaders seek necessary resources to ensure all workers are trained with and strive for balance between training, while maintaining continuous operations.
- Frontline leaders continually seek to ensure competency of the workforce as opposed to check mark compliance where critical training is required.
- Frontline leaders will undertake to ensure that all workers are properly trained and competent to do assigned work.
- Workers know that they will not be penalized for failing to undertake work they are not trained to do or are not competent to perform.
- Workers are observed and assessed while performing work to ensure competency.
- Refresher training is provided to all workers on a defined frequency or when necessary to ensure the competency of the workforce.
- Selecting the best for mentoring and developing followers.
- Promoting job-shadowing for building competency for critical roles.

Frontline supervisors/leaders must continually consult the business unit or facility training and competency matrix to identify gaps in the competency of workers. This gap analysis must also be confirmed by field verification as workers are observed in practice. Efforts must be made to continuously develop the skills and competency of all workers. In today's lean business models, creative methods must be applied to ensure that all personnel are adequately trained and competent. Technology must be leveraged to ensure that people receive training in a timely manner and frontline supervisors and leaders must ensure that no untrained personnel are allowed to perform safety-critical work at any time.

Engage and Motivate Workers to Do the Right Things

Maccoby (2010) identified 4Rs for motivating workers. These are as follows:

1. Responsibilities—when workers find their "responsibilities are meaningful and aligns with their values" (Maccoby, 2010, p. 60), they are motivated. According to Maccoby, motivation is highest when responsibilities "stretch and develop skills" (Maccoby, 2010, p. 60). The reality is that people are motivated by the things they enjoy doing. Hence, when workers are placed in jobs that do not align with their skill sets, they become frustrated and can be easily demotivated. As a consequence, frontline supervisors/leaders must

place workers in roles consistent with their passion, skills, and competence.

2. Relationship—good relationships with peers, supervisors, and leaders tend to motivate workers. As a consequence, frontline supervisors/leaders must actively seek methods and opportunities to develop collaborative relationships with workers. This can be a challenge for many leaders from autocratic cultures. Collaboration among such cultures is generally weak and so relationships tend to be primarily transactional between workers and frontline supervisors and leaders. The organizational challenge, therefore, is to ensure that all frontline supervisors/leaders are properly trained in principles of collaboration and on how to develop collaborative relationships. Collaborations start with removing the word *no* from the frontline supervisor's/leader's vocabulary and replacing it with other more engaging phrases like *let us hear more about this concern* or *what are your thoughts on how we can fix these issues.*
3. Rewards—"appreciation and recognition are the kinds of *rewards* that strengthen motivating relationships" (Maccoby, 2010, p. 61). Research has shown many times over that a simple thank you, a pat on the shoulder, and a letter of recognition are powerful motivators. Where recognition is concerned, such recognition must be timely, relevant, and appropriate, or else the motivational impact is eroded.
4. Reason—a worker's understanding of how his work fits into the big picture provides a powerful motivator to this worker to continue doing his work. Clearly articulated reasons for working safely provide all workers with powerful motivators for continuing to work safely.

Behaviors that may indicate engagement and motivation of workers by a frontline supervisor/leader in the oil and gas industry include the following:

- Workers are consulted and encouraged to provide input on how work is to be completed. Where input is not considered, feedback to workers is provided in a fair and impartial manner on the reasons for pursuing alternative options.
- Genuine care and empathy for workers.
- Treating all workers fairly and with respect.
- Rewards are shared among workers who perform well. Workers have opportunities to determine who gets rewarded from whom does not within work teams.

Leverage Existing Tools and Workforce to Ensure That Work Is Conducted Safely at All Times

Many tools are available for ensuring the safety of workers in the workplace. Frontline supervisors/leaders must be able to leverage them in a sustainable manner to continually improve the way safety is managed at the frontline. The use of work permits, check sheets for verification and control, training matrixes for verification of worker competency, and risk matrixes for correctly identifying the risk exposures, and developing the appropriate risk mitigation plan are only a few of the many available tools.

Other important tools for ensuring the health and safety of all workers include the consistent use of SOPs, work practices, and adherence to site and corporate policies and standards. These tools have undergone much iteration of improvements and are intended to protect the health and safety of all workers. Depending on the scale of operation, some organizations may have specialized support organizations for training and development, communication and EH&S that can be leveraged for improving personnel and process safety throughout the organization.

Core Skills of Frontline Supervisors and Leaders

Some may argue that the core skills of frontline supervisors/leaders are buried in technical competence. The authors argue that frontline supervisors/leaders must be team builders, motivators, influencers, communicators, facilitators, and mediators, and can prioritize properly and can manage change on an ongoing basis. Essentially, the core skills of the frontline supervisors/leaders are no different from that of a senior leader. Many of these skills will be used to varying degrees on any given day in the lives of frontline supervisors/leaders.

Frontline supervisors/leaders are the custodians of health and safety in the workplace. As a consequence, organizations have a responsibility for preparing frontline supervisors/leaders so that they may proactively respond to the health and safety needs of all workers. The principles of situ-transformational leadership provide organizations with opportunities for developing required skills to recognize the maturity state of the worker and to respond to worker needs with transformational leadership behaviors as discussed earlier in Chapter 6.

Training the Frontline Supervisor/Leader

Selecting the right frontline supervisor/leader requires the right balance of technical competence and leadership traits defined in the skills identified in the previous section. However, when workers are moved into supervisory/leadership roles, there are organizational responsibilities to ensure that they are set up for success. Training should ideally be divided into stages to assist the frontline supervisor/leader as he/she develops in competency. Training stages are as follows:

1. Onboarding—refers to the period when the worker is new to the role. This is ideally a prejob period exercise. The frontline supervisor is not yet performing supervisory work in the role. Table 12.1 provides development details for frontline supervisors/leaders that are new to the role.
2. Mentoring—refers to a period of development postonboarding where the supervisor/leader is working in the role but under the guidance of a mentor assigned to support and assist the new supervisor/leader in the role. Table 12.2 provides development details for frontline supervisors/leaders during the mentoring stage of development.
3. Developing in role—the frontline supervisor/leader is now competent in the basic requirements of the role and is allowed to function alone on an ongoing basis. Continuous guidance from his/her leader is required to ensure success. Table 12.3 provides development details for frontline supervisors/leaders during the developing in role stage.
4. Sustainment—the frontline supervisor/leader is now fully competent in the role. Continuous improvements are occurring and the frontline supervisor/leader is able to lead the workers, consistent with the organizational values and behaviors. Table 12.4 provides development details for frontline supervisors/leaders during the sustainment stage of development.

It is important to point out that frontline supervisors may originate internally to the organization or may be hired from external resource pools. The duration of each stage of training may vary based on the following:

1. Initial skills and competency of the frontline supervisor/leader
2. The complexity of the work being undertaken at the site
3. The hazards and risks associated with the worksite
4. The maturity status of the follower workforce
5. The level of ambiguity and changes occurring at the worksite
6. Greenfield versus Brownfield versus continuous steady-state operating facility

TABLE 12.1

Development Details for Frontline Supervisors/Leaders during Onboarding

Stage of Development	**Onboarding**
Duration	**Prejob (First Week)**
Learning objectives	*Leadership goals* • Good appreciation of SMS, safety culture, mindset, and its importance to the organization • Clear understanding of how his/her role fits into the SMS and the protection of health and safety of all workers • Is aware of location and can access all policies, procedures, and information related to the SMS. Knows where to get help if necessary • Clearly understands standards to be set or maintained and safety leadership behavior expectations • Recognizes and embraces the organization safety culture and values • Can recognize regional/country/business unit cultural sensitivities • Understands and demonstrates respectful and professional workplace behavior requirements *Regulatory and technical* • Has reviewed and is familiar with all corporate policies particularly those related to the health and safety of all workers • Has completed all business unit-specific required safety-critical training in preparation for entering the worksite (e.g., H_2S awareness, confined space entry procedures, lockout/tagout procedures) • Understands corporate standards and business unit-specific critical practices
Learning outcomes—what does it look like (capabilities)	*Leadership* • Confident and planned approach to acquiring full knowledge required to lead workers in role within 120 days • Clearly understands and demonstrates the organization's core values (health and safety and business) • Can locate and access reference materials/tools with little difficulties • Can demonstrate the organization's safety leadership behavior expectations—trustworthiness, respectfulness, and character • Can identify and respond to cultural sensitivities unique to the role, business unit, or facility and region *Regulatory and technical* • All required safety training to enter worksite completed and testing verification of competency established
Methods of training	• One-on-one with leader, competent delegate and/or subject matter experts • E-learning • Leverages existing tools and resources

TABLE 12.2

Development Details for Frontline Supervisors/Leaders during Onboarding

Stage of Development	**Mentoring**
Duration	**On the Job (120 Days)**
Learning objectives	*Leadership goals* • Competency in organization's safety leadership behaviors, including teambuilding, communication, relationship management, respectfulness, trustworthiness, and cultural awareness • Leverages existing safety tools and resources. Competent in accessing and applying them • Strives to develop self and identifies gaps in the skills and competency of direct reports • Can consistently interpret and apply business unit's operating and critical practices • Consistent application of corporate business and health and safety standards *Regulatory and technical* • Can lead regulatory inspection and compliance requirements • Supports the organization's emergency preparedness and response plan at a leadership level of competence
Learning outcomes—what does it look like (capabilities)	*Leadership* • Demonstration of the organization's leadership behavior consistent with expectations • Delivers consistent safety messages aligned with the corporate safety vision. Demonstrated use of SMS tools designed to support the corporate safety targets • Demonstrates engagement and involvement in addressing field safety concerns • Incorporates health and safety development in discussions with all workers • Visible at the frontline supporting and guiding workers in working safely and consistent with procedures and corporate values • Communicates and demonstrates corporate values • Able to access all resources available to perform his/her job • Demonstrates competence and compliance in application of business unit safety-critical practices and corporate policies *Regulatory and technical* • Has developed plans for closing gaps identified in self and workers under supervision for regulatory and business unit compliance • Demonstrates competency in key business unit and corporate SMS tools such as risk matrix, training matrix, and contractor prequalification • Identifies appropriate/inappropriate behaviors relating to the technical aspects of the business (challenge the process and identify improvement/deficiency areas)

continued

TABLE 12.2 (continued)

Development Details for Frontline Supervisors/Leaders during Onboarding

Stage of Development	Mentoring
Duration	**On the Job (120 Days)**
Methods of training	• One-on-one with leader, competent delegate and/or subject matter experts • E-learning • Job shadowing • Capability check list—gap analysis • Leverages existing tools and resources

Source: Data from Suncor Energy Inc., Copyright of Suncor Energy Inc., approval from Suncor is required to reproduce this work.

TABLE 12.3

Development Details for Frontline Supervisors/Leaders during Developing in Role

Stage of Development	Developing in Role
Duration	**365 Days**
Learning objectives	*Leadership goals* • Consistent application of principles of the SMS and demonstrates commitment to the organization's safety leadership behaviors and values • Growing expertise in the use of tools designed to enable the SMS • Becoming comfortable in leadership roles bringing synergy to the team and creating a motivated workforce • Demonstrated application of leadership behaviors designed to reinforce the safety culture of the organization *Regulatory and technical* • Takes proactive measures to respond to regulatory and technical requirements of the job • Participates in emergency response drills and demonstrates competence in responsibilities for personnel, environmental, and property preservation (in the order provided) during emergency response training
Learning outcomes—what does it look like (capabilities)	*Leadership* • Demonstrates leadership behaviors for motivating others to higher levels of performance—understands his/her role in leadership accountabilities to the SMS and safety vision of the organization. Demonstrates leadership and take-charge positions for ensuring work is performed safely • Consistent application and use of SMS supporting tools. Knows where to locate tools and resources available to support leading safety (leverages existing tools and resources) • Establishes clear safe work practice expectations and consistently seeks to improve safety at the worksite

TABLE 12.3 (continued)

Development Details for Frontline Supervisors/Leaders during Developing in Role

Stage of Development	**Developing in Role**
Duration	**365 Days**
	• Proactively recognizes and responds to all unsafe work situations and working conditions • Runs effective toolbox meetings with teams • Seeks to develop followers • Manages conflicts and responds well to interpersonal and team dynamics • Flexible and accommodating to personnel needs. Demonstrates genuine empathy and care for people • Understands team, communicates and adapts communication style accordingly. Promotes an environment that supports open dialogue regarding safety opportunities • Can present to followers and senior leaders safety message in a consistent manner. Can coach and mentor followers on safe work practices and behaviors for the business unit or organization • Promotes individual consideration, motivates and engages individuals and team to work safely *Regulatory and technical* • Understands and consistently applies regulatory requirements and critical practices and reviews responsibilities in work plans; works to close training gaps that may exist; ensures that all mandatory safety training is complete for self- and direct reports • Demonstrates his/her knowledge and responsibilities for emergency response • Seeks opportunities to improve critical practices, SOPs, and the technical aspects of the business (challenges the process and seeks continuous improvements)
Methods of training	• Classroom based • Case presentations • External audits • Check lists and QA documentation completed • Leverages existing tools and resources—especially environment, safety & social responsibility (ES&SR)

Source: Data from Suncor Energy Inc., Copyright of Suncor Energy Inc., approval from Suncor is required to reproduce this work.

Table 12.5 provides a template for allocating training time both in terms of days and percentage of overall training period for each stage of the frontline supervisor's/leader's development for either new or experienced supervisor/leaders. This is primarily a stewardship tool that is adjustable for different frontline supervisor's/leader's roles.

TABLE 12.4

Development Details for Frontline Supervisors/Leaders during Sustainment

Stage of Development	**Sustainment**
Duration	**Ongoing Process after First-Year (Gap Closure and Capability Growth)**
Learning objectives	*Leadership goals* • Consistent application of principles of the SMS and demonstrates commitment to the organization's safety leadership behaviors and values • Consistent use of tools designed to enable the SMS • Comfortable in leadership roles bringing synergy to the team and creating a motivated workforce • Consistent application of leadership behaviors designed to reinforce the safety culture of the organization • Sets increasingly higher safety standards and holds team accountable for improvements in safety performance • Continuously communicates with workforce on safety expectations and standards • Motivates workers to higher levels of safety performance • Trains followers on best practices in safety *Regulatory and technical* • Takes proactive measures to respond to regulatory and technical requirements of the job • Participates in emergency response drills and demonstrates competence in responsibilities for the emergency response plan
Learning outcomes—what does it look like (capabilities)	*Leadership* • Workers automatically respond to the SMS requirements and safety standards of the organization, business unit and facility while working • Workers are motivated and want to be here. Workers are prepared to produce at higher levels of performance because they feel valued and their health and safety are looked after at work • Safety performance is on the rise. Fewer incidents—near misses and accidents • Workers actively seek training to develop skills necessary for their jobs • Workers feel comfortable talking to their supervisors/leaders in bringing safety concerns forward • A proactive approach to addressing workplace hazards • All workers wear PPE and workplace reflects good housekeeping standards. The safety culture is felt as soon as you enter the worksite

TABLE 12.4 (continued)

Development Details for Frontline Supervisors/Leaders during Sustainment

Stage of Development	**Sustainment**
Duration	**Ongoing Process after First-Year (Gap Closure and Capability Growth)**
	• Understands and consistently applies regulatory requirements and critical practices and reviews responsibilities in work plans; works to close training gaps that may exist; ensures that all mandatory safety training is complete for self- and direct reports • Demonstrates his/her knowledge and responsibilities for emergency response • Proactively improves critical practices, SOPs, and the technical aspects of the business (challenges the process and seeks continuous improvements) • Personal development plans from 360 feedback, attends safety and technical seminars, conferences, forums • Professional development training
Methods of training	• Performance reviews with leader • External audits • Check lists and QA documentation completed

Source: Data from Suncor Energy Inc., Copyright of Suncor Energy Inc., approval from Suncor is required to reproduce this work.

TABLE 12.5
Stewardship Tool for Frontline Supervisor/Leader Training

Stage of Development	Duration of Training	Type of Training	Estimated % of Time and Days of Training (in Parenthesis) Required (Varies from the List Provided Earlier)			
			New		Experienced	
			Inside	Field	Inside	Field
		Classroom	0	0	0	0
Onboarding	First week (Prejob)	One-on-one with leader/designate	40(2.0)	10(0.5)	20(1.0)	30(1.5)
		E-learning	50(2.5)	0	50(2.5)	0
	% Total time training (training days over 5 working days)		90(4.5)	10(0.5)	70(3.5)	30(1.5)
		Classroom	25(3)	0	33(2)	0
		Job shadowing	8(1)	50(6)	0	33(2)
Mentoring	On the job (120 days)	E-learning	8(1)	0	17(1)	0
		One-on-one with leader/designate	8(1)	0	17(1)	0
	% Total time training (training days over 5 working days)		5(6)	5(6)	3(4)	2(2)
		Classroom	40(2)	0	50(2)	0
Development	On the job (365 days)	Job shadowing	0	0	0	0
		E-learning	20(1)	0	25(1)	0
		One-on-one with leader/designate	40(2)	0	25(1)	0
	% Total time training (training days over 240 working days)		2(5)	0	1.7(4)	0
		Classroom	50(3)	0	50(3)	0
		Job shadowing	0	0	0	0
Sustainment	Going forward (yearly)	E-learning	17(1)	0	17(1)	0
		One-on-one with leader/designate	33(2)	0	33(2)	0
	% Total time training (training days over 365 working days)		1.6(6)	0	1.6(6)	0

Source: Data from Suncor Energy Inc., Copyright of Suncor Energy Inc., approval from Suncor is required to reproduce this work.

13

Shared Learning in Safety

Is shared learning an area of underexploitation in safety? The authors believe that shared learning in safety provides numerous opportunities for improving the health and safety of any workplace. Many organizations are concerned about the legal implications of sharing information within the organization, particularly as it relates to incidents and workplace safety. While this is a genuine concern among organizations, leaders must find creative ways to ensure that knowledge generated from safety incidents is effectively shared across the organization to avoid repeat incidents.

When legal implications are introduced into the process, there is a general hesitance by senior leadership to share information. Furthermore, some cultures would not even consider sharing information at all regardless of its implications for improving the health and safety of the workforce. The challenge for the safety professional is the need to develop an adequate process for shared learning that convinces senior leaders that knowledge relating to safety incidents is being shared in an organized manner that does not place the organization at risk. In this chapter, we shall explore how shared learning in safety can be achieved to the benefit of the organization.

Why Is Shared Learning Important in Workplace Safety

Among safety professionals, there is a belief that 90% of incidents are predictable and avoidable, and 80% are repeated. If these numbers are correct, then it follows from good logic that learning from all incidents can help to reduce repeat workplace incidents. Indeed, when we consider the potential impact of shared learning in safety, on the cost of workplace incidents (US$120 billion to $240 billion and approximately US$170 billion annually), it appears to make good sense to do so.

Aside from the cost-saving implications, studies show that workers want to give more—a 10-year study conducted by DuPont found that 96% of accidents were the result of unsafe actions by employees going beyond their limits, rather than unsafe conditions (Pennachio, 2008). Liberty Mutual (2009) also indicated that 24% of all workplace injuries are a result of overexertion. Pinker (2010) advised that when work–life balance is neglected in the Canadian workplace, catastrophic health-related outcomes can amount to Can$25 billion annually.

This desire to do more must be supported in the organizational interests of efficiency. However, the goal should be to get the work done safely and without incidents. When we consider all of the conflicting messages in the workplace, of high incident cost, increasing demand on the workforce, and high numbers of repeat and avoidable incidents, the opportunities are enormous for shared learning in safety. Table 13.1 highlights the opportunities that are available for shared learning in safety.

Where shared learning is limited, there are significant numbers of repeat incidents of the same types within the same geographical areas and industry and across industries. Table 13.2 highlights the prevalence of repeat fatal incidents in the province of Alberta where shared learning in safety is limited. There should be no limits as to how far we go to protect our workforce. Pennachio (2009) advised the then U.S. Secretary of Labor Elaine L. Chao in an address to the American Gas Association Safety Leadership that health and safety programs in industry are not only a legal requirement but as we have indicated earlier it makes great business sense too. According to Pennachio, Secretary Chao also advised, "No price can be placed on the most important benefit, and that is to see that every worker returns home safely to their loved ones at the end of each work day" (Pennachio, 2009, p. 27). Shared learning in safety provides us the opportunity to better protect our most important assets without compromising business competitiveness—our human resources.

TABLE 13.1

Shared Learning: The Last Remaining Low-Hanging Fruit

Area of Safety Management	Exploited	Unexploited
PPE	√	
COPs, SOPs	√	
Engineering controls	√	
Regulations	√	
Collaborative relationships with stakeholders	√	
Training and development	√	
Internal sharing of learning		√
Industry sharing of learning		√
Cross-industry sharing of learning		√

TABLE 13.2

Repeat Fatalities in the Province of Alberta, Canada (2004–2008)

Incident Type	2004	2005	2006	2007	2008
Inadequate purging—work preparation—hazard identification	Nil	Nil	1 Explosion from welding the outside of a tank containing hydrocarbons	Nil	1 Explosion resulting from welding at the top of an empty hydrocarbon tank
Improper lock out tagout/energy sources—hazard identification	3[a] Separate instances of contact with live electrical sources	Nil	1 Overhead electrical line strike 14.4 KV	2 Two separate incidents of contact with live electrical contacts	2 25 KV overhead line strike. Contact with live wires
Overexertion	1 Cardiac arrest—fatality	Nil	Nil	4 Cardiac arrests from overexertion	4 Cardiac arrests from overexertion
Working from heights—falls—hazard identification	5 Five separate incidents for fall-related fatalities	6 Six separate incidents for fall-related fatalities	Nil	3 Three separate incidents for fall-related fatalities	4 Four separate incidents for fall-related fatalities

Source: Derived from OH&S Canada.

Note: Numbers (1–6) define the numbers of workplace fatalities from the Incident Type identified.

[a] Fatalities.

Kaizen in Safety

Toyota may be considered the godfather of Kaizen—an approach generating continuous improvements through focused attention and subsequent sharing of the knowledge generated across the organization. When Kaizen is applied to safety, the following characteristics are essential:

1. Worker engagement and involvement in workplace safety are critical for success.
2. The focus is on improving various elements of the SMS on an ongoing basis.
3. Improvements in safety are incremental.
4. Through repeated efforts, continuous improvements are generated.

Kaizen in safety is essential for long-term success in safety and it starts first with learning from events followed by a systematic approach toward sharing this learning.

Some may argue that there is an adequate amount of industry and cross-industry (between industries) sharing of safety information. Nevertheless, our experiences have shown that within industries and across industries, sharing of knowledge in safety is disorganized and indiscriminate. Additionally, there is no consistent format for sharing information related to safety incidents. More importantly, when sharing does occur, the true knowledge generated from the safety incident is often clouded by gory details and inconsistent messaging. Figure 13.1 highlights the possibilities for improvements in workplace safety when shared learning is exploited in an organized manner.

Can we successfully leap-frog Kaizen in safety to a more advanced level from current levels? The authors argue that significant improvements in safety can be generated from organized sharing of the learning derived from incidents within an organization. In this way, we learn from the internal experiences of the organization and avoid repeating the same incident in future. Where organized external sharing is occurring, we also learn from the experiences of our peers, thereby avoiding incidents experienced by our peers. The benefit from shared learning is a shortened process to take the organization further along the learning curve while avoiding the ill consequences of learning from internal incident. The strategy is to develop a process that meets stakeholder's needs for focused sharing of knowledge while avoiding liability concerns.

When one considers the common themes of Kaizen as it relates to safety, all of these themes are occurring to some extent in many organizations. Where world-class leaders in safety are concerned, we expect that these themes would be advanced across the organization. Real success in Kaizen in safety is derived from an effective IMS that allows for effective root cause

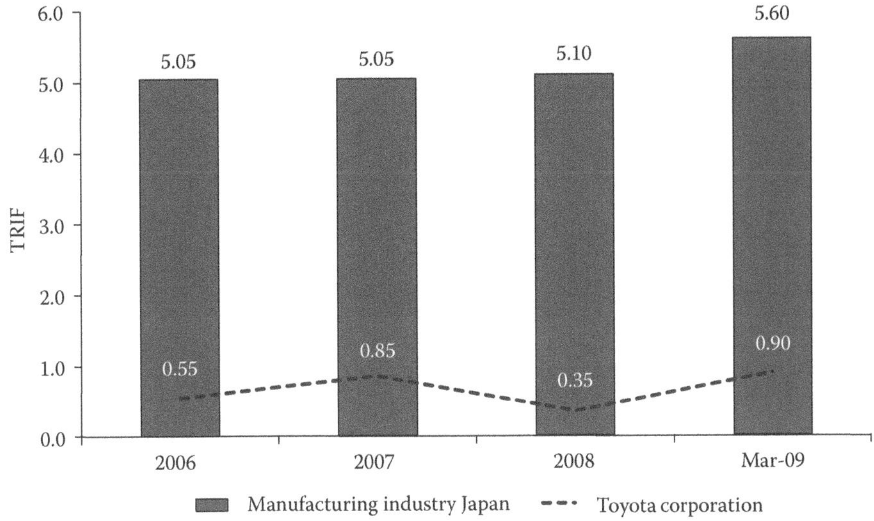

FIGURE 13.1
Recordable injury frequency, Toyota versus manufacturing industry, Japan. (Derived from Toyota Annual Reports.)

analysis such that measures can be activated to prevent repeat incidents at the very foundational level.

Internal Sharing of Learning in Safety

As discussed earlier, sharing of safety knowledge generated from incidents is largely limited because of liability concerns among senior leaders. We also identified that real success in shared learning comes from developing a strategy that meets the needs of all stakeholders. Let us explore a process that can be used for full exploitation of shared learning across the organization. This process shown in Figure 13.2 may be scalable to the size of the organization and is intended to facilitate consistency in transmission of learning across the organization.

The key elements of this process are as follows:

1. Risk ranking of incident: This helps in segregating which event (incident) requires sharing of learning. An organization may adopt the position that all high- and medium-risk events must be investigated and learning must be generated and shared. Others may consider sharing learning only from high-risk events. Flexibility and discretion are recommended to avoid overwhelming the

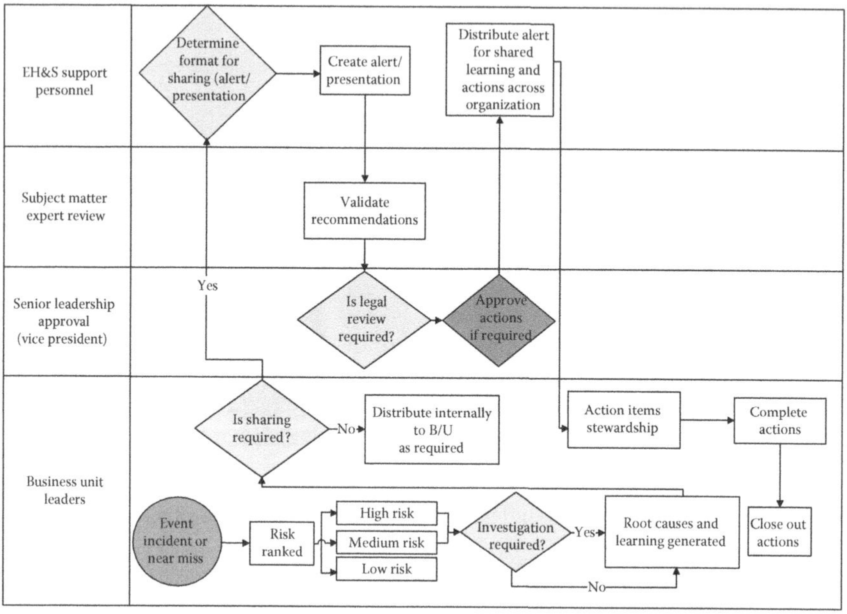

FIGURE 13.2
Internal process for shared learning in safety.

capacity of the organization to learn. Additionally, other business activities that are generating change in the organization may lead to frustration and weak uptake from the learning from an event process.

2. Stakeholder engagement and involvement: in the example provided, multiple stakeholders are engaged and involved in the process. Some of these stakeholders may be centralized to the organization while others may still be decentralized and reside within the various business units of the organization. Engagement and involvement generate ownership and ultimately buy-in. This is a highly recommended requirement for success in shared learning in safety, across the organization.
3. Quality control: The use of EH&S support personnel (e.g., communications experts) in determining the type of communications required (alerts/presentation) is very useful in creating the right messages for maximum impact among a diverse adult workforce. In a workplace dominated by two distinct generations, shared learning in safety must cater to both workgroups in order to generate the desired value. As a consequence, shared learning in safety must be delivered in a format acceptable and accessible by both workgroups. Similarly, the

use of subject matter experts for reviewing the content of alerts or presentations generated is an essential requirement for credibility of information and ensuring that the chances of repeat incidents are reduced. The engagement and legal resources, prior to sharing information, also help in removing liability concerns within the organization from sharing of knowledge generated from an incident.

4. Control of information distributed: Senior leadership approval prior to distribution helps in determining whether or not the knowledge generated from an incident is intended for information only or requires action by various stakeholders. Where actions are required, senior leadership approval suggests that resources required for supporting the required actions will be made available to the business units.

While our discussions on shared learning have focused primarily on incidents and near misses, this process is much adaptable to other proactive measures toward safety when sharing of technical and environmental alerts is considered. Preventative maintenance, technical knowledge, and environmental expertise may generate knowledge for sharing before events occur. By creating alerts or presentations in a similar fashion, proactive corrective actions can be undertaken before events occur.

Adopt a Consistent Format for Shared Learning

Where shared learning is concerned, it is imperative that a consistent format for communicating messages is adopted. Alerts cater for individual or group learning. Individual learning is achieved when alerts are posted to notice boards or are accessible by workers from individual work stations from an electronic database. Group learning takes place when alerts are discussed during toolbox talks, prejob meetings, and FLRAs. The authors recommend that alerts should be limited to a single page arranged in a simple manner for easy reading and deciphering of messages. The following key sets of information should be included in an alert:

1. Title.
2. Date and time of incident.
3. Wherever possible, a graphic or picture of the incident should be provided. A single picture tells a thousand words and helps the end user to better appreciate the severity of the incident or potential severity of a near miss.
4. Originator of the alert (business unit, etc.).
5. A short description of what happened in the particular incident.
6. The root causes identified for the incident.

7. Key learning outcomes generated from the incident.
8. Recommendation/action items if required.
9. Subject matter expert contact details (email address and telephone contact) in case additional information is required by those tasked with actions or who may need further information in interpreting the knowledge provided in the alerts.
10. An alert tracking number to ensure control on follow-up actions recommended.

Figure 13.3 is a sample alert generated for shared learning across a large commercial oil and gas producer. Figure 13.3 is a sample template of an incident alert template for sharing learning. Alerts are essentially simple one-page communication tools filled with information regarding incidents. Alerts can be classified under the broad headings of safety, environmental, health, or technical.

In its simplest form, an alert at its very minimum should convey the information provided in Figure 13.4.

When presentation formats are required, it follows that more details are necessary to ensure that full information is transmitted to the organization. Presentation format may require expert delivery of what was learnt from an incident. Presentation formats for shared learning provide opportunities for group learning and interaction. A well-designed presentation format in addition to much of the information carried in the alert template must also include a discussion section that prompts the learning group to consider where, within their business unit or operations, gaps may exist that can lead to similar incidents. Presentations should provide a summary of the final investigation report from an incident. The following should be included in the presentation:

1. Title page inclusive of the date and time of the incident.
2. Highlights of the incident.
    a. Description of what happened.
    b. Impact of the incident: includes fatalities, dollar impact to business, damages to facilities, and mitigating circumstances that prevented further severity of the incident.
3. Wherever possible, graphic or picture of the incident should be included. A single picture tells a thousand words and helps the end user to better appreciate the severity of the incident or potential severity of a near miss. In addition, the presenter can provide greater detailed explanations as required.
4. Details of all root causes and contributing factors identified from the incident investigation.

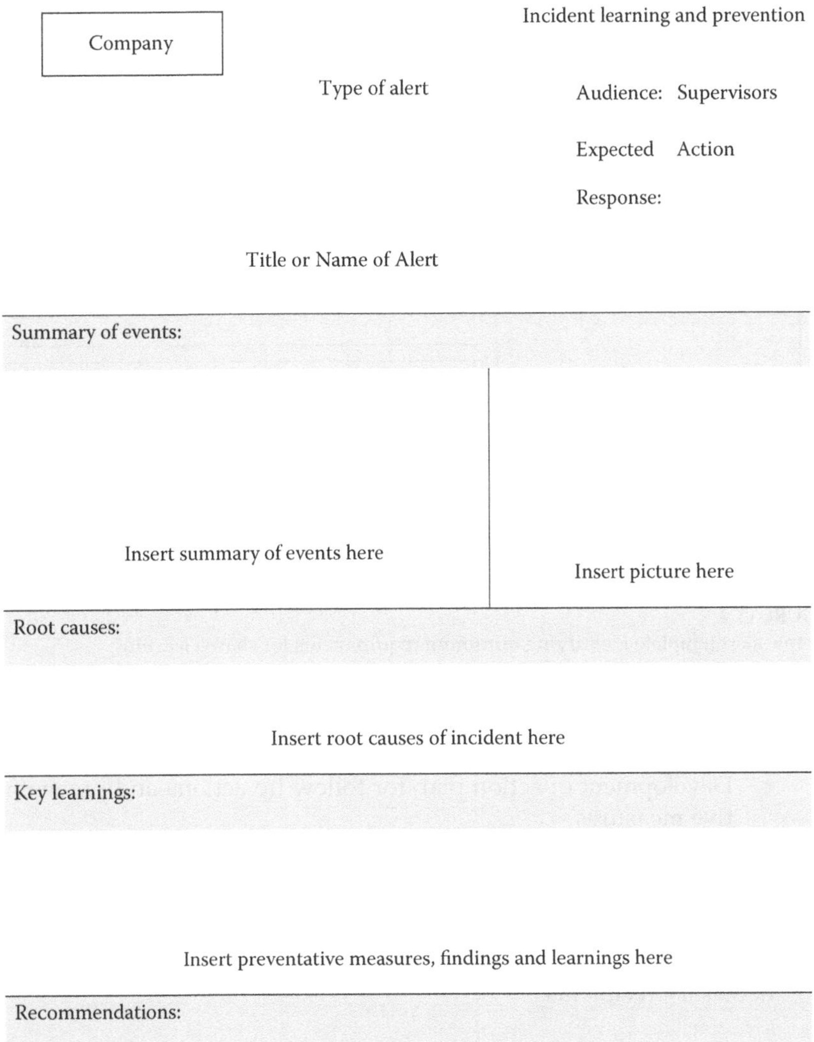

FIGURE 13.3
Sample detailed alert template for shared learning. (Adapted from Suncor Energy Inc., © Copyright of Suncor Energy Inc., approval from Suncor is required to reproduce this work.)

5. Key learnings generated from the incident.
6. Recommendation and generic action items necessary for the prevention of a similar incident across the organization.
7. A discussion page that is designed to generate interaction. Included in the discussion page are the following:
    a. Can this incident occur in your business unit or facility?

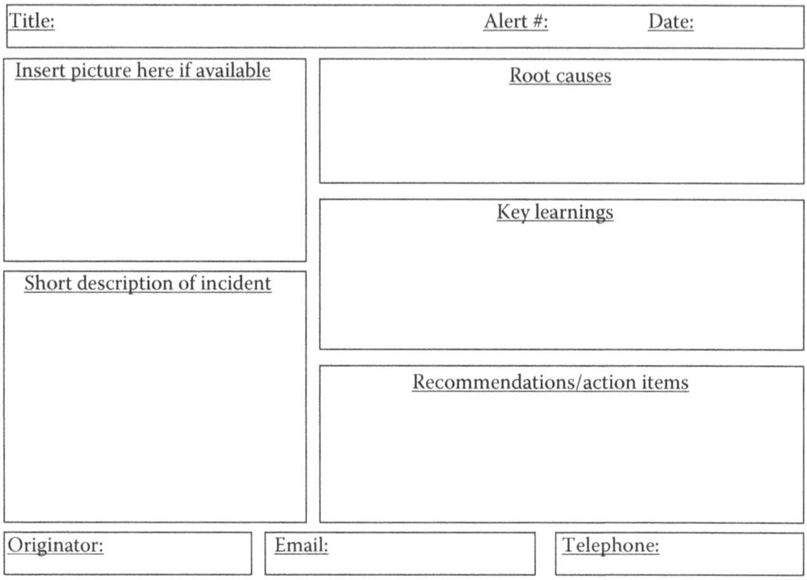

FIGURE 13.4
Sample alert template identifying minimum requirements for shared learning.

  b. List where, and prioritize for action.
  c. Development of action plan for follow-up actions and preventative measures.
  d. If training and competency are required, develop timeline for completion and identify target group for training, trainers, and competency assessors.

8. If a detailed report is available and can be shared, provide access to necessary recipients.

Presentations must be well planned and coordinated such that maximum value is derived by participants from the presentation.

Industry and Cross-Industry Sharing of Safety Learning

With the exception of a few industries, shared learning is almost absent. Unfortunately, in the quest to protect against liability, the very thing that can help protect our most precious assets, shared learning in safety, gets lost in organizational reluctance to share knowledge both within and across industries. Culturally, shared learning appears to be an area many leaders are not prepared to venture into. In spite of this reluctance to share safety learning externally unless regulated, organizations are slowly responding to the benefits of externally shared industry learning in safety.

Traditionally, industry shared learning in safety has been disorganized and generally unapproved by leaders, with safety professionals informally sharing learning via email and other forms of transmission. Organized sharing of knowledge in safety generally occurs during conferences and workshops. However, these events occur very infrequently and are often focused on larger initiatives than sharing of learning from events.

The reality is that many organizations operate across multiple industries. For example, organizations in the oil and gas industry operate within the following industries to name a few:

1. Oil and gas
2. Power generation
3. Transportation
4. Construction
5. Chemicals and manufacturing

Knowledge generated from safety incidents in the oil and gas industry therefore can be very beneficial to many of the industries listed above but is almost always never shared among other industry players beyond oil and gas. The challenge therefore is the need for an established organized process for sharing safety knowledge across multiple industries in an organized fashion.

Shared learning within and across organizations will continue to evolve as organizations recognize the value creation associated from learning within and external to the organization. The first hurdle that must be overcome is that leadership must seek to develop internal processes to remove liability from the sharing process and to better recognize what constitutes competitive advantage and intellectual property. The authors believe that strategies for improving the health and safety of all workers cannot be regarded entirely as intellectual property unless the organization is primarily involved in providing safety to organizations.

The authors firmly believe that shared learning in safety continues to be one of the last remaining low-hanging fruits that must be exploited in an organized manner by organizations seeking to improve safety performance and safety culture. Organizations must first develop and leverage internal sharing processes and capabilities before seeking to share learning externally. However, once the true value of this effort is recognized and appreciated, shared learning in safety will evolve to a major value creation venture for organizations.

14

Safety Training and Competency

Training and competency in safety are critical requirements for preserving the health and safety of all workers in the workplace. There are many types of technical safety courses that vary from job to job, which workers are required to be trained in and demonstrate competency. This chapter focuses on the skills and capabilities that are required to ensure that work is completed safely at all times. The authors will also focus on the challenges to achieving the safety vision of the organization and the training and competency requirements that enable personnel to support and achieve this vision.

Understanding the Business Drivers

The training provided to personnel must provide an understanding of the business drivers for the organization. This is both in the context of stakeholder involvement and sustainable development and growth. Historically, the key business driver of an organization has been profit maximization. Sustainable development today, however, has shifted this focus to value maximization. In this context, value maximization focuses on satisfying the needs of all stakeholders. Value maximization is derived when the business processes are aligned with all stakeholder needs. Figure 14.1 provides a demonstration of this alignment.

The list of stakeholders shown in Figure 14.1 varies with the types of business. Nevertheless, for the most part, value maximization is derived when business drivers are aligned with stakeholders' needs and interests. Training and competency in safety start with this basic understanding among the entire workforce from the frontline worker through senior leadership. When this understanding is achieved, every worker in the organization is now able to appreciate how his/her work fits into the bigger picture.

Understanding and Internalizing the Core Values and Beliefs of the Organization

Great safety performance comes from understanding and internalizing the core values and beliefs of the organization and it starts with the

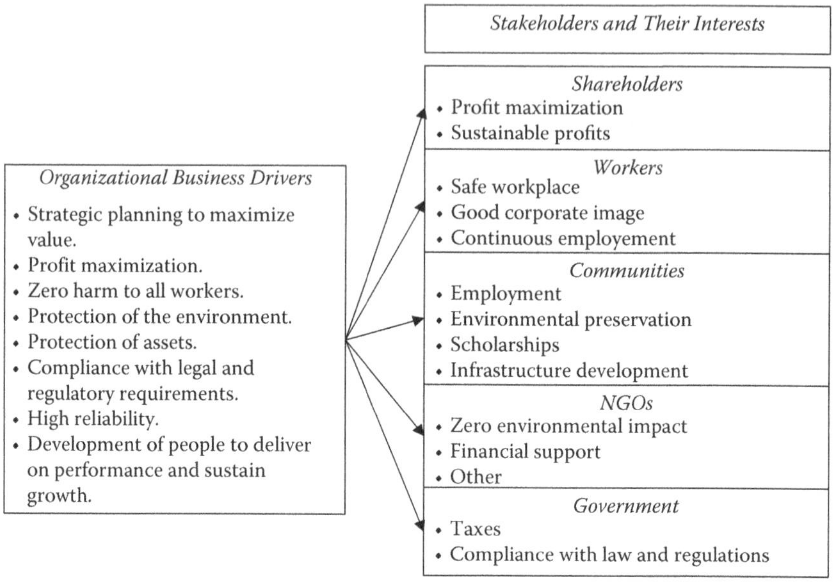

FIGURE 14.1
Business drivers and stakeholder interests alignment.

demonstrated behaviors and commitment of senior leadership. Where safety is concerned, Suncor Energy Inc. advises: "Nothing is more important than protecting ourselves and one another from harm. That means safety is never compromised—for any reason. If we can't do it safely, we don't do it" (Suncor Energy Inc., 2011, para. 3).

The value placed on the organization for safe work requires all workers to work safely. When this value is internalized by all workers, every worker holds each other accountable for safe work with a consequential improvement in overall safety performance. Values and beliefs that support strong safety performance include the following:

1. Strong commitment to the SMS. When an organization has taken the time and effort to develop an SMS, there is a moral obligation of all workers to follow the SMS so that we can protect each other, our environment, and our workplace.
2. All incidents are preventable and avoidable. Strong safety performance starts with the underlying premise that all incidents can be prevented. If we believe otherwise, then we are cursed with the expectation that an incident is always likely to happen at any moment. All incidents are indeed preventable, if we take the time and apply the efforts required to properly analyze the hazards associated with each task and develop mitigating actions for each hazard identified.

3. Working safely is everyone's responsibility and we hold each other accountable for safe work at all times. I am my brother's keeper in the workplace. I look after the safety of myself and my colleagues at all times. I will not do anything that will jeopardize the health and safety of myself or my coworkers. I expect the same behavior from each of my coworkers.
4. Involvement and engagement of workers play an important role in strong health and safety performance. Where workplace hazards are concerned, it is important to engage and involve everyone involved in performing all assigned work. The collective hazards identification, of a group of workers, is much greater and more complete than that of a single worker or a few of the workers among the workgroup.
5. Focus on sustainable development, operation excellence, and high performance from all workers. The leadership must address the long-term implications of business decisions. Striving for business excellence and sustainable development requires focused attention to making the right decisions and allocating resources to support these decisions. For example, investments in training and competency development build the capabilities of workers such that high and sustainable performance can be generated.
6. Ethical and responsible leadership. In all businesses, *doing the right thing* at all times regardless of the consequences of such actions distinguishes a business from its peers. While there may be initial cost implications from such activities, value maximization is ultimately derived and stakeholder support and satisfaction are maximized.
7. Proactive approaches to safety management. A focus on proactive approaches to improving safety performance helps in sustainable development and operational excellence. Stewardship of leading indicators is a proactive approach to managing health and safety in the workplace.

Challenges to Realizing Health and Safety Vision

There are many challenges and obstacles that will prevent businesses from achieving its safety vision. Among the leading challenges to influence realization of the health and safety vision of an organization are the following:

1. Leadership commitment. When leadership commitment to the safety vision of the organization is weak, worker commitment will also be weak. Leadership commitment is measured in demonstrated behaviors that are visible to all workers. Simple behaviors such as

adherence to safety procedures and policies send powerful messages to workers about how important safety is to the organization. The resources allocated to improving health and safety in the worksites are a very powerful indicator of leadership commitment.

2. Management skills and capabilities. Workers look to leaders and managers for guidance and support in completing their daily work and assigned tasks. When management skills and capabilities are weak, there is poor alignment with leadership goals and objectives for safety performance. A skilled and competent management team is essential for driving strong health and safety performance.
3. Communication of the safety vision. The tools for communicating the safety vision today are numerous. Electronic medium, print medium, forums, and team meetings are among the many mediums used today to achieve this goal. While getting the message out to receivers is important, the feedback messages received from workers are perhaps even more important. Feedback tells us how well the message was received, by those we intended to connect with. Communication is also very costly and organizations tend to weigh the cost of communication against message receipt in selecting its communication medium. Among the most powerful communication methods are face-to-face discussions, team meetings, and safety forums. These communication methods provide immediate and direct feedback of the recipient understanding of the intended messages. The communication skills of the presenter are also an important factor in how well safety messages are conveyed to workers.
4. Work tools management, processes, and maintenance. From our earlier discussions, PSM plays an important role in workplace health and safety. As we discussed earlier, PSI is foundational to strong health and safety performance in the workplace. Access to relevant, up-to-date critical work information will, in no uncertain terms, influence the health and safety performance of an organization. More importantly, where contractors are concerned, organizations are protective of its information and often critical work information such as SOPs are not available and accessible to contract workers. Ultimately, these conditions lead to workplace incidents. Until these procedural issues are resolved, possibly through the use of contractor agreements, the safety vision of an organization is in jeopardy.
5. Training and competency of all workers. A common thread among many major incident investigation reports is the training and competency assurance of workers. Training alone is not enough to prevent workplace incidents. Competency on the other hand requires *knowledge*, which comes from training, the *skills and abilities* of workers, and *experience* that helps to further develop these skills and abilities. In any given workforce, despite our best efforts, training

and competency assurance will always be challenged because of turnover, attrition, succession planning, and internal transfers of workers. Furthermore, when contractors are introduced into the workforce, without an effective means for verification of competency assurance, the vulnerability of the workforce to health and safety incidents is further compromised. In many organizations, it may be best to set a target competency assurance of the workforce of say 80% and develop strategies to maintain this level of performance. A competency assurance target of 100% will always be an elusive target in today's global mobile workforce.

6. MOC. Much of our earlier discussions highlighted that incidents occur during periods of change. The way organizations manage change will ultimately affect the safety performance of an organization and its ability to achieve its safety vision.

In spite of these obstacles and challenges, significant improvements in health and safety performance are within our grasps. With strong leadership commitment and a working SMS, organizations have the ability to eliminate many of these obstacles or minimize their impact to safety performance. Key to our success, however, is recognizing them as potential hurdles and proactively working to eliminate them.

Back to the Basics

Managing the health and safety of workers requires that all work be done according to the basic principles of management. These include the following:

Planning

Planning is a critical step in ensuring that work is done safely and the health and safety of all workers are preserved. Where nonroutine tasks and work are to be done, before the work is started, a planning exercise that requires engagement and involvement of workers is an essential process for enhancing safety performance. The planning process is intended to break work into smaller tasks such that hazards associated with each task can be properly identified and mitigated, such that incidents can be avoided. In many instances, this may be regarded as an FLRA or what is known as job safety task analysis.

In many organizations, this may be an informal process whereby workgroups may brainstorm for a moment to identify hazards associated with a particular task and for identifying the hazard mitigation plan and controls. This approach may work well for small tasks that are low risk. Figure 14.2

Tasks	Hazards	Mitigation and controls
Task 1	Hazards 1-n	Controls 1-n
Task 2	Hazards 1-n	Controls 1-n
Task n	Hazards 1-n	Controls 1-n

FIGURE 14.2
Whiteboard template for hazards identification and mitigation.

provides a simple template for the whiteboard FLRA process. This process is informal and can be completed in any work environment.

The goal is to identify all hazards associated with each task and adequately mitigate these hazards. Some may argue that the documentation capability is lost in the whiteboard process. However, digital photography technology allows for a simple shot of the whiteboard to capture all information, thereby creating the due diligence requirements for traceability. This process is ideal for remote operations where computer and other information technology (IT) may be more difficult to sustain. This type of planning is typical of drilling-type operations in remote sites where risk can be high.

For higher-risk nonroutine tasks (such as hot tapping or boiler cleaning), the authors recommend a more formal approach to this planning exercise, where all tasks are documented, hazards associated with each task are identified in a team environment, and the mitigation plans and controls are similarly defined. Documentation is an absolute requirement in all high-risk work. The benefits of documentation are twofold. Documentation makes it easier to share and review the plan with all workers. It also provides a traceable process in the event of an incident.

Figure 14.3 is a sample whiteboard planning exercise completed in the field. A digital picture of the whiteboard field-level hazard assessment is taken and retained for records and documentation.

Figure 14.4 provides a simplified overview of the planning process for hazards identification and mitigation.

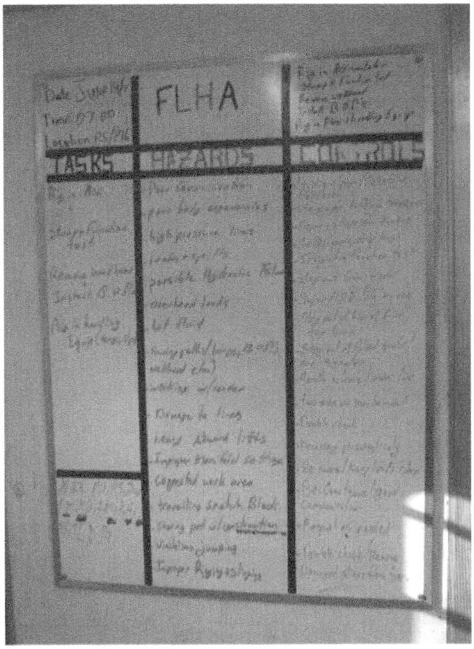

FIGURE 14.3
Sample field completed whiteboard planning exercise. (Copyright of Silverstar Well Servicing Ltd., approval is required to reproduce this work.)

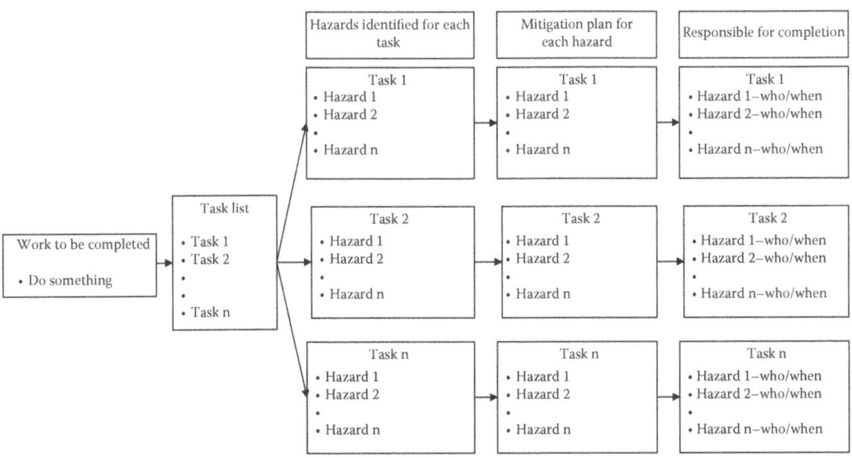

FIGURE 14.4
Planning process for hazards identification and mitigation.

Organizing

Organizing refers to the effective allocation of resources such that work can be completed safely and efficiently. Leaders allocate work based on the skills and competency of all workers and the tasks to be completed. Skilled leaders will engage workers in the organizing process and would allocate personnel based on their strengths and weaknesses, such that all work can be completed without incidents.

Leading

All work requires a leader from whom those in doubt can seek guidance and support. Leading refers to taking charge of the work when conditions may vary from planned. At the frontline, leading may take the form of supervision. Ultimately, leading is the process of taking charge of the work situations and making decisions that are consistent with protecting the health and safety of the workgroup.

Good frontline leadership is reflected in the following behaviors:

1. Demonstrate impartiality and care (empathy) for all workers
2. Ability to delegate responsibility and decision-making authority at the right levels
3. Engagement and involvement of all workers involved in the work and related tasks seeking feedback and providing feedback and guidance where necessary
4. Build trust among workers and team members and to create an environment of trust by demonstrated consistent, fair, and transparent behaviors and by doing the right thing at all times
5. Treating all workers with respect
6. Technical competence and capabilities to guide and support followers
7. Willingness to listen to the recommendations of those doing the work
8. Knowing when to be firm
9. Team building and generating confidence among team members
10. Ability to take charge in crisis situations and bring calm and confidence to fellow workers and those under your guidance

An effective means of leadership demonstrated in the drilling industry that is very applicable to all industries is the concept of video recording and review of personnel performing work in the field. The process works by engaging workers in an improvement campaign whereby workers allow themselves to be videotaped while performing work. Once in the program, workers may be informed or not informed when the taping occurs. This

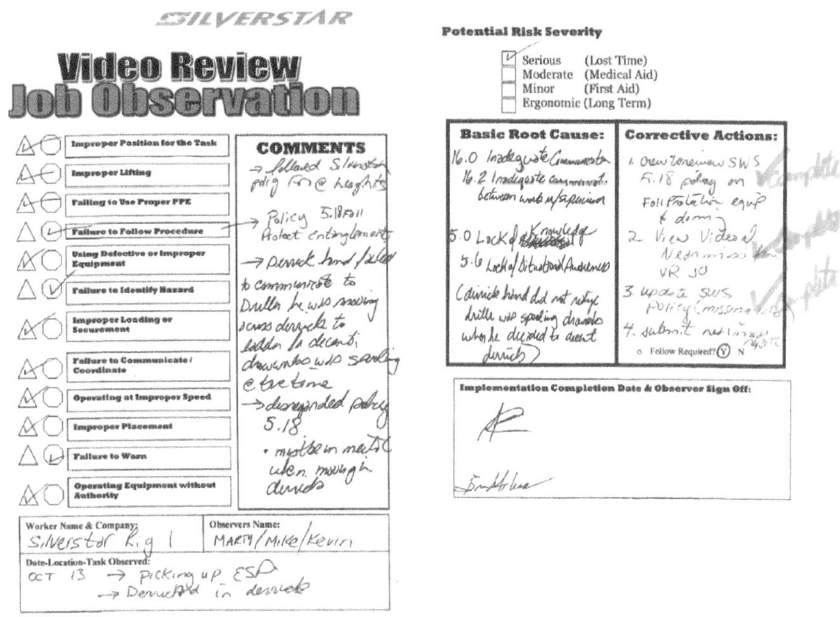

FIGURE 14.5
Sample video review job observation debrief form.

process has been used with great success in changing worker behaviors and engagement in a hazardous work environment.

The videotape is then reviewed with the worker or group of workers and evaluated for positive and at-risk work actions and behaviors. Actions such as in-the-line-of-fire body positioning or failure to follow procedural steps or agreed safe practices are identified and pointed out to participating workers. Corrective and remedial actions such as procedure reviews, training, or procedure upgrades may be undertaken to improve performance and reduce hazards in the workplace. Figure 14.5 provides an example of a video review job observation and corrective actions taken to improve an unsafe action observed in the review. Workers participating in the videotaping exercise and review are generally rewarded for participation regardless of whether or not safe or unsafe actions are identified.

Control

Control refers to making decisions that are consistent with the plan. Controlling work requires that the leader or supervisor understands what is being done and ensuring that tasks are completed in a manner that is consistent with the planning activities outlined in the work plan and FLRA documents. When conditions change, control refers to the leader's ability to work with his team to assess the impact of the changes in working conditions, reassess the new hazards introduced into the work process, and mitigate these hazards such that the health and safety of all workers are maintained. This may occur several times before the assigned work is completed.

Due Diligence Requirements

When major incidents including fatalities occur in the workplace and regulators are called in to investigate, organizations must be prepared to demonstrate due diligence in securing the welfare of all workers. Bill C-45 Section 217.1 of the Criminal Code advises: "Everyone who undertakes, or has the authority, to direct how another person does work or performs a task is under a legal duty to take reasonable steps to prevent bodily harm to that person, or any other person, arising from that work or task."

This means that supervisory and leadership personnel who supervise and direct the work of others have a legal responsibility to ensure the health and safety of all workers under their supervision. They can be held liable for worker injuries, fatalities, and illnesses from negligent behaviors. As a consequence, supervisory and leadership personnel must be made aware of all legal obligations and due diligence requirements in protecting the health and safety of all workers.

When a major or severe incident occurs and regulators are called in to investigate the incident, among the first sets of information collected by this group are the following:

1. The SOPs in use when the incident occurred
2. The competency assurance records of all personnel involved in the incident
3. The work permit and hazards analysis or assessment and mitigating actions conducted before the start of the assigned work

In view of these actions, all supervisory and leadership personnel supervising the work of others must take proactive measures to demonstrate that reasonable steps were taken to prevent harm to all workers.

SOPs

Supervisors and leaders must ensure that SOPs are updated and current, correct and validated, and accessible for use by all workers involved in the assigned work. For all nonroutine work, personnel must use the SOPs provided in performing assigned work. SOPs must be easy to follow and free of ambiguity. All workers involved in the assigned work must be able to understand the SOP. Lutchman (2010) advised that an SOP will typically provide the user the following details:

1. A reference number and title for identifying and searching for the procedure
2. Date on which procedure was written
3. Author of the procedure
4. Person(s) who reviewed the procedure
5. Revision number of the procedure
6. Page numbers of the procedure—normally numbered in the format page *x* of *y*
7. The steps involved in starting up, shutting down, or operating the equipment or machinery
8. Hazards identified and mitigation actions
9. Precautionary measures and PPE required

Occasionally, an SOP may also include a hazard analysis assessment and recommendations for addressing each hazard identified. All steps in an SOP must be followed as outlined in the procedure. When difficulties in following the procedure are encountered, or a step is no longer valid, the MOC process must be followed with an appropriate level or authority available to approve the required changes to the procedure with follow-up training for all personnel involved in using the SOP.

Competency Assurance Records

Competency assurance records extend beyond training records and represent the formal evaluation of demonstrated application of skills associated with the training. Competency is a reflection of the worker's knowledge (derived from training), skills and abilities (developed from practice), and experience (developed from repetition of the skills). A competent assessor is required to observe and evaluate the performance of the worker before they can be officially signed off as being competent. A competency assurance form is provided in Figure 14.6.

The assessor may be required to observe the worker over an extended period and may consider input from the worker peers and direct supervisor when evaluating the competency of the worker. Lutchman (2010) advised that competency assessors are required for two reasons:

1. Competent trainers and assessors provide the best opportunity for well-trained and qualified workers. Well-trained workers provide the best opportunity for reducing the likelihood of incidents,

Employee name:		Operating area / system	
Job title:			
Assessor name:		Assessment date:	

Employee assessment

1. System or area description:		
2. Evidence presented (attach copy where applicable):		
Performed by supervisor	Comment	Feedback and other supporting information
➢ Observation of employee doing task as per critical practice and SOPs. ➢ Assessment of outcome of employees work. ➢ Task discussions and review of questions. ➢ Discussions on responses to simulated conditions. ➢ Work related assignments.		➢ Feedback from peers and other workers. ➢ Team outcomes where the individual contribution is evident. ➢ Record of work or training activities.
3. Employee comments:		
Candidate's signature:		Date:
Assessor's signature:		Date:
To be forwarded to training department upon completion.		

FIGURE 14.6
Competency assessment form. (Copyright Suncor Energy Inc., approval from Suncor Energy Inc. is required to reproduce this data.)

injuries, poor performance, rework, and damage. They are also essential for sustaining strong performance.

2. In the event of an incident or accident, where the competency of personnel is questioned, credibility is enhanced when expert trainers have conducted the training and competency assessment of personnel involved in the incident.

A competency assessment form is required for each area of competence required for the worker and must be supported by a training matrix as discussed earlier in Chapter 4. Once this assessment form is completed, the form must be retained for traceability requirements. Refresher training and assessment may be required over time and the process of competency assessment must be repeated all over again.

Work Permit and Hazards Analysis or Assessment

A work permit that has identified and mitigated all hazards is also required for all nonroutine work. A work permit is essentially a contract between the organization and the worker performing the job. Work permits can be classified into categories based on the type of work being undertaken. Some examples of separate permits are as follows:

1. Cold work permit—required where no flames or fire is introduced when work is being performed in a hydrocarbon-rich or flammable environment.
2. Hot work permit—required where open flames or sparks can be generated from the work in a hydrocarbon-rich or flammable environment. Such work may include welding, grinding, hammering, or even driving a vehicle into a facility where hydrocarbon or where flammable vapors may be present.
3. Ground disturbance permit (also called excavation permits)—required where excavation and ground disturbances are possible. Buried electrical cables, pipelines, and structures must first be located before excavation can occur.
4. Confined space entry permit—required when work is to be performed in areas where access and egress are restricted. Such environments may or may not contain materials immediately dangerous to the health and safety of workers. However, special precautions are required to ensure that the work is done safely. Examples of confined space entry work may include entry into a vessel or tank, entering a sewer or trench, climbing to a platform where access or egress is restricted.

The alternative to separate permits is a single permit that carries sections for all of those identified above so that the appropriate section can be used when necessary.

Typically, the sections of a work permit may include the following:

1. Job scope identification
2. Atmospheric testing
3. Equipment isolation and identification
4. Hazards identification and control
5. Authorization and agreements—work initiation, to be signed by both the performing authority and the issuing authority
6. Authorization and agreements—work closeout, to be signed by both the performing authority and the issuing authority
7. Details of prejob tailgate safety meeting—all workers must attend

8. Record of prejob meeting attendance
9. Repeat atmospheric testing
10. Emergency response information

Training and competency in issuing a work permit are essential requirements. Furthermore, this legal contract is binding and must be retained on file upon completion of the work for a defined period, before it may be destroyed.

Typically, the permit is essentially an identification of the work to be done and the hazards identification and mitigation actions done to ensure the health and safety of workers performing the work. Appendix 3 is a sample work permit highlighting the relevant sections and requirements for ensuring the health and safety of all workers. Hazards identification and risk mitigation is also achieved using the process described earlier in Figures 14.2 through 14.4 for high-risk and critical tasks.

Documentation and Traceability

Due diligence must be demonstrated. This requires documented evidence of actions and activities designed to protect the health and safety of all workers. As a consequence, SOPs, training and competency assurance records, and hazards assessment and mitigation records must be properly documented and easily accessible. In today's digital age, many of these documents can be scanned and stored electronically such that filing rooms and retention of hard copies of data and information can be reduced. Without proper documentation and traceability, the credibility of the earlier discussed due diligence requirements is diminished.

15

Audits and Compliance

Audits and compliance are proactive means for verifying compliance to the organization's SMS and governing standards and procedures. When used properly as a business improvement model, audits and compliance are very effective in generating continuous improvements in business and can move the organization along its safety culture maturity path. Auditors are no longer organizational watchdogs and policemen; rather, they provide guidance to business leaders on gaps and opportunities in an organizational SMS for proactively addressing these gaps and risks exposure to prevent incidents and losses before they occur.

This chapter focuses on audits and compliance as a business process aimed at continuously improving the SMS of an organization.

Avoiding the Blame Game

Safety accidents occur in organizations for various reasons such as miscommunication and employees not adequately trained and lacking basic knowledge and required safety skills (Kletz, 2009). Errors also occur because people deliberately decide not to carry out instructions that they consider unnecessary or incorrect. These are referred to as safety violations or noncompliance. When employees commit a safety infraction, companies can take disciplinary action, which can vary from doing nothing, to verbal warnings or written warnings, to more severe actions such as time off and termination. It has been reported that disciplining employees for safety infractions is an effective tool to reduce incident rates at the third occurrence.

An organization that implements a discipline policy, which includes time off or termination for safety infractions, should realize a reduction in total recordable incident and lost time incident rates. Companies that implement an aggressive discipline policy for safety infractions should realize substantial reductions in total recordable incidents and lost time incident rates in the workplace. With regard to safety audits, it is important for organizations to realize that an increase in the number of safety audits significantly reduces recordable incident rates (Brahmasrene and Sanders, 2009). This was also well demonstrated in Chapter 11, Figure 11.6.

Many employers today blame workplace accidents on workers rather than taking a look at hazardous job conditions or failures in the SMS. Dr. W. Edwards Deming, through his pioneering work on quality, was able to support statistically using root cause analysis that quality is 85% the responsibility of management and 15% the responsibility of employees, which can be interpreted as 85% of the time, problems that arise in organizations are related to inappropriate and ineffective processes, and not to people. By blaming problems on people, organizations are missing an opportunity to improve by tackling the pertinent processes or identifying which element of SMS is not working as intended (Hein, 2003).

Groups and organizations with a rampant culture of blame have a serious disadvantage when it comes to creativity, learning, innovation, and productive risk-taking. According to research, when companies or individuals play the blame game, they lose status in the industry and show decreased performance compared to others who own up to their mistakes. Fast (2010) indicated that creating a culture of psychological safety is the most important thing a leader can do. One example of the *blame game* follows the Deepwater Horizon accident in BP's Gulf of Mexico oil spill where three executives of the companies involved were caught blaming each other for the accident instead of taking responsibility.

BP America blamed its subcontractor Transocean, saying that it was their supposedly fail-safe blowout preventer that failed. The chief executive of Transocean blamed the catastrophe on a cased and cemented wellbore that suddenly failed. Halliburton's global business lines, which did the cementing, tried clearing their name, saying that the company's work was done in accordance with industry standards (Helman, 2010). Arising out of this disaster, BP announced the creation of a new safety division with overriding audit and oversight capabilities across its global operations.

The new organization is designed to strengthen health and safety and RM across the BP group and this goes beyond deepwater drilling. These are lessons for BP and many other organizations relating to the way they operate, the way they organize the company, and the way they manage risk (Industrial Safety & Hygiene News, 2010). According to Baldwin (2001), in some organizations, someone has to take the blame for bad decisions. This phenomenon leaves employees in a position where they always seek to remove themselves from blame rather than achieving results. Baldwin (2001) indicated that blame can be a powerful and constructive force and can be an effective teaching tool that helps people avoid repeating their mistakes. When used judiciously and sparingly—blame can also induce people to put forth their best efforts, while maintaining their confidence and their focus on goals. Indeed, blame can have a very positive effect when it is done for the right reasons. The key then is the way blame is managed, which can influence how people make decisions and perform their jobs, which ultimately affects the culture and character of an organization.

Thus, a precondition for auditing and managing risks is a work environment with a culture that focuses on learning from near misses and adverse events as opposed to concentrating blame and shame and subsequent punishment. When audits are conducted and findings are identified, the principle of reporting should be aimed at closing the gaps identified in a proactive manner as opposed to finger pointing regarding noncompliance. When blame is assigned, the outcome is that future audits result in window-dressing to disguise obvious SMS gaps such that auditors are misguided regarding the actual exposure. When blame is avoided, there is greater engagement and involvement among stakeholders with genuine desires to close gaps and address noncompliance issues.

Audits Support the Gap Closure Process

One component of the SMS is conducting safety audits and assessments, which is a methodical assessment or review of a workplace and worker safety to ensure that legal or industry/company standards are met. The definition of an audit varies from one organization to another and is referred to by a number of different terms such as appraisal, survey, assessment, evaluation, and inspection. The definition of an audit by the American National Standards Institute (ANSI) Z10 is that it is a process for data and information collection followed by objective evaluation to determine compliance to defined business criteria (Fearing, 2008).

Safety regulations from governmental and organizational bodies are instituted in most workplaces but facilities often add site-specific safety requirements to the audit process. An internal safety audit program is used by an organization to evaluate the effectiveness of the procedures and compliance in the use of these procedures to meet safety legislative and other statutory administrative requirements. It is a particularly useful management tool for ensuring that the safety management is being conducted efficiently and properly at all levels.

Basically, audits help employers continuously improve existing health and safety policies, standards, and procedures by identifying risks, hazards, unsafe behaviors, and safety gaps that may not have been addressed. Furthermore, audits help determine whether employees and management are following established safety guidelines, rules defined in these policies, standards, and procedures. Audits also identify and elevate safety concerns that employees may have relative to the workplace safety program management and operation and can potentially generate recommendations that will enhance the existing program.

The ability to identify program strengths and weaknesses and rate an organization's safety program is fundamental to continuous improvement and success, while at the same time, essential for good management. This entails not only identifying opportunities for improvement but also creating processes and procedures to mitigate future risks and losses. In addition to assessing conformance to safe work practices, audits measure the senior management's safety philosophy and attitude (Esposito, 2009). Audits should be continuous and aligned with the day-to-day operations of an organization. An ongoing audit process is a mechanism by which management can obtain measurable and meaningful data about the organization's safety and health programs. In contrast, a single audit is ineffective in that it only provides a snapshot of the overall status of safety and health programs (Brahmasrene and Sanders, 2009, www.nsc.org).

Typically, the three types of safety audits are program, compliance, and management system audits as discussed below (Johnstone, 2001; Esposito, 2009). A determination of audit types will be based on requirements defined within the SMS. Factors relevant to the audits may involve the level of maturity of the SMS, policy initiatives and implementation, and similar issues. The knowledge and understanding of key stakeholders within the organization and further organizational factors (e.g., number of sites, geographical locations, state, national, or international application) may also apply.

Determining whether the audit program should be compliance based or system based is typically a result of a company's philosophy and the maturity of the program. Costs and frequency of an audit are usually driven by a company's concern for being viewed as a good corporate citizen by its stakeholders and the public (Fearing, 2008).

All safety audit comments, recommendations, and corrective actions should focus on these four questions:

1. Does the program cover all regulatory and best industry practice requirements?
2. Are the program requirements being met?
3. Is there documented proof of compliance?
4. Is employee training effective—can and do they apply specific safe behaviors?

Program (SMS) Audit

The goal of an SMS audit is to ensure that the company has designed and follows its own procedures and policies related to managing safety. The SMS audit gauges the strategy and implementation of a safety program, regardless of whether that program is required by regulation. To respond to regulations, management develops appropriate programs. For example, one regulatory requirement is to record accidents on an OSHA 300 log and to do so within 6 days. A written program describes who has the responsibility to get this done, as well as the method one would use to investigate the incident. OSHA, while providing suggestions for investigating an incident, does not regulate how to investigate. Thus, a company must define and distribute the procedure for investigating the accident in order to implement the safety rule or requirement and make it meaningful. An adequate control is a detailed list of procedures that clarifies how employees should perform various tasks, report operating problems, make decisions or take initiatives, and judge work progress.

A control is effective if it properly rectifies problems that employees report. Having done so, the company now has a safety program/procedure in place to respond to the requirement. The program assessment will examine how requirements are implemented, in addition to the compliance specifics themselves. One challenge to a program audit is in knowing what to use as a standard or evaluation criteria. There is guidance, but not much consistency, in professional practice when it comes to what should be included in safety programs, policies, and procedures. Some fundamentals do exist, for example, in any safety program, all procedures must be documented so that consistent communication and implementation are achieved. Procedures usually document the responsibilities and implementation strategies and detail how the procedure will be checked, measured, and audited. Another challenge is in knowing how to keep a safety program current. New facilities, equipment, and personnel often require changes in a program. Effective program management responds to changes that occur in the workplace and will help ensure that change is successfully managed (Esposito, 2009).

Compliance Audit

Compliance audits are performed in response to regulatory requirements and can help determine whether the company is providing a safe and healthy workplace. These audits ensure that both employees and employer abide by regulations. It helps the senior leadership to ensure that the staff conforms to governmental laws and regulations when performing duties (Codjia, 2010a). Audits are also part of corporate responsibility. Compliance audits are also important in corporate business decision-making processes as it helps top executives prevent operating losses resulting from adverse regulatory initiatives, such as litigation, fines, and other punitive sanctions. Fines can vary from $20,000 to as high as $90,000 when an employee gets hurt as a result of noncompliance and this cost is not covered by insurance (Leman, 2010). In one case study, Maryland occupational safety and health issued a citation against Allen Family Foods, a 91-year-old, family-owned poultry processor in July 2010. The citation assessed a $1.03 million fine for 51 violations of occupational safety and health laws at the plant.

Violations included hazards with industrial trucks and deficiencies in machine guarding and PPE (Mook, 2010). Others included missing exit signs, exit route maps not posted as required by law, blocked exits and electrical panels, missing protective eyewear, missing fire extinguishers, or extinguishers not properly marked or located where easily visible (Leman, 2010). An effective compliance audit process also helps department heads detect, measure, and identify regulatory nonconformity risks in operations. Official safety audits are performed by the OSHA in connection with inspection visits, usually to high-hazard work places or companies with elevated accident rates such as construction sites and laboratories.

Federal and local government offices will perform annual safety audits while large companies with health and safety offices regularly self-audit to ensure compliance to OSHA and other internally established standards. These audits will include checks on compliance to emergency response plans, safety equipment, material and substance hazards, and personal safety. More recently, audits have started focusing on compliance to PSM standards and requirements. Auditors will also focus on safety training and accident rations (Codjia, 2010a). In the United States, OSHA has specific requirements, such as annual lockout/tagout assessments or weekly eyewash inspections as well as standards relating to ergonomics, respirator use, hearing conservation, blood-borne pathogens, and use of PPE. OSHA requires emergency evacuation plans for many facilities. Required are written and oral plans for an evacuation; a definition of what constitutes an emergency; all exits and fire extinguishers clearly identified and easily accessible; and a functioning emergency alarm system (Chacos, 2010).

A department or site inspection is performed to look for hazards and the absence of controls. Since incidents may include some at-risk behaviors, inspections often include a review of behaviors and conformance to procedures. A wall-to-wall compliance assessment will typically include three components:

1. Conformance
2. Record-keeping
3. Training and competency assurance

While OSHA does not specifically require companies to conduct compliance audits, compliance requirements dictate certain recordkeeping, programs, and training requirements. A compliance assessment is typically performed annually to review the status of written programs, records (e.g., training outlines, attendance), and performance (e.g., conditions and behaviors in the workplace itself).

Management System Audits

The management system audit is designed to be a complete process to evaluate and validate the effectiveness of management's commitment to compliance, the level of employee involvement, applicable risk control procedures, and the culture of the organization as a whole. Management system audits also examine other business processes, such as accountability and effectiveness of implementation, to determine how well safety is integrated into the organization.

A management systems audit uses a combination of audit techniques such as review of documentation, observation of the workplace, and employee interviews to validate and determine effectiveness. It is an evaluation of process and not just an evaluation of program. A management system-based audit is more nonregulatory but necessary for a successful safety program. This type of audit should include evaluating senior management's support and active participation (policy statement, staff meeting agenda item), employee participation (safety committees, "off-the-job" safety efforts, and so on), inspections and audits, training, contractor management, and emergency response programs. Some of these topical areas have regulatory implications, but collectively they are generally considered system-based programs. The usual examples in this type of audit are OSHAS 18001 and the OSHA Voluntary Protection Program, and another option is the ANSI Z10 standard (Fearing, 2008).

Auditors

Audits can be performed by federal or local government administrations, the company that owns the workplace or specifically hired independent auditing companies to ensure that corporate policies are *adequate*, *effective*, and *in compliance* with industry practices, regulatory requirements, and business and ethical standards. The audits should be performed by a team of safety professionals and SMEs from locations independent (to ensure neutrality and objectivity) of the specific site being audited. Audits may also be conducted by qualified consultants. All auditors must be trained in the standards as well as on a wide range of technical topics from safety, industrial hygiene, and environmental science to more specific subjects such as ergonomics. These auditors should be familiar with both the company program and the various local, state, and federal requirements. Audit team members should review all existing safety program material in advance of the safety audit. It is recommended that all auditors possess:

- The correct level of audit experience
- Comprehensive experience in compliance, including regulatory requirements
- A strong knowledge base and resources, including relevant industry knowledge where applicable
- Proper accreditation for the audit operations required

Audit Check Lists

Audit at a minimum should be performed using industry protocols that set out standards for measuring the effectiveness of an SMS. These protocols are documents that outline the procedure and offer guidance and instructions to assist the auditor in completing the evaluation process. However, whether the audit is performed on a manufacturing industry or on a construction industry, the basic concepts for the evaluation are the same. Each audit includes an evaluation of management systems, program responsibilities, a review of records, observations of physical conditions, equipment and personnel, and interviews with employees throughout the facility. Changes caused by equipment updates, building maintenance, job duties or personnel issues, and emergency response are examined in audits. Auditors will also focus on health and safety training of workers and accident ratios. Extended checks will be performed in high-hazard environments such as construction sites and laboratories. OSHA requires that employers first attempt to eliminate or reduce hazards before requiring employees to use special safety equipment to deal with them. Sites are required to comply with local regulations and with company standards. In cases where there is a regulation and comparable company standard, the site must comply with the most stringent. An integral part of the OSHA standards since 1992, and applicable to industrial processes containing more than 10,000 lbs of hazardous materials, including explosives and pyrotechnics, audits are conducted against a PSM standard. This standard is intended to prevent or minimize the consequences of catastrophic releases of hazardous chemicals, thereby protecting employees and property, the public, and the environment (Turnbull, 2010).

Although many off-the-shelf audits are available and have been developed by SMEs, in some cases, the best choice is an audit designed specifically or customized to a given application, based on what the users want to measure, what they ultimately wish to learn, and how they will use what they learn. Although many audit tools are used in the field, few have been validated using additional scientific methods. A good audit tool should include consistency and accuracy, by examining the reliability (refers to consistency and stability)/validity (refers to accuracy and precision) when applied to workplace safety (Yueng-Hsiang and Brubaker, 2006).

Schedule Considerations

Inspections can be performed on a daily, weekly, monthly, or annual basis. Frequency depends on the nature of the control or how often conditions, equipment, or people change, as well as legal requirements. Employers can decide when to conduct audits, in order to monitor the effectiveness of their policies and to safeguard the working environment for employees. For example, manufacturers usually recommend that emergency stops be tested at the beginning of each shift, which emergency eyewashes and safety showers be inspected weekly. Although audits are not required to be performed with any specific frequency, workers who are unsatisfied with existing workplace health and safety practices can contact OSHA to inspect the premises or they have the right to address safety concerns with OSHA officials privately.

Conducting a single annual comprehensive safety audit can actually hide the facts and hazards that need to be discovered. A better approach would be to schedule various specific safety audits throughout the year. The internal audit schedule should be owned by a person who has authority to manage and review the implementation of the schedule. Senior management should have direct oversight of the process, to allow checks on audit functions and efficiency. The internal audit schedule will provide, as a minimum, the following information:

- The type of audit to be conducted
- The expected time frame in which the audit is to be conducted
- The lead auditor responsible for the audit activity
- Location of the audit
- Time frame for final report

Having identified the types of internal audit activity to be conducted, the organization will need to consider the most suitable audit delivery options and scheduling. These considerations will include one or more of the following points:

- *Mandatory audit requirements*: These may be conducted for inclusion in reporting requirements. The internal audit, in these cases, can also act as "radar" for management, ensuring that compliance issues are under proper scrutiny.
- *Size of the organization and geographical locations*: Consideration should be given to the different types of activities (and risk potential) within an organization, similarity of activities at different locations to evaluate uniformity of application, and geographical locations that may impact on audit schedule delivery. This approach additionally allows for targeting of areas related directly to policy implementation.
- *Identified high-risk potential subject areas*: Any high-risk areas identified as a result of statistical analysis or management review require audit verification to assist in implementing strategic management.
- Changes in legislation or organizational structure that may impact on the capacity to implement safety requirements. Legislative changes usually include a time frame for compliance.
- Resource base (provision of internal or external resources) from which the organization can allocate qualified and competent auditors

to conduct the audit verification activity type and depending on the scope of activity to meet audit needs. The resource base must be able to deliver the necessary standard of auditing to ensure efficiency and compliance.

- *Organizational planning and development cycles*: Planning of audits to align provision of audit reports with review cycles. This planning is a particularly useful management tool that can ensure proper control and timeliness of operational data.
- Prioritized activities (e.g., based on risk evaluation or business need) that will impact on implementation of safety requirements.

Common Problems with Internal Audits

One major issue in relation to developing internal audit schedules is that they either do not meet organizational needs or merely aim to achieve a level of conformance with statutory body requirements of regulators. *This is quite inadequate, and may expose organizations to serious liabilities.* To ensure a fully functional SMS that is capable of dealing with all safety issues and meets the standards of both statutes and major legal claims, the organization must ensure that the internal audit is conducted on a holistic, best-practice basis in which all areas of liability, RM, and safety are properly audited. The SMS must achieve full coverage of all potential liabilities. Conversely, the needs of internal audits may exceed the capacity of the organization to meet the requirements defined therein.

The organization may lack the expertise required to deal with some areas of RM. Any internal audit carried out on this basis will inevitably be inadequate, and can create a risk of serious deficiencies in the SMS. Another key problem is to ensure that audit competency (either internal or external) is at a level to achieve a suitable outcome as defined within the internal audit schedule, for example, there are serious risks in using auditors to conduct compliance audits without the necessary understanding in relation to legislative application of requirements.

Laying the Foundation for an Effective Audit

In response to competitive pressures, companies are being reorganized to facilitate structures that efficiently incorporate all aspects of operations (e.g., production, quality, and safety performance) into a cohesive and profitable process managed jointly by all organizational levels. Safety management has evolved from its technical roots to incorporate the qualities that characterize these transformed organizations. Modern safety management involves the management of people and their behavior, and recognizes the value of cross-level teamwork to improve business systems. The National Safety Council has an increasing body of evidence supporting the finding that comprehensive SMSs are effective in reducing the risk of workplace incidents, injuries, and fatalities.

Effective SMSs must be woven into the fabric of an organization, becoming part of the culture and the way that people do their jobs. Successful SMS share certain attributes, which include leadership from both management and employee representatives to assure that necessary resources are available, technical and operational elements to assure there is ongoing reduction of risk, and cultural and behavioral considerations to maximize improvement by engaging the workforce and fostering collaborative efforts for all to contribute. Aligned to the safety audit, sites must also have a fully implemented environmental management system. This is to ensure that the facility has a sustainable process to ensure compliance with local regulations and conformance to company standards. The National Safety Council model, SMS, includes the following nine elements organized into three key performance areas.

Leadership: Management

- Management leadership and commitment
- System management and communications
- Assessments, audits, and performance measurements

Technical: Operational

- Hazard identification and risk reduction
- Workplace design and engineering
- Operational processes and procedures

Cultural: Behavioral

- Worker and management involvement
- Motivation, behavior, and attitudes
- Training and orientation

Phase One: Safety Audit Preparation

Step 1: 1–2 months prior to the audit, inform all involved managers and supervisors. This warning period provides them adequate opportunities to ensure that all compliance records, documents, and demonstrable proof of compliance to policies, standards, and procedures are available for presentation to the auditors during the audit.

Step 2: Review all past program area audits and corrective action recommendations. This provides auditors the opportunities to understand business unit/area leadership commitment to remedial and corrective actions proposed for finding and recommendations.

Step 3: Review all company, local, state, and federal requirements for the specific program. Become familiar with the document, inspection, and training requirements.

Step 4: Determine the scope of the audit. This can be based on accident and inspection reports and input from various managers. Set a start and stop time and date for the audit.

Phase Two: Fact Finding

A fact finding event is used to gather all applicable information. Auditors should be impartial and unbiased and must make conscious efforts to avoid forming opinions or make evaluative comments during this phase.

A team approach: If a safety audit team is used, make assignments to each person that defines their area of inspection. Ensure that they have the proper program background information and documents.

Safety audit areas: Most audits can be broken down into the following focus areas:

Employee knowledge and competence: OSHA standards require "effective training"—an effective program ensures that employees have the knowledge required to operate in a safe manner on a daily basis. The level of knowledge required depends on the specific activities in which the employee is involved and their specific duties and responsibilities. Generally, managers and supervisors should have a higher level of knowledge than general employees.

This includes practical knowledge of program administration, management, and training. They should be able to discuss all elements of each program that affect their assigned employees. Many programs divide employees into these two groups—authorized employees and affected employees. Authorized employees must have a high level of working knowledge involving hazard identification and hazard control procedures. Determining employee level of knowledge and competency can be achieved through written quizzes, formal interviews, or informal questions in the workplace. In many instances, demonstrated skills and capabilities are required to establish competency and competency assurance.

Written program review: During the health and safety audit, a comprehensive review of the written SMS should be conducted. This review should compare the company program to requirements for hazard identification and control, required employee training and record keeping against the local, state, and federal requirements. Additionally, if applicable, the company insurance carrier should be asked to conduct an independent written program review.

SMS program administration: This part of the health and safety audit review checks the implementation and management of specific program requirements. This section asks these and other similar questions:

- Is there a person assigned and trained personnel to manage the program?
- Are specific duties and responsibilities assigned?
- Are sufficient resources provided?
- Is there an effective and ongoing employee training and competency assurance program?

Record and document review: Missing or incomplete documents or records is a good indicator that a program that is not working as designed. Records are the company's only means of proving that specific regulatory requirements have been met. Record review also includes a look at the results, recommendations, and corrective actions from the last program audit.

Equipment and material: This area of a health and safety audit inspects the material condition and applicability of the equipment for hazard control in a specific program. Examples of audit questions for this area are

- Is the equipment maintained in a safe and reliable condition?
- Is there adequate equipment to perform assigned tasks safely?
- Is the PPE used and stored properly?
- Is the equipment, such as exit lights, emergency lights, fire extinguishers, material storage, and handling equipment, designed and staged to control hazards effectively?

General area walk-through: While safety audits are not designed to be comprehensive, physical wall-to-wall facility inspections and a general walk-through of work areas can provide additional insight into the effectiveness of safety programs. Auditors should take written notes of unsafe conditions and unsafe acts observed during the walk-through.

Engaging the workforce: Engaging the workforce is perhaps one of the most effective means for deriving useful information regarding SMS compliance at a worksite. Worker behaviors and willingness to talk with auditors in an informal manner is an indicator of the business unit/area culture and approach toward audits and in general health and safety.

Phase Three: Review of Findings of the Health and Safety Audit

After all documents, written programs, procedures, work practices, and equipment have been inspected, gather your team and material together to formulate a concise report that details all areas of the program. Remember to focus on the four basic questions mentioned earlier. Each program requirement should be addressed with deficiencies noted. Audit findings must highlight findings that show strong compliance and good work while at the same time showing areas of weak compliance and exposures. As a consequence, auditors must learn to include comments of a positive nature for each element that is being effectively managed.

Phase Four: Recommendations from the Health and Safety Audit

Develop recommended actions for each deficient condition of the program. Careful forethought should be applied to ensure that this is not a process that simply makes more rules, additional record keeping requirement, or makes production tasks more difficult. Examine the manner and means in which the current deficient elements are managed to determine if there is a simpler procedure that can be employed.

Phase Five: Corrective Actions from the Safety Audit

Development of corrective action should involve the managers and supervisor who will be required to execute and steward corrective actions and recommendations. Corrective action priorities are established based on the level of hazards and risks exposure determined from the audit findings. All corrective actions should be assigned a completion and review date. Records of completed corrective actions should be reviewed through the normal management chain and then be filed for use during the next audit. Table 15.1 provides a simple template for stewardship of assigned corrective actions with sample findings and corrective actions. The key to success is to prioritize corrective actions properly, provide adequate resources, and ensure that reasonable time is given for completion of assigned responsibilities and tasks.

Phase Six: Debrief and Publish the Safety Audit Results

It is essential to let all supervisors and the manager know the basic findings and recommendations. At the end of any audit, it is very important to debrief with the entire team and business unit/area personnel who participated in the audit so that they may have a preliminary insight into findings and recommendations. Acknowledge also those departments, managers, and supervisors who are properly executing their responsibilities. After a few audits, everyone will want to show up on the plus side of the results, making both the audit process and the work of the audit team much easier.

TABLE 15.1
Corrective Action Item Tracking and Stewardship Process

Audit Finding	Corrective Actions	Assigned to	Due Date	Completion Date
• No processes for ensuring workers are trained and competency assessed	• Establish a training matrix for determining trained versus untrained personnel and upcoming needs for refresher training • Develop process for verifying competency and documentation of sign-off process	Operations manager	June 30, 2011	Open
• Poor compliance to MOC standard. MOC standard not accessible to workers	• Determine methods for ensuring MOC standard is available to employees and contractors	Maintenance director	June 30, 2011	Open

Resource Allocation Based on Risk Exposure

The higher the level of risk faced by a firm, the greater the need for investments in resources in order to adequately minimize the risks. The allocation of resources is based upon risk exposure that is ascertained through the audit and compliance process. A risk score calculator is used to determine the level of risk. This process involves the defining of consequence, exposure, and probability through the use of a risk matrix that allows the prioritization of risks from high to low risk. Organizations with high-risk processes, technologies, and/or environments need their employees to be more safety conscious than those operating in lower-risk industries. Resource allocation is necessary because of the increasing demand for services and rising costs to provide those services. To manage hazards effectively, greater vigilance and problem solving are required. In addition, the further the company is behind in safety practices from industry leaders, the greater the incentive for the firm to improve underlying variables that affect longer-term sustainability (Mol, 2003).

There are many associated hazards and risks in any workplace; therefore, a system must be put in place, based on the RA, to ensure that duties may be carried out safely and that workers involved in any ongoing process are safe. Considering the wide range of hazards and risks in the workplace, it is necessary to include these in a comprehensive management system of the company. Allocating sufficient time and resources through ensuring training and competence of staff, putting in place safe systems of work based on an appropriate RA, and effective communication between management and staff are key issues (Boyle, 2002; European Agency for Safety and Health at Work, 2009).

Another resource allocation that needs to be considered is in relation to procurement procedures, which must be put in place to ensure that the necessary tools and PPE (along with the necessary training and care of this equipment) are available for safe maintenance. During the procurement of new machinery and buildings, ease of access for performing maintenance should be considered: risks during maintenance can be minimized or even eliminated through good design of work equipment, availability of relevant tools, and information from the supplier or manufacturer. Whenever a company or an organization makes changes to the physical environment of the workplace, or buys new equipment, it is important to ensure that those changes or purchases are also suitable for the diversity of the workforce. Also, organizations are increasingly outsourcing their maintenance activities, which means that the procurement and management of contracts between companies have a strong impact on OSH. Maintenance carried out by a contractor has to be well integrated into the ongoing activities of the company to safeguard the safety and health of all workers involved. Good practice examples, where the needs of both contractor and the host company are taken into consideration, include "good neighbor schemes," "safety passports," and induction procedures (European Agency for Safety and Health, 2009).

During the procurement process, in addition to competency and communication, the issues of cultural and language differences have to be considered, in the case of migrant workers, as well as issues resulting from the precarious employment of some subcontractors. Management commitment and safety culture are essential for safety and health at work in general. Management commitment may be the single most important determinant of the safety culture of an organization. It determines the resources (time, people, and money) allocated to safety and health and produces higher levels of motivation for health and safety throughout the organization.

Audit all Facets of the RM

OHSAS 18001 is the internationally recognized assessment specification for occupational health and safety management systems. OHSAS 18001 has been designed to be compatible with ISO 9001 (Quality) and ISO 14001 (Environmental), to help organizations meet health and safety requirements in an efficient manner. An area most often forgotten is the way contractors are managed within the organization. Most leaders in the past believed that when work is contracted out, risks and responsibilities are transferred to the contractor and the organization is no longer responsible or accountable for incidents associated with contract work.

Contractor safety management is perhaps one of the greatest opportunities for improving workplace health and safety since experience has shown that injury frequency of contractors is 2–3 times that of employees. In addition, contractors are injured on average 3–4 times more frequently than employees are. As more and more companies are forced to become more competitive, downsize, and subcontract workers employed to perform jobs within the company in order to augment internal capacity and technologies, the impact will result in a negative safety performance. There are many factors that cause poor contractor safety, including

- Owner and contractor's attitudes toward health and safety.
- Differences in approaches to health and safety.
- Owner's failure to communicate expectations about health and safety to the contractor.
- Focusing exclusively on prime contractors and overlooking subcontractors.
- Owner SMS and compliance to OHSA requirements are substandard.
- Contracts that are confused about owner's safety expectations. Some owners may up front establish priorities as follows: health and safety, quality, cost, and finally schedule.

To improve management of contractor safety, in addition to addressing each of the pitfalls listed above, an important initial step is to ensure that safety is included in the hiring criteria. Careful contractor selection, a clear statement of work, identification of the hazards the contractor will encounter, and audits of the contractor's performance (either random or on a schedule) are key in ensuring contractor safety. Facilities should choose contractors on the basis of work experience and ability to complete the job in a safe fashion, not just based on the bid for the project (Ayers, 2007).

Effective prequalification of contractors and having an overview of the contractor's documentation, including liability insurance trends and health and safety performance before work begins, are also key to success. Contractors need to ensure that not only this documentation is in order but also that their own workforce has the appropriate level of training and competency assurance. When the hiring process begins, the owner needs to perform a prequalifying interview of the contractor based on elements such as their injury statistics and their ability to identify hazards. The owner should specify basic safety requirements in the contract document, the basic components are required in the contractor's safety program, and the owner needs to address safety in the design phase. The owner assigns at least one safety representative to the project with high authority, with responsibilities clearly defined, and regularly monitors the contractor's safety performance.

Once this has been established, communication is absolutely essential between both parties to ensure that everyone understands expectations, to provide regular status updates, and that everyone is being held accountable. For some companies, this may translate into audits and inspections on the part of the employer to keep an eye on whether the agreed-upon procedures are being followed. Proper training is at the heart of every safe workplace, and ensuring that education is in place when on-site contractors are used for a project is crucial for both the client and the contractors themselves, according to safety experts. All projects involving contractors should be managed through a risk-management rating system. By using a safety rating system, you will give each project a risk rating of high, medium, or low. Then, for projects at each level of risk, you can identify what safety systems you require of the contractor and spell out how they should interface with your own safety systems. For example, at each different risk level, you can have different standards for what a contractor must show to prequalify for a project or how much oversight or feedback on the project you demand. For any firm trying to adjust to a large influx of contract employees, the key to success is the management's ability to apply high safety standards to employees and contractors alike. Although such actions may not guarantee an injury-free environment, they lay the foundation for minimizing risk.

All owners, regardless of the type and size of their projects, should recognize that they have a responsibility for construction safety. Safety should be integrated into the overall project objectives. By realizing the significance of construction safety to the success of their capital projects and overall reputation, owners should make a firm commitment to construction safety. This includes making a commitment of funds to enhance safety. Contract projects can be managed in a manner that emphasizes mutual responsibility, zero injuries, minimal incidents, and heightened safety awareness among all involved—workers, owner-operators, contractors, and the community.

Leadership

A senior manager should actively participate in the internal audit to ensure currency of management information and proper critical organization-wide scrutiny of administrative and operational functions. The right attitude about compliance must come from the top (Leman, 2010). Several factors need to be considered when using the audit as a primary measure of safety performance. To be successful, the audit should address positive programs and activities (best practices), as well as program gaps, recommendations, findings, and local attention items. Another cornerstone for success is for the final report to be a clear and accurate evaluation of the overall environmental health and safety (EHS) program and having the results sent to the right people in the organization, including senior management (Fearing, 2008).

The audit process assumes that there are defined safety elements and sub-elements that all companies must possess in order to achieve good safety performance. Audits, however, do not measure human factors such as trust and workplace norms. Organizations seeking to achieve world-class safety cannot do so unless these human factors contributing to safety performance are measured and addressed. One of the best ways of doing this is to employ alternate measurement tools such as safety perception surveys in order to tap into the collective knowledge of employees. Such organizations actively seek information from their workers on how things can be improved. Now, more than ever, employee opinions and perceptions play a key role in the success of an organization.

As companies start to realize that the solutions to most of their problems are already well known to their stakeholders, more and more companies will see the value in soliciting opinions and perceptions from employees, customers, and constituents. Many of the factors that affect quality, productivity, and service will also affect safety. Safety perception surveys are a key tool for understanding these factors. If an organization's current culture fosters unsafe behaviors and procedural shortcuts, a perception survey can help identify why they are happening. If current corporate culture influences employees to work safely even when the boss is not around, then these organizations have achieved a key prerequisite of a world-class safety system (Ryan, 2010).

16

Auditing the Safety Management System

Auditing the SMS of an organization is essential to determine if policies and procedures in place are adequate to protect the health and safety of workers, public, and the environment. An effective audit should clearly identify all weaknesses of the SMS as well as provide recommendations for enhancement. Generally, audits should be performed using industry protocols that set out standards for an effective SMS. However, many organizations may hold themselves to higher standards than industry performance and will often audit against internally developed standards for performance improvement opportunities.

Audit protocols are documents that outline the procedure and offer guidance and instructions to assist the auditor in completing the evaluation process. Various audit protocols have been developed in specific industries, offering standards for measuring the effectiveness of the safety program as outlined in Chapter 15. However, whether the audit is performed on a construction company or on an oil and gas industry, the basic concepts for evaluation are the same.

Gap Analysis and Identification

Gap analysis is a business process designed to enable an organization to compare its actual compliance or performance with the organization's standards relative to requirements of the standard. Gap analysis is sometimes spoken of as *the space between where we are and where we want to be.* Often conducted by an unbiased team with representation from within and external to a business unit or area, the process involves the physical comparison between what is actually done in the business relative to the requirements of the standard. The process involves

1. Developing a table that highlights the requirement of the standard for compliance
2. Identifying business unit or area actual performance relative to each of the standard requirements
3. Determining gaps in compliance

4. Prioritization of gaps based on risk exposures to the organization
5. Developing gap closure strategies and actions

These help to provide the organization with insight into areas that have room for improvement. As a result, internal assessment and audits among other tools can be used to identify gaps or areas of noncompliance. A gap analysis can be conducted for all standards and procedures. Gap analyses provide proactive opportunities for continuous improvements in compliance and safety performance management.

Internal Assessment

Internal assessments are used to evaluate an organization's strengths and weaknesses. It can be considered as a formal process to identify opportunities for improvement, potential risks, liabilities, and pitfalls. From a health and safety standpoint, an internal assessment may be a predecessor or a component of performing an RA for given activities and/or operations within a company. An internal assessment can also take the form of an internal audit. Such audits seek to determine the degree to which an organization conforms to the requirements of a specification, standard, or to its own organizational requirements. The results of the audit compare current activities or performance to the benchmark of standards that the organization seeks to establish in its operations. It also records observations, important milestones/efforts, and can be used to identify opportunities for improvement.

One of the reasons why internal auditors question individuals is to assess their knowledge of a particular system or process that is used by their company that they may be directly or indirectly exposed to. If the proper implementation of a particular benchmark or system is not present, it may be a challenge to determine whether or not employees are effectively utilizing the system and/or are knowledgeable of their roles in meeting the required specifications. The U.S. Occupational Safety and Health Administration (OSHA)'s official policy toward voluntary self-audits is not to ask to see the audit report. However, if a voluntary self-audit was performed and violations were immediately addressed prior to the visit of an OSHA inspector, then the self-audit report can be used to show good faith and may possibly help avoid a citation (Safety Compliance Letter, 2006).

An example of an internal audit tool is the management awareness and action review system (MAARS) used by Johnson & Johnson (J&J) as a global process for the identification and elimination of safety risks (Johnson & Johnson, 2009). MAARS provides a framework to identify and proactively

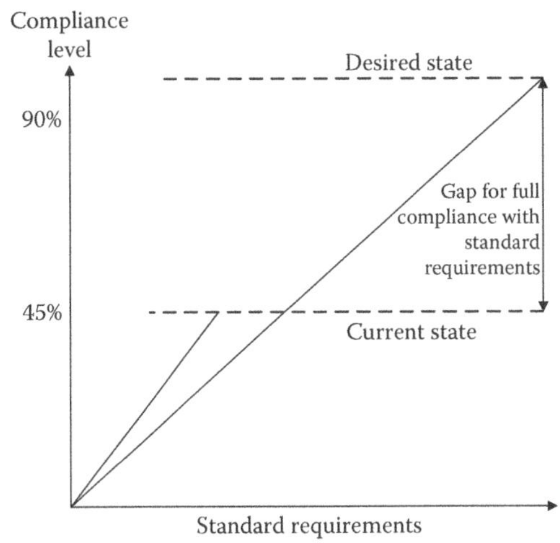

FIGURE 16.1
Gaps between current and future states from gap analysis.

manage issues beyond compliance with regulatory requirements. A complete and accurate self-assessment is critical because it

- Assures a solid foundation for the development of complete and accurate management action plans (MAPs). A sample template for MAPs is provided in Figure 16.1.
- Provides a systematic mechanism that assures proper alignment of priorities and deployment of resources.
- Helps eliminate losses associated with environmental, safety, sterilization, and quality risks.
- Facilitates earlier product approvals.

The criticality of the self-assessment prompted J&J to select it for improvement when a Six Sigma (DMAI^{2}C) Green Belt project revealed that only 45% of their facilities were conducting complete and accurate self-assessments. Using voice of the customer feedback, four areas were identified as significant drivers for effective self-assessments:

1. Skills of the worker performing the self-assessment
2. The quality of tools used in the self-assessment
3. Leadership and management commitment
4. Organizational culture toward continuous improvements

The MAARS process complements J&J's partnership with its management to establish a continuous learning and improvement culture at each of its global facilities.

TABLE 16.1

MAP as a Business Tool

	Date :						
	Company:			Country:			
	Prepared By:			Signature:			
	Approved By:			Signature:			
PROCESS EXCELLENCE MANAGEMENT ACTION PLAN:							
Ref #	Process / Category	OFI Description	Action Steps	Owner	Due Date	Status	Business Impact[a]

[a] Business impact may be stated in financial terms or in non-financial terms, such as improved cycle time, defect rate, employee satisfaction, customer satisfaction, improved competencies and/or capabilities of processes, people, etc.

The MAP as a business tool was a very effective resource for supporting the improvement culture in J&J (Table 16.1). This tool will facilitate the evaluation of each opportunity as to its relative impact and urgency, and facilitate the allocation of the appropriate resources relative to the potential business benefit. It also provides a mechanism for business leaders to monitor and follow up on a regular basis the progress being made on priorities until projects are completed and the benefits achieved. It may also be used as a periodic communication device for creating organizational excitement around reaching the next level of operational excellence. Improvements of the self-assessment tools have also occurred using electronic access to the quality assessment tool and initiatives to combine the self-assessment tools for EH&S into one web-enabled system.

A similar process is used in the energy and oil and gas industries. Internal self-assessment processes are used to compare actual performance at any point in time, relative to requirements imposed by the organization upon itself in the form of standards. Workers within a business unit or area will perform this self-evaluation on an ongoing basis (generally annually) to determine gaps and opportunities for developments relative to the standard requirements. When gaps or improvement opportunities are identified, these are prioritized, resourced, and auctioned out to responsible individuals or groups for closure. A tracking mechanism similar to that defined in the MAP business tool is used for stewardship. To perform an effective internal assessment, the steps outlined below are recommended:

- Identify the specific standards and work instructions for which the gap analysis is to be conducted.
- Develop the gap analysis check list for this standard.
- Nominate and support the audit team.
- Forewarn stakeholders regarding the self-assessment process.
- Schedule an audit and conduct field compliance audit and assessment.
- Engage stakeholders in performing the self-assessment and perform the audit to identify the current gaps.
- Develop report and action plan from the gap identified in the analysis. Prioritize actions based on risk exposures.
- Develop future action plan to ensure sustained improvements.
- Conduct follow-up self-assessments and audits to provide objective unbiased assessment and support.

Gap closure identifies the steps that can be used to eliminate the gaps in a company's progress and development. In order to solve the gaps, issues are included in a MAP and management assigns resources to address these issues. Root cause analysis is used to correct gaps. For training gaps, a training plan is established annually and communicated to the relevant departments detailing the type of training, employees required, and a tentative date for completion. Selecting a successful strategy to close the gaps is both a critical and a difficult decision to make. This is dictated by the customers and competition of the organization. Customers can be unreasonable and want the world, but if it is physically impossible, then attempts to satisfy them will drive an organization out of business. If the competition is capable of satisfying such requirements, then companies will have no choice but to ensure the improvement.

DuPont states that the use of surveys can help organizations determine "best and worst sites and make comparisons to uncover critical truths" (DuPont, 2008, p. 3). Once established, greater attention can be paid to poorly performing sites to determine underlying values and beliefs that may be contributing to the performance and can therefore be acted upon. DuPont (2008) proposes that benchmarking safety culture within the organization and extending the benchmarking process to external organizations in the same industry or across industries are also beneficial in improving organizational safety performance and culture. The success of benchmarking lies in the response to the question: *Are we happy with our safety culture and with our health and safety performance?*

External Benchmarking

One method used by organizations to gauge effectiveness of safety performance improvement is to set benchmarks by comparing data for various facilities, processes, and practices within an industry. Benchmarking is the process comparing organizational performance with industry peers as a means of identifying improvement opportunities. The key aspect of benchmarking is that it allows the performance of a company to inform improvement activity or affirm excellence (ReVelle, 2004). According to Love and Dale (2007), there are three main types of formal benchmarking as discussed below.

Internal Benchmarking

Internal benchmarking involves benchmarking between business units or functions within the same group of companies. In this way, internal best practices and initiatives are identified and shared across the corporate business. Internal benchmarking helps to standardize business practices and processes within the organization and to remove silo effects wherever they may exist.

Competitive Benchmarking

Competitive benchmarking provides a comparison with direct competitors, whether of products, services, or processes within a company's market. It is often difficult to obtain data for this form of benchmarking as by the very nature of being a competitor the company is seen as a threat. However, where health and safety are concerned, barriers of protectionism and secrecy are slowly being eradicated in the interest of collective improvements to safety performance and protection of workers, environment, and assets.

Functional/Generic Benchmarking

This form of benchmarking involves analyzing outside organizations that are known to be best in class. This is the comparison of specific processes with "best in class" in different industries, often considered to be world class in their own right. Functional relates to the functional similarities while generic looks at broader similarities of businesses, usually in disparate operations. It is usually not difficult to obtain access to health and safety information from other organizations to perform this type of benchmarking.

For companies that have performed internal benchmarking and want to investigate new ways in which to improve performance of their internal processes, functional/generic benchmarking can produce significant improvements. Many companies believe that their processes are as efficient as possible, but quite often, the efficiencies are limited by the knowledge within the company. The external benchmarking process takes a company outside of its own industry and exposes them to different methods and procedures. There are a number of steps in a formal benchmarking process and these may be grouped in five phases discussed below.

Phase 1: Planning

This is an essential stage in the benchmarking process as is essential to any plan development. It addresses the what, who, and how requirement of the process. Planning answers the following questions:

- What is to be benchmarked?
- To whom or what will we compare?
- How will the data be collected?

Phase 2: Analysis

This phase must involve a careful understanding of current process and practices, as well as those of the organizations being benchmarked. What is desired is an understanding of internal performance on which to assess strengths and weaknesses. Some general questions that may be asked include

- Is the safety performance of the organization better than our performance?
- If yes, why is this so?
- By how much is the organization performance better than ours?
- What best practices are being used now or can be anticipated?
- How can their practices be incorporated or adapted for use in our organization?

Answers to these questions will define the dimensions of any performance gap: negative, positive, or parity. The gap provides an objective basis on which to act and close the gap or capitalize on any advantage your organization has.

Phase 3: Integration

Integration is the process of using benchmark findings to set operational targets to create change. Integration involves careful planning to incorporate new practices in the operation and to ensure that benchmark findings are incorporated in all formal planning processes, including communication of action plans and findings, to all organizational levels, to obtain support, commitment, and ownership.

Phase 4: Action

This phase involves converting benchmark findings, and operational principles based on them, to specific actions to be taken. Use the creative talents of the people who actually perform work tasks to determine how the findings can be incorporated into the work processes. Any plan for change should also contain milestones for updating the benchmark findings, and an ongoing reporting mechanism. Progress toward benchmark findings must be reported to all employees.

Phase 5: Maturity

Maturity will be reached when best industry practices are incorporated in all business processes, thus ensuring superiority, and when benchmarking becomes an ongoing, essential, and self-initiated facet of the management process. Benchmarking becomes institutionalized and is done at all appropriate levels of the organization, not by specialists.

While benchmarking is extremely beneficial in creating strong and compelling reasons for change, there are disadvantages to organizational benchmarking. Table 16.2 highlights some of the advantages and disadvantages of benchmarking in health and safety (Table 16.2).

Another opportunity for benchmarking in an organization is in setting out the key objectives and targets for safety management and creating an incentive structure for employees, which drives good safety performance, balancing both leading and lagging indicators, and capturing both tangible and intangible factors (Boardman and Lyon, 2006). Too often, whether due to lack of time or information, organizations find themselves picking measures that fit into a generic mold as opposed to identifying essential drivers of performance and this approach does not help much, since *you get what you measure.*

If you are not measuring the *right things,* then the ability to impact performance and outcomes will be beyond reach. Safety performance management is multidimensional. No single measure provides an overriding indication of an organization's success or failure in managing work-related

TABLE 16.2

Advantages and Disadvantages of External Benchmarking

Advantages of Benchmarking	Disadvantages of Benchmarking
• Measures strength and weakness of health and safety management systems and provides opportunities for improvements • Engine for strategic changes and improvements • Creating new business development strategies • Increasing efficiency and productivity • Motivating employees	• Time consuming and expensive • Process that needs expert knowledge • High risk • Requires significant resources of company • What may work for one organization may not be effective in another

risk. The appropriate set of measures varies by organization, based on its fundamentals of vision, mission, values, goals, and objectives (Higgins and Hack, 2004). Some questions to answer are

- What measures will determine success by the organization's standards?
- What are the few vital measures that will define how the organization is performing against these goals and objectives?
- Will the measures drive the behaviors required?
- Can valid and reliable data be collected?

It is a matter of determining the few critical measures that reflect specific goals that will impact on outcomes (not giving in to the temptation to measure everything) that depict organizational health. KPIs must be specific to the business drivers and linked to the overall strategy. Some KPIs include leading indicators (such as measures of culture and measures of the integrity and performance of management systems) and lagging indicators such as standards of control for principal risks (such as incidents, near misses). Zero injuries and zero illnesses is a goal of many organizations in order to prevent workplace accidents and illnesses.

Part of BHP Billiton's performance incentive scheme like many other large high-risk organization operates on a set of health, safety, environmental, and people-related targets, arranged around a corporate scorecard. A combination of lagging and leading indicators is used. Examples of lagging indicators include the number of environmental incidents and safety figures and failures of safety critical instrumentation or alarms. Lagging indicators of process safety performance suggest corrective actions only after an accident. Leading indicators on the other hand, attempt to measure some variable that is believed to be an indicator of future safety performance so that the desired safety outcome (zero incidents) can be achieved. Implementation of action plans to improve performance is an example of a leading indicator.

Six Sigma is a culture that entails total management commitment and a philosophy of excellence, customer focus, process improvement, and continuous measurement, and continues to be a driver in J&J's way of doing business and achieving operational excellence (Saldana, 2004). For J&J, improvement methodologies such as Six Sigma, Design Excellence, and Lean provide a systematic and data-driven approach and are key drivers of competitive advantage. In J&J, the Health & Safety (H&S) organization believes that Six Sigma is critical to ensure that they continue to meet and exceed customer needs, reduce risk, and drive operational excellence.

In a Six Sigma culture, there is strategic alignment of vision, mission, goals, and objectives throughout the organization. Metrics (the critical few) that demonstrate results and focus on the key outcomes are carefully selected. Goals are set for each critical metric. Performance against goals and targets is tracked over time to identify improvement opportunities. Data are displayed in the form of a "dashboard" or "scorecard" and reviewed frequently to understand the underlying causal factors of change. The focus is to more effectively manage business processes, improve them, and deliver outstanding business results.

DuPont (2008, p. 2) advised characteristics of leading indicators that generate real improvements in safety including the following:

"Simple, close connectivity to the outcome/results" (p. 2)

- "Objectively and reliably measurable" (p. 2)
- "Interpreted by different groups in the same way" (p. 2)
- "Broadly applicable across company operations" (p. 2)
- "Easily and accurately communicated" (p. 2)

Halama et al. (2004) of the Construction Owners Association of Alberta listed 300 leading indicators for consideration of which the following 10 were identified as the top leading indicators. Consistent with DuPont's recommendations on characteristics, these indicators can be allocated a quantitative measure for stewardship and tracking purposes. They also meet the SMART requirements, discussed in Chapter 3.

Leading indicators

1. Presence of an effectively functioning behavioral-based observation process.
2. Leadership inspection and observation process in place and working.
3. Presence of a functioning near miss/near hit reporting system.
4. Employee perception surveys conducted and baseline data developed for sustained and continuous improvements in EH&S.
5. Prehire screening and drug and alcohol testing of all workers conducted.
6. Presence of a functioning and effective contractor prequalification process.
7. Leadership visibility stewarded and reflected in active management safety participation—such as tours/walkabout/written communications.
8. Supervisor safety activity evaluated.
9. Presence of a functioning hazard ID/analysis process.
10. Demonstrated proof that FLRA is conducted prior to new work and at the start of each shift (Halama et al., 2004, p. 26).

Halama et al. also recommended the following lagging indicators for consideration:

Lagging indicators

1. Recordable injury frequency
2. Lost time frequency
3. TIF
4. Lost time severity
5. Vehicle accident frequency
6. Workers compensation costs
7. Property damage costs
8. Accidents investigated (Halama et al., 2004, p. 26)

Figure 16.2 provides a simplified tracking tool for stewarding both leading and lagging indicators. Figures 16.3 and 16.4 illustrate sample dashboard measurements in order to achieve safety compliance.

Best-Practices Identification and Alignment

A best practice can be defined as any technique, method, process, activity, incentive, or reward that can be regarded as more effective at delivering a particular outcome than any other technique, method, process, and so on when applied to a particular condition or circumstance. Best practices are practical techniques gained from experience that organizations may use to improve internal processes and that

- Has been shown to be effective in the prevention of workplace illness or injury
- Has been implemented, maintained, and evaluated (demonstrated value and clear results)
- Is based on current information
- Is transferable and can be leveraged by other companies

According to the Canadian Centre for Occupational Health and Safety (2005), best practices cover a wide scope such as

- Global—such as hazard assessment, incident investigation, health and safety management programs, and evaluation.
- Industry specific—such as those developed by industry associations.

Hours worked	Jan	Feb	Mar	Apr	May	Jun	Jul	Aug	Sep	Oct	Nov	Dec	Total	Target
Employees														
Contractors														
Leading indicators (employees and contractors)														
Percentage of personnel onboarded and oriented														
No of formal safety meetings this month (employee and contractors)														
No of toolbox talks held this month (employees and contractors)														
No of safety alerts/bulletins/communicated this month to all workers (employees and contractors)														
No of joint (owner/contractor) field safety inspections conducted by site leaders														
No of visits conducted by owner senior leaders														
No of visits conducted by contractor senior leaders														
No of joint owner/contractor leaders visits														
No of near misses														
No of vehicle inspections conducted														
No of at risk work behaviors observed and corrected (PPE etc)														
No of at risk conditions observed and reported for remedial actions														
Total														
Lagging indicators (employees and contractors)														
No of first aids														
No of medical aids														
No of recordable incidents														
No of disabling injuries														
No of personnel on restricted work weeks														
No of loss time hours														
No of fatalities														
No of process safety incidents														
No of MVI's														
No of environmental incidents														
TRIF														
Total														

FIGURE 16.2
Tracking tool for leading and lagging indicators.

Leading indicators (employees and contractors)	On target	Attention required	Behind
Percentage of personnel onboarded and oriented			
No of formal safety meetings this month (employee and contractors)			
No of toolbox talks held this month (employees and contractors)			
No of safety alerts/bulletins/communicated this month to all workers (employees and contractors)			
No of joint (owner/contractor) field safety inspections conducted by site leaders			
No of visits conducted by owner senior leaders			
No of visits conducted by contractor senior leaders			
No of joint owner/contractor leaders visits			
No of near misses			
No of vehicle inspections conducted			
No of at risk work behaviors observed and corrected (PPE etc)			
No of at risk conditions observed and reported for remedial actions			
Lagging indicators (employees and contractors)			
No of first aids			
No of medical aids			
No of recordable incidents			
No of disabling injuries			
No of personnel on restricted work weeks			
No of loss time hours			
No of fatalities			
No of process safety incidents			
No of MVI's			
No of environmental incidents			
TRIF			

FIGURE 16.3
Easy to follow EH&S performance dashboard.

- Hazard specific—such as workplace violence or working alone.
- Task/procedure specific—such as patient lifting in health care, framing in construction, cooking in a restaurant, and so on.

Sources of best practices include

- High-performing companies (award-winning, reports on positive interventions)
- Government agencies (if performed by industry or advisory groups)
- Researchers (case studies, evaluation of interventions, systematic reviews)
- Industry associations (industry recommended practices)

Best practices in health and safety are the outcome of shared learning in safety as discussed in Chapter 13. When knowledge is shared, a wider pool of resources becomes available for improving the practice. Successfully iden-

Leading indicators (employees and contractors)	Target	Caution	Needs attention
Percentage of personnel onboarded and oriented	>95%	75–80%	<75%
No of formal safety meetings this month (employees and contractors)	4	2	<2
No of toolbox talks held this month (employees and contractors)	100	70–80	<70
No of safety alerts/bulletins/communicated this month to all workers (employees and contractors)	10	6–7	<6
No of joint (owner/contractor) field safety inspections conducted by site leaders	4	2	<2
No of visits conducted by owner senior leaders	8	4–6	<4
No of visits conducted by contractor senior leaders	8	4–6	<4
No of joint owner/contractor leaders visits	8	4–6	<4
No of near misses	0	2–4	>4
No of vehicle inspections conducted	1/week/vehicle	1/month/vehicle	1/quarter/vehicle
No of at risk work behaviors observed and corrected (PPE etc)	0	1–4	>4
No of at risk conditions observed and reported for remedial actions	0	1–2	>2
Lagging indicators (employees and contractors)			
No of first aids	0	5–10	>10
No of medical aids	0	1–2	>2
No of recordable incidents	0	1	>1
No of disabling injuries	0	1	>1
No of personnel on restricted work weeks	0	1	>1
No of loss time hours	0	8–16	>16
No of fatalities	0	0	0
No of process safety incidents	0	0	0
No of MVI's	0	1	>1
No of environmental incidents	0	1	>1
TRIF	0.4	0.4–0.5	>0.5

FIGURE 16.4
EH&S critical outcomes performance dashboard.

tifying and applying best practices can reduce business expenses and improve organizational efficiency (Hill, 2000). While benchmarking can be used to ensure safe physical conditions of a workplace, benchmarking of BBS programs can also aid in improving safety culture in organizations. Many examples of successful systems exist; among these are the safety training observation program (STOP) that assists companies in curbing the negative behavioral patterns of employees from a safety perspective, who operate mainly in high-risks areas. For the program to be disseminated and implemented effectively throughout the company, it is imperative to obtain commitment and involvement from all senior management. The STOP program is based on the regular participation of all management (supervisors, managers, directors, vice president, and president), recording negative observations of tasks being performed by employees on a STOP card.

Opportunities for improvement, as well as good practices, are loaded onto a web-based database for frequent management review and follow-up actions. Not only does this keep all employees actively involved in promoting safe working behaviors but it also drives continuous improvement as well. Additionally, a focus on near-miss reporting, which works best with a reward/nondisciplinary policy, is perhaps one of the most effective tools in truly preventing serious injuries and fatalities, whether they are behavioral-based issues or not.

STOP findings are easily trended and can provide useful information regarding the health and safety culture of the workforce. Figure 16.5 shows several STOP parameters that have been trended to help leaders determine where concerns with the organization's SMS may exist. When STOP cards

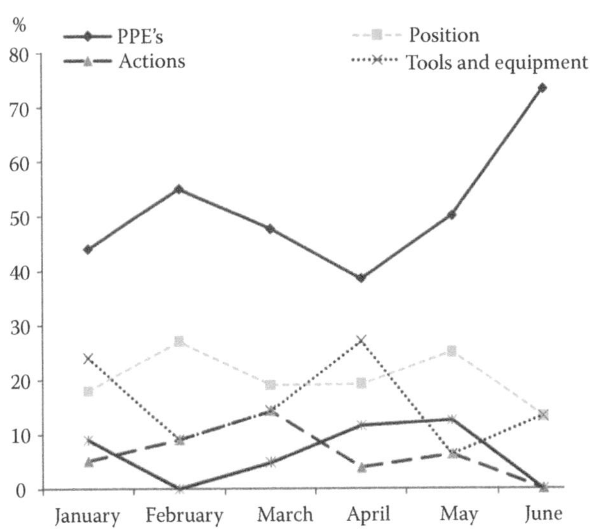

FIGURE 16.5
Trended data derived from STOP cards observations.

are submitted, findings can be categorized and expressed as a percentage of all observations. Greatest effectiveness of the STOP card is derived when a reporting culture is sustained and findings are shared with the entire organization on an ongoing basis.

Industry Leaders and Peers in Safety

Creating a culture that promotes safety and strong business performance depends on what leaders do and how they influence others. No one can be an effective safety leader, or credibly communicate the importance of safety efforts, without genuinely valuing the safety of others. What is important is not what you say you value but what you actually value—the ethics manifested in your personal behavior. These values directly influence safety decision making, interactions with subordinates, the priority placed on safety, and the ways in which you measure success. A sure sign that a leader does not "get it" is when he defines safety purely in quantitative terms or dollars and cents. Leaders try to motivate others by talking about the cost of poor safety in terms of dollars or loss of productivity. Leaders are effective at motivating others when they genuinely understand what safety means to them as human beings and appreciate the awesome responsibility they have in running a business and doing so without harming those who do the work (Spigener, 2009).

Great leaders care enough about employees and customers to do what is necessary to protect their health, well-being, and safety. Companies that have been titled as industry leaders are committed to safety; they collaborate with their employees on safety at all levels and encourage employees to utilize their safety knowledge in all areas of their lives. Developing safety leaders in your business can mean the difference between marginal and world-class results. According to Maxson (2010), if you want to develop successful safety leaders, then you must consider the following in your business practices:

- Talk safety
- Walk safety
- Do safety

Building communication between the boardroom and the shop floor will be an important facet of all safety leaders. Executives know that if they can win the hearts and minds of workers, they will not only have great results in safety, but it will reflect well on their bottom line. Apart from having detailed SMSs composed of safety-focused policies, programs, and procedures, other activities that support a continuous focus on safety include

- Internally generated weekly safety tips e-mailed to all employees that address various safety topics that are applicable on and off the job
- A brief "safety moment" provided by someone prior to all meetings
- A monthly "safetygram" newsletter on a specific safety topic with an associated quiz
- "Safety tip of the week" automatically posted on the company's intranet
- Lunchtime "safety brown bags" on various safety topics
- Monthly office safety committee meetings attended and led by employees and supported by local management
- Employee participation in safety-focused events such as "safety day" and "STOP" where office operations shut down for a portion of the day to focus on planned safety events and activities
- All work activities requiring the generation of a job hazard analysis developed in a team approach with project management, supervision, and employees who will be performing the work
- Sharing lessons learned and best practices through an innovative process that e-mails applicable learning content to employees

Reconfiguration of the Organization to Achieve World-Class Safety Performance

As we continue into the twenty-first century, successful companies must tailor their business model to continually increase efficiency for proven revenue-generating processes by reducing costs associated within the areas of operations, technology, manufacturing, and logistics. Associated with these efficiencies are many challenges that are faced by the safety, engineering, and management communities. Risks and the potential for catastrophic loss are dramatically increasing as technology advances at an ever-increasing rate. Focusing on economics must never be detrimental to safety, and operational efficiency standards, which remain at the forefront of all operations. The public demands a high level of safety in products and services; yet, in the face of world competition, the safety effort must be timely and cost-effective. Studies that examine safety performance show that organizations that perform well in safety tend to perform well across the board. A knowledgeable and competent workforce is necessary to sustain efficient "cost-savings-driven" operations, while looking to safety practices and performance (Bell et al., 2008).

The safety field recognizes that companies with world-class safety have a certain culture in which people do not engage in unsafe acts that result in an actual loss. Because most injuries result from human actions, the key battle in the safety war is training of employees, which is an important tool in changing the minds, and subsequently the behaviors, of the workforce. Training is a vital part of the communication process in order to change the behavior and culture of workers. It is culture that determines if a group can internalize safety by doing the right thing when no one is looking. A solid, dynamic, and responsive safety culture transcends any written program, training gap, program gap, or unexpected event. Consistent with the definition of many writers, the safety culture of an organization is the outcome of shared values and beliefs. Hughes and Ferrett (2009) suggested that positive safety cultures are reflected in mutual trust and shared perceptions regarding health and safety as well as high confidence in effective preventative measures.

J&J, a healthcare company with over $62 billion in annual sales, is considered one of America's safest companies. A core value is that safety is a very essential part of J&J's business functions and a measure of organizational excellence (Smith, 2003). J&J uses high-performance work systems that assume that employees are a primary source of competitive advantage and that workers are capable of continuous improvement and will perform at higher levels if they are motivated to do so. J&J achieves this by encouraging practices such as participative decision making, providing high-quality training, and sharing information.

By treating workers with respect and as capable and intelligent individuals, organizations like J&J will find that workers will be more committed to the organization and more trusting of management, which will result in improved performance. Bombardier launched a North American initiative called "world class manufacturing" that has propelled its environmental, health and safety (EHS) performance forward. WCM is not only about reactive measures (i.e., injury rates), but the safety culture has evolved with increased near-miss reporting, employees coaching other employees on safe behaviors, and a true ownership, responsibility, and pride for all aspects within each work area. Each of the steps in the journey has criteria that must be met and maintained in order to progress to the next step. The "qualified" and "performing" steps focus on workplace and organizational aspects, while the "excellence" and "world class" steps involve improving workplace culture. Annually, each North American site must achieve established levels of performance. Initially, the driving force was the commitment and support of upper management, but as sites progressed into the final steps, employees took pride in achieving their goals in the previous steps and assumed responsibility for accomplishing the final steps (Leemann, 2007).

World-class organizations in safety can be summarized under three umbrellas listed below:

Vision: "World-class" organizations define and communicate safety as a core business value, rather than a priority subject to changing conditions. This value applies everywhere, to everyone, and in every activity and decision, at all times.

Performance: "World-class" organizations set out to eliminate, or at least minimize, workplace injuries and illnesses, then establish a strategy to instill a prevention-oriented culture ("how we do things around here"). Making safety and health an essential measure of business success requires a sustainable commitment by management leadership at all levels ("walking the talk"), allocating sufficient resources, and reinforcing safe behavior while not tolerating at-risk behavior. It is accepted as fact that a safe operation is a productive, reliable, and profitable operation.

Systems: "World-class" organizations strive to define and use best practices to manage safety and health through formal systems integrated into every aspect of the business. Effective planning, proactive risk control, performance monitoring, and corrective actions based on lessons learned are constantly promoted. Such processes typically include employee participation, hazard identification and RA, competence, safe operations, integrity of facilities, contractor management, product stewardship, MOC, incident investigation, emergency management, and compliance assurance (Leemann, 2007).

The common attributes of companies with superior EHS performance according to MacLean (2003) include the following:

- Organizations must be proactive in its response to safety concerns.
- Organizations must be committed to improving health and safety in the workplace.
- Transparency and openness regarding health and safety are essential. Information must be shared in a timely manner.
- Organizations must seek to meet internal and external health and safety goals.
- EHS issues are well integrated and embedded into core competencies of workers as is ethical behaviors.
- Leaders must seek continuous improvement at every step of the way.
- They must be aggressive and vigorous in their approach to removing workplace hazards.
- Leadership behaviors must demonstrate consistency and accountability.
- Leadership is progressive and shares its vision of the future with all followers and stakeholders.
- Organizations must be recognized as a socially responsible corporate citizen.

- Beyond compliance and contractual agreements with labor or the community.
- There must be focused attention to meeting both spirit and intent of the law.
- Reduced life-cycle impacts/supply chain.
- SMART performance targets for results.
- A well-developed RM and mitigation culture.
- Strive for performance better than peers with leading indicator metrics.
- Fully implemented environmental and SMSs.
- Focused attention on a prevention-based philosophy.

Safety Audit Programs in the New Millennium

Safety audit programs have undergone continual changes to meet the needs of the corporation, operating groups, and stakeholders. Audit programs have continued to change and evolve as regulations, demands of the public, technology, and management philosophies have changed. Regulatory bodies have increased the number of regulations and penalties associated with violations over the past several years. Numerous nongovernmental organizations are pressing for more accountability to the public. Additional liabilities of operating any business venture have greatly increased as the public finds creative avenues to prosecute corporations for behavior viewed as unacceptable.

Often, programs have been modified after major disasters such as the BP Texas incident or to conform to changes in governmental policies. Management system audits implemented to assure policies were being followed and that appropriate controls were in place have evolved from an informal internal review process to one, which in some countries, is now stipulated by international bodies. Compliance audits brought about by a change in government expectations have moved away from outside auditing firms to being managed by facility personnel. Public recognition or the desire to obtain certification has moved to the stage where international organizations are now requiring accredited auditors as well as certifications, to demonstrate to business partners, regulatory agencies, and customers, that they are employing adequate systems to manage not only safety but quality and environmental issues.

Some organizations have even progressed from compliance to management systems and to self-governance. Several forms of audits have disappeared while other methods have come into use; however, auditing continues to remain a key factor in measuring a company's ability to effectively conduct business. While the strategic goals have remained constant throughout time, the programs have continued to evolve to reflect both internal and external forces. The key goals for the safety auditing programs have been to

- Manage risks effectively
- Improve compliance
- Increase awareness and understanding of hazards
- Improve public image
- Provide assurance to executive management
- Assist the facility manager to operate efficiently and safely

The audit approach a company uses is very personal. The approach chosen depends on many considerations such as regulatory environment, actual and perceived liabilities of the company, the culture the executive management chooses to promote, the performance and capabilities of the business units, safety metrics used in the company, the safety issues faced by the company, the interest level of stakeholders, public, and the competition.

Successful companies generally find their comfort zone and continually adapt that zone to suit their specific yet changing needs. The challenge is to constantly drive for a program that adds value to client organizations. This, in turn, requires a conscious effort to challenge everyone involved in the program to constantly drive for improvement (Johnstone, 2001). Thus, audit programs will be influenced by regulatory requirements, published industry standards, best operating practices, and influences from stakeholders. Management systems audit, compliance audits, and certification audits will continue to evolve in response to external and internal forces.

17

Emergency Management

At some point in time, every organization will have an emergency of some sort and will necessarily have to respond to this emergency. As a consequence, having an established emergency management process that is able to meet the needs of the organization when emergency arises becomes very beneficial to the organization. For large complex organization, regulations require the existence of a well-maintained IMS within the organization.

IMS generally becomes necessary when unplanned events occur. Nevertheless, all organizations seek to operate its businesses in such a way that unplanned events are avoided. However, in the event of an unplanned major incident, organizations must be equipped and ready to respond to the event in a way that satisfies stakeholder needs. To achieve this goal, organizations must establish an IMS plan and team, train personnel, conduct drills, resource the IMS team, and always be ready for an unplanned event.

History of IMS

IMS evolved from the incident command system (ICS) that was originally developed in 1970 as a result of shortcomings identified in response to wild fires that killed 16 people, destroyed 772 buildings, and cost $233 million (1970 US) dollars in Southern California (Ministry of Environment, British Columbia, 2002). The ICS was designed to assign personnel and coordinate tasks during emergency operations. ICS was considered a rational management system whereby all activities and communications flow through one centralized authority, the *incident commander.*

The ICS was developed after repeated incidents of poor interorganizational coordination that resulted in communication, logistical, and coordination problems as multiple organizations responded to the same incident as the size and scale of the incident grew (Abdallah and Burnham, n.d.). Shortcomings identified from these responses included the following:

1. Having uncommon radio frequencies, signals, and codes—this leads to poor interagency communication.
2. Lack of common terms—when agencies did talk, they often misunderstood each other.

3. No effective or functional command system—each agency operated on the luck and personality of its leaders. In some situations, the operational effectiveness depended on which leader or chief was working that day.
4. Insufficient methods for giving out resources effectively.
5. Poorly defined ways of responding to disasters—there were no standard guidelines. How each response related to other functions depended upon individual interpretation (Abdallah and Burnham, n.d.).

The ICS approach to incident management provided the much-needed coordination and control required to effectively manage incidents. Almost two decades after the inception of the ICS, it evolved into the IMS process that provided a more holistic approach to managing incidents (Abdallah and Burnham, n.d.).

IMS became an "all-risk, all agencies, co-ordinated system for managing humanitarian emergencies" (Abdallah and Burnham, n.d., p. 10–13). The IMS process has over time resulted in greater response coordination and effectiveness and provides opportunities for more effective resources allocation and accounting. Today, there is very little mention of ICS as many organizations have adopted IMS as the emergency response model for addressing unplanned events.

Why Should Organizations Have an IMS?

Is an IMS necessary? It is costly to maintain and one may argue that we have not had a major incident ever in our organization, so why bother? Some of the costs incurred in maintaining an IMS include the following:

1. Maintaining a dedicated and fully equipped incident command center in a ready state for responses
2. Training personnel in a single forum, often several times per year. Attrition and turnover of personnel increase this cost as new workers must be similarly trained and competency assessed
3. Travel and accommodation costs associated with geographically distributed organizations
4. Conducting drills and scenario responses to maintain competency

Despite these costs, an effective response to one single major event can easily justify the existence of an IMS process and team.

An IMS is essential for three compelling reasons discussed below.

Legal Compliance

Maintaining a functioning and active IMS is a legal compliance issue. In North America, the following regulatory bodies require organizations to maintain an effective and functioning IMS. In Canada, the following regulatory bodies require an IMS: Canadian Environmental Protection Act (CEPA), Occupational Health and Safety Acts (OH&S), and Sarbanes-Oxley (SOX). In the United States, regulators such as OPPC Oil Pollution Acts (OPA-90), Environmental Protection Act (EPA), and SOX also demand that organizations maintain an effective and functioning IMS.

Good Business Sense

There are many instances of IMS at work that has demonstrated the value and benefits of an IMS. Responses to the Texas City BP incident in 2005 and the more recent Gulf of Mexico oil spill in 2010 are examples of IMS at work. Multiple fires and explosions in the oil sands of Alberta have also tested the validity of an IMS response and existence with great success.

In the absence of an IMS, responses to incidents are poorly coordinated and can result in significant business impact. Among plausible business impacts are the following:

1. Collapse of entire business
2. Significant damage to image and brand from a poorly managed IMS response
3. More damage, extended outage, and repair requirements, and ultimately greater recovery cost implications
4. Loss of market share and credibility
5. Worker turnover and employee flight

While this list is not exhaustive, the impact to business can be tremendous. An effective and functioning IMS can mitigate some of these impacts for unplanned major incidents.

Corporate Social Responsibility

Many communities depend on organizations for employment, infrastructure development, scholarships, and community developments benefits. When organizations collapse from unplanned incidents, the impact on communities can be devastating. Organizations also have a social responsibility for preserving the environment and communities within which they operate. Table 17.1 highlights the impact on communities from unplanned events and incidents.

Many smaller incidents have had very significant impact on organizations. Needless to say, the retention of an effective and functioning IMS is in the best

TABLE 17.1

Community Impacts from Selected Unplanned Major Events

Incident	Impact
April 20, 2010 Gulf of Mexico oil spill—BP (National Commission on the BP Deepwater Horizon Oil Spill and Offshore Drilling, 2011)	• Worst oil spill in U.S. history. Almost 4.9 million barrel spilled (Powell, 2011) • Eleven fatalities and several serious injuries • Significant impact to fishing industry along the U.S. coastline • Reduction in tourism to the impacted regions along the west coast of Florida and other popular tourist's destinations around the Gulf coast • Projected cleanup cost of >$40 billion dollars
Exxon Valdez Prince William Sound, Alaska, spill of 1989	• 11 million gallons of crude oil spilled covering thousands of miles of pristine waters and coastal areas • Major impact to fish and wildlife in the region • Significant impact to the livelihoods of communities in the region • Approximately $2.1 billion cleanup cost • Criminal fine and civil restitution of ~$1.0 billion
Bhopal—Union Carbide India Limited (UCIL) pesticide release incident December 2–3, 1984 (Mishra et al., 2009)	• World's worst industrial disaster—30 square miles affected • 2500–6000 killed with debilitating impacts on more than 200,000 people • More than 500,000 registered victims that survived the tragedy • 30–40 tons of a toxic gas, methyl isocyanate released (Mishra et al., 2009)

interest of any organization because organizations have a social responsibility to all stakeholders for value maximization and for maximizing public welfare.

Types of Events Requiring IMS Responses

As discussed earlier, IMS is an "all-risk, all agencies, co-ordinated system for managing humanitarian emergencies" (Abdallah and Burnham, n.d., p. 10–13). As a consequence, the IMS process is very effective for managing catastrophic and major incidents of the following types:

- Business continuity challenges arising out of the following:
  - Pandemic outbreaks
  - Fires and explosions in process and operating facilities
  - Process safety management incidents such as spills
  - Workplace fatalities and severe injuries
  - Work stoppages arising out of strikes and bargaining failures

    - Worker hostage incidents/kidnappings
    - Marine/mine incidents/collapses
    - Train derailments
    - Dangerous goods occurrences
- Natural disasters:
    - Avalanches
    - Earthquakes
    - Flash floods/tidal waves
    - Hurricanes/storms/tornados
- Public security and social unrests:
    - Blackouts/brownouts
    - Building/structure collapses
    - Bomb/terrorist threats
    - Political unrest/riots
    - Fuel/resource shortages

This list can be expanded to include incidents of varying sizes and scope.

Organizing Response Structures

For many large, geographically distributed organizations, emergency response management may be tiered for best performance. Tiered response is also tied to the scope and scale of the incident. For example, within the oil and gas or energy industries, an organization may have several facilities located at different locations across the world. As a consequence, emergency response from a head office location may be impractical for responding to a small-sized plant emergency that is localized with localized impact. However, when an incident has the potential to impact the organization's entire business, the scope of the incident is greater and a much more coordinated and supported response is required.

From the authors' experiences, in many instances, three tiers of responses are established. These are as follows:

Tier 1 Response

Tier 1 response refers to emergencies that can be managed by the facility or site leadership personnel and emergency management team to take control of an incident and to manage the incident such that the overall impact of the incident is minimized. Generally, Tier 1 responses are specific and aimed entirely at containment and control of the specific incident such that danger to personnel is removed and damage to the environment and facilities (equipment and machinery) is minimized. The aim for Tier 1 responders is to directly address the incident and stop the event from proceeding further.

Under the leadership of the on-scene incident commander, Tier 1 responders will be addressing the following in the response:

1. Controlling access to the site of incident
2. Physical fighting of fires, spill containment, air monitoring, rescue and recovery efforts, and all immediate and frontline work required to prevent further fatalities, damage to the environment, and damage to equipment and machinery in that order
3. Managing mutual aid support and coordination
4. Local communication with community groups

Tier 2 Response

A Tier 2 response is activated to support the Tier 1 responders when the scope of an event is larger than local impact. The potential for brand and image to be impacted is real and timely intervention is essential to support the efforts of Tier 1 responders who are addressing the physical requirement of the event. Tier 2 responders will generally gather at the organization's command center and under the guidance of a Tier 2 incident commander, shall provide support to the Tier 1 response team in dealing with nonimmediate response events and support requirements.

Tier 2 response team will support the efforts of the Tier 1 response in the following ways:

1. Providing subject matter health and safety expertise in responding to hazards generated from the incident
2. Providing support to the on-site security team in controlling access to nonauthorized personnel and media groups
3. Providing support to external stakeholders seeking information about families and friends who may be involved in the incident
4. Interfacing with media groups seeking information regarding the incident
5. Addressing stakeholder management requirements such as suppliers, customers, and communities
6. Providing accounting and financial support to the team such that all required resources used in the response are properly accounted for
7. Planning support to ensure that proactive strategies are developed to not only contain the incident but also to assist in the recovery of the facility

8. Providing technical and operations support to the on-the-ground response teams in terms of process fluids, isolation points, and other hazards that may be involved in the process
9. Logistical and supply chain management support to ensure that all materials required by the on-scene response team are provided in a timely manner such that the incident is brought under control very quickly and recovery can be initiated

Tier 3 Response

Tier 3 responses are the collective efforts of Tier 1, Tier 2, and Tier 3 response teams in an overall crisis management strategy for the organization. The goal is to avoid any crisis through timely intervention and strategies. Tier 3 responders generally include senior executives of the organization whose principal role is to mitigate damage to organizational image and branding. Tier 3 responders are governed by the incident commander (generally the CEO or designated senior executive of the organization) who is updated based on new information being generated from the Tier 1 and Tier 2 response.

Activating the Response

The key to an effective emergency response is communication. Once an incident has occurred, a sequence of events is required to generate an effective response. Table 17.2 shows the sequence of events and required responses and actions. Upon activation of the Tier 1 response, it is important that the on-scene incident commander alert the Tier 2 response team such that standby support is available as quickly as possible. In a similar fashion, the Tier 2 incident commander should also notify the Tier 3 response team so that they are available in the event the event transitions into a crisis situation.

Organizational Structure and Key Supporting Roles

Abdallah and Burnham (n.d.) provided a typical IMS structure comprising an incident commander supported by a management staff and the following organizations:

1. Administration
2. Planning

TABLE 17.2

Emergency Response Stage and Response Actions

Response Stage	Response Actions and Behaviors

Stage 1—Activation	• Initiate communications
	  • Call for help
	  • Sound alarm
	  • Notify control room
	• Tier 1 response team activated
	• Tier 2 response team notified
	• Tier 3 response team notified via Tier 2 incident commander

Stage 2—Reactive and tactical response	• Reaction
	  • Team activation
	  • Assessment
	  • Accounting for people
	  • Assessing the hazards
	• Team response
	  • Consulting response procedures and contingency plans
	  • Activating skills practiced in drills
	  • In the trenches responses for containment and control
	  • Recovery
	• Tier 2 team support
	• Tier 3 team action and crisis management actions

Stage 3—Proactive and strategic response	• Project management
	  • Assessing the damage
	  • Developing the immediate short-term recovery plans
	  • Developing the strategic long-term recovery plans

3. Logistics
4. Operations

A typical IMS structure described by Abdallah and Burnham (n.d.) is provided in Figure 17.1.

Many organizations have adapted this model to meet their own specific business needs. Suncor Energy Inc. (2011) have evolved this organizational structure as shown in Figure 17.2. The command staff provides relevant expertise to support the response and advises the incident commander as subject matter experts. Success also depends on the ability of the response team to *get big fast.* A key command staff person often overlooked is an IT support, which can be invaluable in today's IT business and communication environments.

Consistent with the organizational structure provided in Figure 17.2 are the roles and responsibilities of each organization, individual, or group that

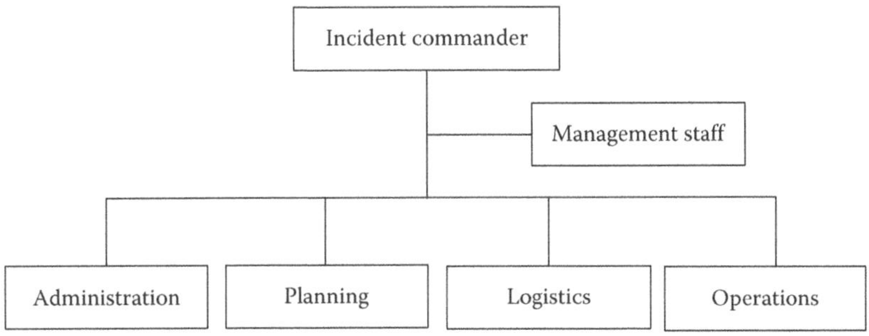

FIGURE 17.1
Typical IMS organizational structure. (Adapted from Abdallah, S., and Burnham, G. (n.d.). *Public Health Guide for Emergencies.* The Johns Hopkins and Red Cross/Red Crescent. Retrieved January 16, 2011 from http://pdf.usaid.gov/pdf_docs/PNACU086.pdf.)

makes up the emergency response team. Table 17.3 shows the roles and responsibilities of the stakeholders involved in emergency responses.

When a Tier 2 response has been activated, it is highly recommended that all personnel associated with the Tier 2 response be called out to provide support. Should the response be effectively contained and managed by the Tier 1 response team, it is easy to scale down and return personnel back to their main duties. In all instances, getting big fast provides a major advantage of having the necessary resources to expeditiously provide required and necessary support to the Tier 1 team in managing any response.

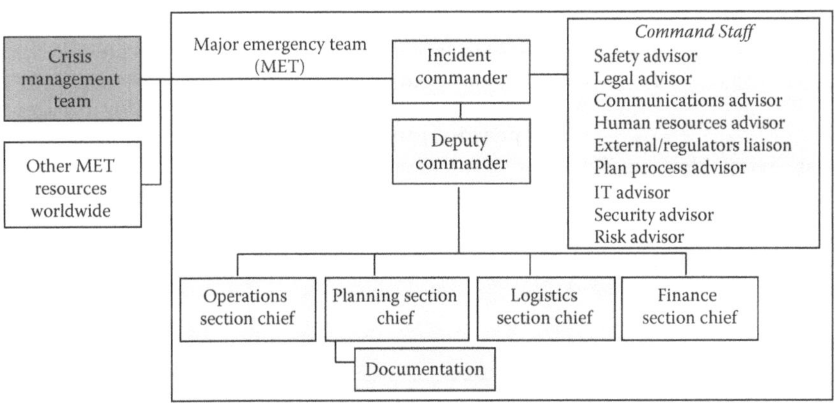

FIGURE 17.2
IMS organizational structure for a large oil and gas producer. (Adapted from Suncor Energy Inc. 2011. *Sustainable development.* Retrieved January 09, 2011 from http://www.suncor.com/en/responsible/302.aspx. Copyright of Suncor Energy Inc., approval from Suncor is required to reproduce this work.)

TABLE 17.3

Roles and Responsibilities of Command Staff

Command Staff	Roles and Responsibilities
Plan process advisor	• Provides guidance to the response team in terms of the emergency response plan. The plan process advisor is a subject matter expert in managing the emergency response
Communications advisor	• Provides guidance in timeliness, types, and quality of media releases provided to the public and other stakeholders
Legal advisor	• Provides legal support to the response team regarding regulatory concerns and other major liability concerns
Safety advisor	• Provides response oversight and SMS policy enforcement
Human resources advisor	• Coordinates handling of personnel matters regarding employees and their family members. Provides support also in stress debriefing and employee assistance
External/regulators liaison	• Manages relationships with external stakeholders and regulators
IT advisor	• Provides support to ensure all technological software and hardware are functioning properly to support the emergency response
Risk advisor	• Provides guidance to the response team regarding risk management of the many alternatives that may present themselves during the actual response
Operations section chief	• Operational and technical support are available to provide help to the on-site response team regarding location of control equipment, process contents, and other predisposing conditions that may be generated as the emergency evolves. For example, when fire impinges on a vessel that may contain a hazardous compound, details regarding controls to minimize catastrophic outcomes are provided by this team
Planning section chief	• Provides support regarding short-, medium-, and long-term planning essential for stabilizing operations and returning the business to preincident conditions as quickly as is possible
Logistics section chief	• Organizes and coordinates delivery of services and supplies to support the emergency response. Logistics essentially manages service providers to best support on-scene response needs
Finance section chief	• Finds ways and means for paying for all requirements to support the emergency response in a way that avoids delays. Finance also provides purchasing support cost monitoring and control

Source: Suncor Energy Inc. Copyright of Suncor Energy Inc., approval from Suncor is required to reproduce this work.

Managing the Response

When an incident occurs, all stakeholders are interested in ensuring that the emergency is properly managed. Each stakeholder seeks to ensure that their interests are best represented during the management of the emergency.

During an emergency response, the following sets of behaviors are essential for success:

1. Appear to be and be in control.
2. Tell the truth at all times.
3. Consult with communications experts for key messages to be communicated. Avoid listening to the lawyers during media releases.
4. Avoid the blame game.
5. Communicate frequently and demonstrate genuine empathy.
6. Documentation.

Appear to Be and Be in Control

Being in control of the emergency is an essential criterion for instilling confidence in the hearts and minds of stakeholders that the emergency is being handled properly. Regardless of how difficult the emergency may be, it is essential that the incident commander not only appears to be in control of the emergency but must be in control of the emergency. This is a very difficult requirement, particularly if there are fatalities involved in the incident.

Being in control requires that the incident commander appear to be calm and collected and present information to stakeholders, such that it is coherent and makes sense to stakeholders. As a consequence, when media releases are generated, messages must be crafted such that they convey caring and concern for stakeholders, they must provide information regarding actions taken or being taken, and they must also provide a perspective on the unfolding incident.

When a leader appears to be in control of an emergency, such behaviors create trust and confidence among responders and encourage them to become even more vigilant in their efforts. On the other hand, if an incident commander appears to have given up, this discourages responders and they, in turn, are likely to give up in similar fashion. A trainer in emergency management once suggested that during an emergency, an incident commander must appear like a duck in a turbulent stream: *unruffled on the visible top, but paddling like hell beneath the surface.*

Tell the Truth at All Times

Stakeholders are generally seeking information related to their specific interest during an emergency response. In situations where the incident commander does not know the answers to concerns raised by stakeholders, saying *I do not know* is quite possibly the best response rather than guessing. Lessons of the 2010 Gulf of Mexico oil spill cannot be underestimated regarding the volume of oil spilled. Early estimates provided by BP regarding the volumes of oil being spilled in the Gulf of Mexico were 1000 barrels per day at the start of the incident (Graham et al., 2011), 5000 barrels per day (Gillis,

2010), followed by almost daily changes to the estimated volumes spilled with a final estimate of almost 60,000 barrels per day (Graham et al., 2011). Final estimated volume spilled into the Gulf of Mexico was >4 million barrels of oil (Graham et al., 2011).

With each revision of the daily estimated spill volume, the credibility of BP in its response to the disaster was decreased. Trainers in emergency management will often suggest that it is always best to provide stakeholders with the near worst-case scenario and to perform better than what had been communicated. When stakeholders are prepared for the worst-case scenario and performance is better than that earlier communicated, there is often a sense of relief among stakeholders. Telling the truth during emergency management that is based on credible information is the best approach to providing updates to stakeholders.

Avoid Listening to the Lawyers during Media Releases

During an emergency response, there is always a tendency to have the corporate lawyers handy to seek to limit liability. Lawyers tend to provide language that says a lot with very little substance during emergencies and crisis situations. This language is exactly what stakeholders do not wish to hear. More importantly, legal language tends to ignore the human aspects of communication. As a consequence, all media releases and communication messages should be generated by the corporate communications team and vetted with the legal organization for liability exposures.

Avoid the Blame Game

During an emergency response, there should be no allocation of blame to any stakeholder. Until an incident is fully investigated and root causes are identified, it is almost impossible to determine where failures may have occurred to lead to the incident. Allocating blame during an incident is, therefore, premature during the emergency response and should be avoided at all cost. Instead, all parties must come together in a united front to manage the emergency and to eliminate hazards to personnel, the environment, and facilities.

A united front generates confidence among stakeholders and sends the message that interests are aimed at maximizing collective welfare as opposed to the welfare of one or a select few stakeholders. Once the emergency has been controlled and all hazards are removed, the organization can then seek to identify liable stakeholders and respond accordingly.

Communicate Frequently and Demonstrate Genuine Empathy

Perhaps one of the most important activities undertaken during an emergency management response is communication. Communication must therefore be done frequently as new information becomes available; it must be timely, accurate, and done with demonstrated empathy. During an emergency, multiple stakeholders are interested in how their interests have been affected by the emergency. To highlight a few, families of workers are interested in their well-being, suppliers are interested in business interruption impacts, environmentalists are interested in impact to ecology and the environment, workers are interested in their jobs, and shareholders are interested in profitability impacts. In view of this, frequent communication is required to provide information to interested stakeholders.

Communicated messages must therefore be crafted as follows:

1. Messages must address concerns and demonstrate sensitivity and caring to each stakeholder group.
2. Messages must be simple, clear, concise, and free of ambiguity.
3. Messages must be accurate.
4. Messages must convey actions taken as the emergency unfolds.
5. Messages must be filled with empathy and care for those affected by the incident.

Care must be taken, therefore, to identify all stakeholder groups and their specific interests in the incident so that communication regarding their interests may be provided.

In recognition of the importance of communication during an emergency response, it is important, therefore, that whoever does the communication and media releases stick to the script and to demonstrate empathy while doing so. What may appear insignificant to you may be the single most important thing to another stakeholder. The many quotes of BP's CEO, Tony Hayward during the 2010 Gulf of Mexico oil spill shall forever remain as lessons to be learned by incident commanders in emergency responses management. A few of Tony Hayward's communication errors are as follows:

1. In apology to the residents along the Gulf Coast of the United States, Tony Hayward said, "We're sorry for the massive disruption it's caused to their lives," … "There's no one who wants this thing over more than I do, I'd like my life back" (para. 2). Tony Hayward's comments appeared insensitive toward the families and loved ones of the 11 workers who lost their lives during this incident.
2. When response workers became ill, Tony Hayward questioned the source of their poisoning suggesting food poisoning as the source of illnesses related to temporary camp accommodations and concentration of people. His assertions were in contrast to the finding of health experts who suggested the source of illnesses were related to chemicals and overwork associated with the spill. Once again

empathy was absent for those trying to address the spill situation, that all stakeholders wanted an end to.

3. Tony Hayward said, "The Gulf of Mexico is a very big ocean. The amount of volume of oil and dispersant we are putting into it is tiny in relation to the total water volume." With the lessons of the Exxon Valdez spill and the Bhopal disaster still fresh in the minds of many stakeholders, these statements appeared insensitive to them.
4. Tony Hayward said, "I think the environmental impact of this disaster is likely to have been very, very modest." Similar to the comment in (3), for many stakeholders who depended on the Gulf of Mexico for their livelihood, such as fishermen, tourism industry along the coast of Florida, restaurants, and small businesses who were already feeling the impacts of the fallout of the spill, these statements were very insensitive and incorrect.

Tony Hayward was eventually relieved of his job as the CEO of BP in 2010 and will continue to provide a case study for emergency management personnel for decades to come.

Documentation

Documentation is a critical component of any emergency management response. Documentation serves to provide a historical perspective of all actions taken in response to the evolving situations associated with the emergency incident. Transcripts of the documentation process must be retained such that a complete compilation of the event can be made in where liability and legal implications exist.

During the response, a well-documented process allows for identification of all key tasks that are assigned among response members. Once documented, tasks can be prioritized such that they can be properly addressed in a timely manner to support the on-scene response. Documentation makes the recall of events, assignments, actions, and the rationale for such actions transparent and clean when a debrief of the event and response is undertaken.

18

Safety Culture Maturity

Health and safety play a very important and critical role in any business today. Businesses in all types of industry, including primary, secondary, and tertiary industries with upstream and downstream operations, have differing risks but the consequences are similar in terms of potential loss of life and injuries, fires and explosions, environmental damage, and negative company image impacts. In the United Kingdom, for example, when it was first introduced, the Health and Safety at Work Act 1974 (HSAWA 1974) was probably one of the clearest and most practical frameworks for health and safety in the world (British Safety Council, 2005). In fact the Health and Safety at Work Act 1974 can be considered to have had a great influence on the European Health and Safety Directives today. The act, like most health and safety legislation all over the world, is binding and thus it is critical to follow the law or suffer the consequences of fines and possible prosecution. This aspect cannot be ignored when looking at compliance.

Internationally, there are various organizations that are important in setting high-level guidance that have played an important role in setting minimum standards. One such organization was the ILO that was established with the growing concern for social reform after the First World War, and as such any reform would need to be done at an international level. The ILO has a tripartite structure in which a dialog between employers, employees, and governments is encouraged in a functional approach to promote and realize standards and the fundamental principles and rights at work; to create greater opportunities for women and men to secure decent employment; to enhance the coverage and effectiveness of social protection; and to strengthen tripartism and social dialog (ILO, 2008).

There are other agencies that also play an active role, such as the World Health Organization, a United Nations agency that promotes technical cooperation for health among nations, carries out programs to both control and eliminate diseases, and strives to improve quality of life. As such, it gives worldwide guidance on health issues; sets global standards for health; cooperates with governments in strengthening national health programs, and develops/transfers appropriate health technology, information, and standards (WHO, 2008).

Another agency that does similar activities at the European Union is the European Agency for Safety and Health at Work, which started activities in 1994. It is involved in the collection and dissemination of information to all the member states of the European Union. Others include the European Trade Union Institute, the Health and Safety Executive and Commission of the UK, the OSHA of the U.S. Department of Labor, the U.S. National Institute for Occupational Safety and Health, and so on (British Safety Council, 2005).

Legal Significance of Health and Safety at Work

The importance of health and safety is directly related to the principles of the prevention of loss. This is applicable in almost any business no matter what operations are involved and regardless of the level of risk. The reasons for this are simple. An employee at any capacity requires some degree of protection when it comes to his/her health and safety. There is thus a legal aspect to health and safety at work. In the United Kingdom, for example, that legal protection to all (including employer, employee, contractors, public, etc.) is provided through the HSAWA 1974. However, regardless of there being legislation with respect to health and safety under common international law, principles such as the duty of care, reasonable care, and protection of all are expected. This is similar in many parts of the world, including the Middle East, Singapore, United States of America, and Canada, to mention a few.

The management of risk concerns itself with the prevention of loss, preventing being negatively affected by issuance of enforcement, improvement, or prohibition notices from the local authority having jurisdiction for health and safety. Prevention also helps organizations avoid punitive action in which criminal courts can impose fines, compensatory lawsuits in civil courts leading to compensation claims, and imprisonment for breaches of legal duties and which can affect companies or individuals and thus their operations and reputation.

Accidents and incidents cost companies money directly or indirectly. The indirect costs of an incident can be estimated as being up to 30 times the direct losses on an incident (Bird et al., 2003). Certain insurance protects employers and these include employer's liability, public liability, workman's compensation, fire and perils, and so on. It is to be noted that losses cannot always be recovered for issues such as lost production time, loss of highly trained personnel, impacts on employee morale, productivity and time, resources spent investigating the incident, and so on. The Health and Safety Executive in the United Kingdom estimates that for every 1 GBP of insured loss there is an estimated uninsured loss of between 8 and 36 times that which is insured (HSE-96, 1996). We also discussed earlier the staggering numbers in the tens and hundreds of billions in Canada and the United States.

Health and Safety at Work in High-Risk Business: Case Studies

As discussed, the importance of health and safety is directly related to principles of loss prevention. The oil and gas industry is a hazardous industry and handling of petroleum products has intrinsic dangers that can lead to catastrophic losses of human life and private/public assets. It is worthy to note that many of the researchers and safety culture scientists (e.g., Hudson et al. 1996; Reason, 2000; Cox et al., 2006; Cooper, 2000/2001/2003) who have written about behavior-based safety, safety climates, and cultures all seem to agree that the most notable incident in industrial history was the Chernobyl disaster in 1986 in Ukraine, which changed the outlook on safety. It was not only the safety culture of the disaster but the subsequent immense damage and impacts it had on the whole of Russia and Europe that made it so important to study. It is also worthy to note that some have also studied the differences between safety behaviors for land-based and offshore employees (Tharaldsan et al., 2006) and employees expected to travel long distances as part of their occupation (Peters, 2008).

Obadia et al. (2006) states that after the Three-Mile Island nuclear disaster in the United States in 1979 and Chernobyl, safety became a paramount concern for the nuclear industry worldwide. Furthermore, they explain that the International Atomic Energy Agency and its development of the safety culture approach through an advisory board and in its *Summary Report on the Post-Accident Review on the Chernobyl Disaster* INSAG-1 (1986); that safety culture became a necessary characteristic to reach the safety levels that were required to improve and maintain it. So, as can be seen, the "quantification" of the safety culture was an aim quite early on. The authors interestingly talk about the organizational performance improvements through the implementation of total quality management (TQM) principles and systems. They state that within hazardous industries where safety is of fundamental importance, integrated safety management must be part of the strategic policy of these organizations besides TQM.

Safety in the nuclear industry has traditionally been very strong indeed, but as noted by Hudson et al. (1996), after the very thorough investigation of the Chernobyl disaster, five out of seven human action contributors to the incident were deliberate deviations from written rules and instructions as opposed to human errors resulting from slips, lapses, or mistakes. In the aftermath of another disaster, the explosion of the NASA Challenger space shuttle, Hopkins (2006), in his commentary on the organizational study conducted by Vaughan in 1996 on the 1986 incident, states quite implicitly a study of organizational culture and how it contributed to the disaster, and the elements such as the *normalization of deviance, the culture of production,* and *structural secrecy,* and she showed how this actually contributed to the outcome.

In fact, in her work with NASA on her cultural organizational study, she views safety culture as ethnographical and like a social historian or anthropologist. This is a very valuable piece of work that took >9 years to complete.

Olive et al. (2006) noted that following the Challenger explosion, the climate at NASA was strongly oriented toward improving safety; however, there was a problem as the underlying culture did not adequately promote the importance of placing safety as a *value* and in fact the safety culture degenerated to such a dangerous degree that the Colombia disaster resulted. It was the nonintegrated safety organizational culture that NASA was suffering from. The safety culture characteristics the authors described included *commitment, communication, resilience,* and *flexibility* as well as a prevailing *attitude of constant vigilance.*

In a similar way, many years later, the Colombia space shuttle disaster made NASA once again look at the social organizational safety culture to help in the identification of the issues that were raised. Hopkins (2006) determines with reference to the Colombia Accident Investigation Board Report, 2003, that NASA had the lacking elements of a high-reliability organization, although technologically NASA was considered as the most reliable of industries. The elements that he noted were

1. Commitment to a safety culture
2. Ability to operate in both a centralized and decentralized manner
3. Recognition of the importance of communication
4. Avoidance of oversimplification
5. Awareness of success
6. A commitment to redundant systems

Parker et al. (2006) provide a safety organizational classification or categorization of maturity levels of *pathological, reactive, calculative, proactive,* and *generative* cultures. These different cultures are described below.

Pathological Culture

A pathological culture can be described as a safety culture in which workers are truly not engaged in any way whatsoever with EHS or safety. They will not care about any process implementation that will exhibit any kind of unsafe behavior as long as they are not caught. This can naturally start right at the top of the organization, and that is where it is most serious.

Reactive Culture

A reactive culture is one where safety and EHS systems are implemented after a certain event, usually an incident or accident of some magnitude. Such a culture only changes as a reaction to an antecedent with usually negative outcomes.

Calculative Culture

A calculative culture is one that is very process- and systems-driven. This kind of culture is usually satisfied with meeting the minimum requirements.

Proactive Culture

A proactive culture is one that can be described as more intelligent in that it looks at performance and incidents and studies through various methods on how to avoid incidents in a very proactive way. It usually has a rigorous cycle of reviews and risk committee recommendations.

Generative Culture

A generative culture is one that is not only proactive but fully integrated in every sense of the word. The organization's leadership, systems, and people are all driven intrinsically by a true and unquestionable belief in safety implementation borne through understanding and covenanted conviction. Both trust/confidence building as well as sharing of knowledge openly through communication and structured training in the organization are required to reach a higher level in terms of organizational maturity (Horizon Terminals Ltd, 2007). In the oil and gas business, in recent years, BP's Texas refinery explosion and the oil storage terminal in Buncifield at Hemel Hempstead have all had very serious implications. The industry was progressively getting safer, apparently because of the technological advancements in engineering safety systems and training of staff and contractors.

In the BP Texas refinery, an explosion in the isomerization unit killed 15 people and injured more than 170 persons. The explosion, in fact, was in one of the largest refineries in the United States. Although the level alarms failed to operate as they should have, there was in the management and process oversight, a twofold failure in following the procedures. This led to a loss of process control and lack of supervisory and management oversight during the start-up operations, although it was known that start-up of units after turnaround were critical operations (CSB, 2007). It was also noted in the initial investigation that the operators did not use their experience and knowledge effectively. Interestingly, in the Chemical Safety Board (CSB) investigations, the three main concerns cited with this incident, which was the worst disaster in BP's operational history, were

1. The effectiveness of the SMS at the BP Texas City refinery
2. The effectiveness of BP North America's corporate safety oversight of its refining facilities
3. A corporate safety culture that may have tolerated serious and long-standing deviations from good safety practice

The Baker Panel (Baker et al., 2007) that was established as a high delegation investigation team looked into this disaster in a different way to the CSB. The Baker Panel identified that one of its main objectives was to focus on how the organizational culture (values, beliefs, and underlying assumptions) influenced the corporate structure and was affected through the management philosophy, processes to affect the control or management of process hazards in these refineries. Against this background, the investigation noted the following:

1. A good process safety culture requires a positive, trusting, and open environment with effective lines of communication between management and the workforce, including employee representatives. Apparently at Texas City, Toledo, and Whiting, BP has not established a positive, trusting, and open environment with effective lines of communication between management and the workforce, although the safety culture appears to be improving at Texas City and Whiting.
2. BP has not instilled a common, unifying process safety culture among its U.S. refineries. Each refinery has its own separate and distinct process safety culture.
3. Although the five refineries do not share a unified process safety culture, each exhibits some similar weaknesses.
4. There is a lack of operating safety culture and discipline, tolerance for serious deviations from safe operating practices, and apparent complacency toward serious process safety risks at each refinery.

The Baker report (Baker et al., 2007) recommended, besides many other recommendations, on PSM and incident investigation, that stakeholder involvement and the development of positive, trusting, and open process safety culture would be beneficial in all the refineries operated in the United States. The report also stated that other companies with similar gaps should seek to improve their safety culture, approaches to PSM, and corporate oversight processes for overall collective industry improvements. Such actions were likely to create an industry that was safer and would incur fewer incidents and loss of lives.

In BP's response to the incident, Mogford (2005), in a very detailed BP lead investigation report made public to the safety practitioner's domain, pointed to failures in both leadership and SMSs, reflected in the following ways:

1. A work environment that was characterized by resistance to change
2. A work environment void of trust
3. Low motivation
4. Weak sense of purpose
5. Unclear supervisory and management behavior expectations

6. Weak enforcement of rules and consequences for failure to follow rules
7. Low empowerment and weak worker engagement

In the BP report, five cultural issues were raised and they are summarized in Table 18.1. The BP Texas City incident is perhaps one of the most important incidents in recent oil and gas industrial history although today it is dwarfed by the 2010 Gulf of Mexico spill.

Of equal significance to the Texas City BP incident was the Buncefield disaster in the United Kingdom. This incident was even more significant in terms of the disruption to business and fuel supplies. This facility provided finished refined products to the smaller depots, retail stations, and the airports in London, and the incident disrupted many businesses. Fortunately, there were no fatalities in this incident primarily due to the day of the incident—a Sunday morning with only two people on the plant.

Most of the root causes were clearly failings from the process control and additionally, the lack of adequate containment systems and response preparedness for an incident of this magnitude. However, to focus on the organizational and behavioral aspects of the incident investigation findings, part of the recommendations of the Major Incident Investigation Board in the incident investigation report (MIIB-HSE, 2006), the high-reliability organizations can be described as those with seven common elements, given in Table 18.2.

The rig had a *lack of management commitment to safety* that was displayed by poor RAs, clear conflict between production and safety, a superficial SMS, and a safety culture that was described in the Lord Cullen-led public enquiry as pathological with several gas leaks occurring before the incident and management not taking any serious action and investigation (Cullen, 1990). This, in turn, created a degraded safety culture in which the employees were encouraged to just keep production up. In fact, Cullen (1990) argued that it was the *poor safety culture* within the operating company that was an *important determinant of the accident,* and stressed that it was fundamental to the investigation and enquiry.

Sadly, after the investigation, it was clear that the actions or in fact the lack of timely shutdown actions of the Claymore rig that continued to pump immense amounts of gas/oil fuelled the Piper Alpha fire contributing significantly to the calamity. The Claymore rig manager *was not empowered to shut down production even after receiving Mayday messages* that were followed by radio silence (after the control room in Piper Alpha was disintegrated in the explosions) until he contacted the mainland office management to obtain clearance to shut down production. His shutting down sequence was undertaken after more than 40 min of receiving the Mayday message but only after losing hope of getting through to the mainland and under serious persuasion of the rig's shift controller.

There are many other reasons behind the incident; these include *absence of backup communication systems,* the *fire pumps being placed on manual* while

TABLE 18.1

Summary of the Five Cultural Issues Identified

Culture Issue	Explanation
Business context	The investigation identified the lack of clearly defined and broadly understood context and business priorities for the Texas City site as the first cultural issue for the incident. The investigation team was not able to identify a clear view of the key process safety priorities for the site or a sense of a vision or future for the long term
Lack of safety priority	The investigation identified the second cultural issue as the lack of safety and basic operations as a priority through the operations on the ISOM. Good safety is delivered through good line operations, underpinned by the right safety culture and values. The quality of basic operations had declined to the level where real safety interventions were necessary to ensure that the right actions were being taken. Evidence of this was that shift change overs were inadequate, procedures were not followed, and line managers were unaware of operations that were underway. There was no evidence of comprehensive and consistent business plans to reduce site risks
Unclear accountabilities/ inadequate communication	The third cultural issue is unclear accountabilities and inadequate communication due to a complex organizational structure that has been modified frequently without improving the required behaviors. The Texas City facility is a large, complex site with multiple levels within the organization, apparently to address the span of control across such a large site. This organization has many interfaces requiring clear accountabilities and good communication, both horizontally and vertically throughout the organization. In reality, the investigation team found examples of a lack of accountability, unclear roles and responsibilities, and poor communication with employees tending to work within silos. This in turn, created confusion around some of the many interfaces. As a result, the working environment is cluttered with many processes, committees, and so on, such that it is relatively easy to lose sight of the basic fundamental requirements for safe efficient operation
Inability to see risk	The fourth cultural issue is the inability to see risks and, hence, toleration of a high level of risk. This is largely due to poor hazard/risk identification skills throughout management and the workforce, exacerbated by a poor understanding of process safety. Although some effort had been expended to raise awareness and understanding in the early 1990s, when OSHA promulgated the PSM rule, this basic training had not been effectively refreshed over the intervening years. There was no ongoing training program in process hazards risk awareness and identification for either operators or supervisors/managers
Early warning system	The fifth and final cultural issue is the lack of a holistic early warning system for process safety exposures. The site has numerous measures for tracking various types of operational, environmental, and safety performance, but no clear focus on the leading indicators for potential catastrophic or major incidents. Numerous audits had been conducted at the site in line with regulatory and corporate requirements, but had generally failed to identify the systemic problems with work practices uncovered by this investigation. Vertical communication was poor Many KPIs used, but not transparent or useful for loss of containment, showing recordable injury frequency improvement Audits were process focused and did not gain verification of action

Source: Adapted from Mogford, J. 2005. Fatal accident investigation report; Isomerization unit explosion, *Final Report*, BP Texas Refinery, Texas City.

TABLE 18.2

Seven Elements of a High-Reliability Organization

Element	Safety Culture Description
Leadership	Leadership from the top. *Staff must see and hear the right messages*, with the right tone and level of commitment. *Managers must walk the talk*, be seen to listen, and to value other's insights. Need to *encourage no blame/open relationships*, promote the control of risk and accident prevention, recognize *safe behavior, and challenge unsafe practices*
Clear accountabilities	A demonstration that process safety is being managed as part of the business process at all levels, *with clear accountabilities*. Important to develop and concentrate on key process safety performance indicators (learn from near misses/precursor events, and avoid major incidents)
Awareness	To make RA real and dynamic, *ensuring that the staff understands the links between hazards and risks*, and the control measures that are in place to control them (the barriers to failure)
MOC	A robust MOC approach to capture, and document, real-time plant and operational issues so that today's plant and *operating envelope are properly understood by those that "need to know it"*
Sustainability	Sustainability, with the business focusing on long-term performance, with investment and maintenance decisions especially focused on the longer term. Also, must maintain *an intelligent customer capability inside the business*
Competency	*Well-trained and competent people* at all levels of the organization—and in sufficient numbers to address steady-state company operation, periods of change and emergency situations and the infrastructure to ensure sustained competency
Safety values	A learning organization that *not only values and encourages learning* from its own experiences but *looks beyond itself for lessons*

Source: Adapted from Major Incident Investigation Board (MIIB). 2006. Bunciefield. *HSE (07/06)/CIO, Third Investigation Progress Report.*

no remote activation was possible, and so on. However, as described by the investigators, the Piper Alpha rig was in many ways bound to fail sooner or later due to the number of *unsafe physical and cultural deficiencies*. The Piper Alpha disaster claimed the lives of 228 men and ironically the only men who survived were those who did not follow the safety instructions to go to the heli-deck but those who took the 300 foot drop into the freezing North Sea waters! (BBC, 1991).

Incident Frequencies and Extent of Maturity of a Health and Safety Culture

There are various relationships established through data analysis between incident frequencies and accidents. If we consider the human losses hierarchy as a fatality, then a permanently partially or totally disabling injury, then a temporally partially or totally disabling injury, then a minor injury, and then a near-miss incident, then we must consider that the elimination of minor accidents or near-miss incidents would work toward eliminating any serious accidents. Furthermore, the near-miss incidents have underlying, unsafe conditions, unsafe acts, and/or unsafe behaviors. Thus, in theory, eliminating unsafe behaviors and acts would most probably also eliminate unsafe incidents. Al Hajiri (2008) confirms this by discussing how the unsafe acts have underlying unsafe behaviors and changing the behaviors will thus eliminate near-miss and accident incidents.

However, even this simple concept needs the workforce to be engaged accordingly. To be engaged they also need to be well informed and trained effectively. The culture of safety at work as extensively explained and highlighted with examples in this chapter needs to be developed to eliminate all kinds of incidents, both minor and major. As such, a more mature safety culture within an organization would help ensure that the organization and its operatives are safer as they are more aware to prevent incidents and behave in such a way to compliment productivity at work with safety (Harvey et al., 2001).

The UK Health & Safety Executive (HSE) explain the need for efficient management strategies to control costs of incidents (HSE, 2006). Chamat (2008) insists that although engineering out hazards is a very critical and important loss-prevention strategy for an organization, focusing, studying, and improving safety behaviors of the workforce have immense impact as he states that 80–95% of accidents are triggered by unsafe behaviors that can be measured (by survey) and evaluated (by observation).

Ajabnoor (2007) confirms that in the industry as a whole, near-miss reporting is the first step and tool to prevent accidents, and that although all incidents should be prevented, focusing on where the highest risks are (in terms of consequence and probability) is how resources can be used most wisely to prevent loss. He focuses on work performance as the main reason behind both unsafe conditions and acts. Al Marzoqi (2003) places much emphasis on the importance of near-miss reporting and in-vehicle monitoring systems to study behaviors as tools to prevent incidents. Al-Sulaby (2008) states that in fleet management, the management of company drivers and contractors besides in-fleet performance monitoring systems and training, near-miss reporting can also form part of the leading safety KPI.

Impact of Trust and Employee Engagement on Maturity of an EH&S Culture

The safety culture and safety climate have been already defined and described in the earlier sections. The HSE-UK developed through its research a regulation and the safety culture components, and interesting information is provided relating to human failure that is well addressed (HSG-48, 2006; HSG-65, 2000). Many practitioners and academics have studied this subject, such as Conchie and Dinald (2006), Cooling (2007), Cox et al. (2006), Cooper (2001, 2003, 2005), Haukeild (2005), and HSE (2005).

People can cause or contribute to accidents (or mitigate the consequences) in a number of ways. Although people do not purposely cause incidents (except for malicious acts), they generally make errors that lead to incidents due to wrong decision making just before the incident. This action is triggered generally by a job that they want to do and due to the lack of assessment of risks, lack of training, and understanding/appreciation of the impact of an action, or an incident. At times, a lack of safe design of equipment, lack of effective maintenance, poor worker competency, and substandard procedures may lead to an incident (HAS, 2004). The control of incidents is best achieved by effectively avoiding unsafe acts or unsafe conditions. These unsafe acts and/or conditions cause incidents. Mitigation is also possible through the use of good and sensible preactions or preventative actions. Thus, the degree of loss can be reduced even in case of an incident in effective preplanning and being prepared for incidents with action procedures and emergency training.

Anderson (2008) stresses on the consistent communication of a strong personnel belief in safety and the creation of an environment that encourages people to provide feedback, all feedback and not only on safety issues. He also advises that the measurement, communication, and rewarding of progress is fundamental to the perpetuation of the process of the development of a safety culture while continuously reinforcing the fact to all that the management and the workers should share a transparent working organizational culture not limited to the above safety issues. Al-Kudmani (2008) also supports this in his analysis of cultural barriers to safety in saying that a workplace culture that promotes employee distrust of instruction from a management is one that cannot develop a productive safety culture. He also states that some cultural factors such as working groups of people who believe that being brave and by-passing safety standards to improve production goals are actually, even if not explicitly communicated by management, favored by management. This is furthermore compounded by the factors of reinforcement through not taking serious counseling steps to prevent this behavior. Ignoring such behavior is automatically considered by workers to mean that what they did was not so bad after all and in fact acceptable by management. Here, the role of the direct line supervisor in taking the right steps in time becomes fundamental to success. For instance, Al-Sulaby (2008) stressed the significance of rewarding and penalizing drivers when it came to demonstrating safe and unsafe behaviors, respectively, on behavioral adjustment.

The consequences of human behavior can be immediate or delayed. Thus, active failures are described as immediate consequences of actions undertaken, such as by drivers and operators. The active failures usually have a direct and immediate impact on health and safety. Latent failures are made by persons who are remote in space or are removed by time such as managers who are at other locations or designers who have well finished and gone after handing over the plant or equipment, and so on. Latent failures, therefore, include poor or ineffective plant design, ineffective training, and inadequate supervision as well as other aspects such as maintenance and the definition of roles and responsibilities (HSG-48, 2005b).

Fleming (2000) noted that although organizations may be at different levels on various elements of a safety culture, which he describes in a 10-element management system he refers to as the safety culture maturity model (SCMM), deciding which level should be achieved for each element depending on the organization or the installation being evaluated. The 10 elements of the SCMM are noted as

1. Management commitment and visibility
2. Communication
3. Production versus safety
4. Learning organization
5. Availability of safety resources
6. Participation
7. Shared perceptions about safety
8. Organizational and leadership trust
9. Industrial relations and job satisfaction
10. Training

This model can be used in the categorization of the questions asked in safety culture surveys (Ghanem, 2009). Thus, the components of this maturity model are used to establish if any of the components are actually missing as well as the prevalence of each component. The SCMM can only be evaluated in organizations that have an effective technical and SMS already in place (Fleming, 2000), one in which technical failures are not causing the majority of the incidents; the company is compliant with the relevant health and safety laws and one in which the organization is driven by the real determination in the prevention of incidents/accidents and not fear of prosecution.

The *opacity problem* is well defined by Reason (2000) when he explains that at times certain organizations and countries speak of the remoteness of a similar incident that has taken place in another organization or cannot happen to them due to the perceived more advanced technological or cultural status. This he describes as a failure to realize that defenses, particularly defenses-in-depth, can create and conceal dangers as well as protect against them. When this ignorance (by the company management) leads to the collective belief in a security of high-technology systems or more superior safety system, this takes on a cultural significance. Thus, in high technologically reliable organizations, the reliance on the multiple-backup safety systems can lead to operator failure to prevent incidents due to (a) heavy reliance on the backup systems to prevent incidents; (b) the concealed latent conditions and (c) the added complexity of the systems caused after a while by by-passing of the redundant safety systems (say at times of testing such as what happened in Chernobyl) with an assumption of the independence of such systems from one another.

The trust factor is one of the cornerstones of any successful safety culture and, in fact, a management system. The behavioral processes work effectively when a high degree of trust exists between management and the employees (Barrett, 2000). In fact, behavioral processes implemented effectively inspire trust and it is a consistent journey in which managements have to maintain their attention and feedback to employees with respect to their suggestions and contributions and can be seen to "walk the talk" or "practice what they preach."

Leaders are very much part of the culture as are the workers and it is very much possible that at times the culture may be better found and driven from the lower levels or more operating (coal-face) parts of the organization as financial pressures and outside influences do not distort the real picture as much (Reiman and Oedewald, 2004). In fact, given that leadership gives enough tangible support through forums such as the safety committees and cross-functional multilevel safety inspections and audits, a greater trust will be built and information feedback systems will be more effective in building an effective safety culture (Burns et al. 2006).

The development and sustainability of a working culture that can be described as being a healthy safety culture of a high-reliability organization such as those working in high-risk industries is based on event reporting, learning at both the individual and organizational levels, as well as the perceived need for a *just* culture. Cox et al. (2006) argues that within the study sites investigated, individuals should be encouraged to take responsibility for safety within the organization and to subsequently develop a questioning and challenging culture. Of course, as also quoted, Clarke (2006) highlighted the importance of communication that is founded upon mutual trust. Cox et al. (2006) also believe that there is an impact of trust and safety cultures on learning organizations in this context.

Harter et al. (2006) using their meta-analysis framework (Gallop Organization) for employee engagement studies of organizations demonstrated that engagement generated better performance as shown in Table 18.3.

As such, engaged employees are more likely to be more beneficial to the organization as a productive and safe resource. This could also mean that without an engaged workforce a safety culture would be difficult to develop and maintain in a sustainable manner.

Even with a BBS program implemented using tried and tested models from world-class consultants, the program fails to yield results at times

TABLE 18.3
Business Impact of Employee Engagement

Employee Engagement	Business Impact
50% employee engagement (i.e., higher employee engagement by being in the top two quartiles out of all the organizations surveyed)	38% higher success rate on productivity 27% higher success rate on profitability 44% higher success rate in health and safety 56% higher success rate with reduced absenteeism

Source: Constructed from Harter, J. K. et al., 2006. Q12 Meta-Analysis. *The Gallop Organization.*

Asseri (2008). The level of unsuccessfulness not only depends on many factors but also includes that the organization was not actually ready to accept such a program, lack of trust between management and staff, and the defective communication systems at work. As such, we can conclude that the BBS programs must be instituted after certain fundamental and functional organizational issues have been actually addressed and it should not be assumed that the investment in time, resources, consultants, and organizational energy will necessarily yield results once a BBS program has started. Thus, if the organizational systems are dysfunctional to start with, the BBS program will inevitably fail. The consistency of management support is also a key issue as workers monitor the seriousness of management on a continued basis and make their own conclusions about their motivations. Trust is built over time, and in the Bahraini refinery BAPCO, Ahmed (2008) attributes the success of their 2-year-long BBS project to the level of involvement of employees with the BBS champions who were chosen from the line and from within the operations. The level of trust was built up over time and the employees *buy-in* to the process that improved with time, not only making the process a success but also enhancing the sustainability of the process.

Cultural Variation: Relationship between Employee Cultural Outlook versus Organizational EHS Performance (National and Organizational Culture)

Much of the discussion so far has been regarding safety cultures, developing safety cultures, and their importance to an organization. It is thus important to understand that an organizational culture is heavily influenced by workforce national cultures as well. Hofsteade (1980) argued that differences in worker motivation, organizational design, and leadership styles are the outcome of deeply seated values and beliefs—culture. In his review work on leadership, he argued that what is common between some of the most prominent leadership theories at the time is that a common theme advocated was of participating subordinates into the manager's decisions. In his discussion on *industrial democracy,* he points out that an important factor in the level of participation is very much dependent on the level of *power distance* within the working and national culture. Power distance is the degree of acceptance of a hierarchical power difference between one level and another, for example, supervisor and driver. Hofstede in 1996 presented various orientations of national cultures and the bearing they have upon organizational or corporate cultures. He discussed five main dimensions. These are summarized and defined in Table 18.4.

Taking from the aspects described in Table 18.4, the scores were developed from Hofstede's work in 2001 on cultural dimension of cultural value safety culture scores in which a major number of employees (100,000+) were surveyed from over more than 70 countries over a period of 7 years (Francesco & Gold, 2005). It is very interesting indeed as it actually means that the cultures with a greater power distance would expect superiors to act autocratically and not necessarily consult with them. In the United States, a certain degree of consultation is expected of superiors by subordinates, yet when managers behave autocratically this is also accepted. This is a very important point that was discussed in greater detail in Chapter 4.

Dapo (2006) insisted that based on the 2002/2003 world-class HSE performance of the Petroleum Development Oman Company, workplace engagement was one of the most critical elements of the HSE program's successes. The safety briefing packages were introduced in three different languages and much of the safety communication messages were also pictorial with the use of employee safety prompt-cards. The scheme was also used across the contractor community, educating and engaging more than 16,000 contractor employees in a total workforce of 20,000. In his analysis of safety cultures and their relationship with organizational cultures based on a major incident investigation study, Hopkins (2006) admitted that one source of organizational culture is derived from societal influence.

Motivation in Health and Safety Culture

To work safely and continue to do so in a sustained manner requires motivated employees. It can be argued that management in an organization that focuses on production and productivity and neglects working safely can create a working environment that encourages workers to work unsafely by virtue of the focus it emphasizes on production as opposed to safe production (Cooper, 2004). Mearns et al. (2003) argued that in recent years the realization that the reliability of the complex work systems within an organization to achieve operational goals safely depended greatly on social structures as well

TABLE 18.4

Hofstede's Dimensions of Cultural Values

Dimension	Definition	Examples for Clarification
Individualism/ collectivism	The tendency to take care of oneself and one's family versus the tendency to work together for the collective good	Collectivism heavily accents the group and values the harmony between group members. Individualism-type orientation is more focused on self-development, career progression, personal rewards as well as personnel rights and freedoms. The United States has an individualistic culture, whereas the Japanese have a more collectivistic culture
Power distance	The extent to which an unequal distribution of power is accepted by members of the society	To what extent do employees automatically defer to the wishes and decisions of their superiors? In North America, a lesser power distance exists when compared to the Indian subcontinent and as such employees are less likely to blindly follow instructions from managers in North America as compared to the Indian subcontinent
Uncertainty avoidance	The extent to which members of the society feel threatened by ambiguous situations and have created beliefs and institutions that try to avoid these	Employees with a high level of uncertainty avoidance rather receive very clear and specific instructions and directions from managers. Greeks, Portuguese, and Belgians exhibit generally very high uncertainty avoidance in their working cultures as opposed to the Chinese and Americans who exhibit less
Masculinity/ femininity	The extent to which highly assertive masculine values predominate (acquisition of money at the expense of others) versus showing sensitivity and concern for other's welfare and the quality of life	The emphasis on money and the acquisition of wealth is very important in more masculine societies such as the Japanese, whereas the Scandinavian society bears greater emphasis caring between workers, relationships among people and a greater balance between family and life (Lewis, 2007)
Time orientation	Called confusion dynamism and is the ability to pursue long-term and general goals versus short-term gain an advantage	Preparing for the future and the value that is associated with savings and the merits of persistence are definitely more prominent in certain societies more than others. Some Far Eastern cultures such as the Japanese, Chinese, and Hong Kong nationals have a greater long-term orientation whereas French, Russian, and West African societies have a shorter-term orientation

Source: Adapted from Buchanan and Huczynski, 2004. *Organizational Behavior* (5th ed.), *Prentice Hall-Financial Times*; Newstorm, J. W. and Davis, K. 1997. *Organizational Behavior: Human Behavior at Work* (10th ed.) International Edition, McGraw-Hill.

as the technical arrangements. Ruuhielehto et al. (2005) suggested various strategies and explained that it was not only the safety programs that were important in as much as getting people motivated to be involved was!

There are many different motivational theories that are applicable to the safety culture at work, including Maslow's hierarchy of needs; X–Y theory on the intrinsic motivation to work; Herzberg's motivation–hygiene theory; McClelland's theory of needs in which power, achievement, and affiliation are important drivers; goal-setting theory; job design theory; reinforcement theory (confirming what behaviors management want and do not want to see!); equity theory; and expectancy theory (Janeeh, 2005).

Reporting on the factors that affect people's job attitudes in 12 separate investigations, Herzburg (2003) shows that in general, achievement, recognition, work itself, responsibility advancement, and growth were the top intrinsic motivators leading to job satisfaction for employees whereas company policy and administration, supervision, relationship with supervisor, work conditions, and salary were the main hygiene factors leading to serious dissatisfaction with the job.

Livingston (2003) reaffirms that one person's expectations shape another person's behaviors and as such a manager who expects more from his employees will probably get higher results and a manager who has lower expectations will have employees with a lower performance output. Furthermore, it is argued by Livingston (2003) that a manager's belief of themselves will be reflected in the performance of their staff as they influence how they view and treat their staff. Therefore, if a manager does not believe he can impress upon his staff to behave and be safer, he will have great difficulty in maintaining a low-incident or incident-free workplace or organization.

Janeeh (2005) made reference to the most critical motivational theories on safety behavior of the workplace and stated the *goal-setting theory* in which we set higher goals for employees and push them to achieve a better level of safety, the *reinforcement theory* in which the employee is rewarded for doing better in terms of safety and in which he/she is reprimanded for unsafe behaviors, and the *equity theory* in which the employees felt that they are treated fairly throughout the organization when it came to safety behaviors were the most significant in the aluminum smelter plant in which he worked and observed employees.

Levinson (2003) explained that to motivate people successfully, leadership must address the issue of mutual satisfaction. He argues that this is one of the main failings in the motivation of employees whose personal goals are very different and thus misaligned with organizational goals. Staff self-motivation thus only occurs at the point of convergence of organizational and personal goals. An individual's personal aspirations must underlie the significant incentive power in which performance-based objectives and targets drive! As such, if safety is not a value that is important to the individual and the organization, it would be difficult to develop a sound safety culture within an organization.

Physical and Physiological Stress and the Health and Safety Culture

In much of the work conducted in recent years, positive correlations between safety and employee wellness have evolved (e.g., Palmer, 2007; Flint-Taylor, 2007; Horseman, 2008). Although the amount of empirical data is very limited and the impact of safety behaviors and cultural factors is not considered in many of the wellness studies based on data collected from predominantly transient working populations and those workers who fundamentally work away from their families in the Middle East, it is worth exploring some of the associations between stress, wellness, and their impacts on developing an organizational safety culture at work.

Horseman (2008) stresses that evidence suggests that a healthier workforce equals a safer workforce. A workforce kept healthy and functionally fit decreases the risk of job injuries, disability, absenteeism, and compensation claims. These, in turn, have a positive impact on general wellness and also contribute to the bottom line through both reducing loss and increasing productivity. Elfarmawi (2008) states that, in fact, stress is the root cause for accidents. From his work experience as an HSE team leader investigating incidents, he says that in the majority of cases stress is either ignored in investigations or otherwise considered as an immediate contributing factor. He categorizes stress in organizations as being caused by nine potential factors (i.e., sources of stress) presented in Table 18.5.

Burton et al. (2005) studied the impact of employee sports center/fitness facilities access and the level of improvement on absenteeism. Although they have commented that little validated data exist on this, in their study that used work limitation questionnaires to determine short-term disabilities (STD), they concluded that there was an association between worksite fitness center user employees and worker productivity and reduced absenteeism from work due to STD workdays lost.

In an extensive and to a great extent unique work conducted by Parkes (1993), shift patterns were an important factor with tiredness and sleep problems compounded by the changing over of shifts from day to night and vice versa. This caused anxiety that was raised as concentration levels needed to be high at all times to the hazardous work being undertaken. The study findings also indicated a difference between onshore and offshore working personnel with regard to anxiety levels that were found to be higher in the latter group.

This indicates greater stress on the mental health of workers in the offshore industry who face greater physiological pressures. This could also be attributed to some extent to more isolation of the individual from family and friends while at work and similar to the short-term effects of those who are involved in lone-working occupations. Similar commentary was made by other authors (e.g., Duff et al., 1994; Bucklow, 2001).

TABLE 18.5

Organizational Stress Leading to Accidents

Dimension	Explanation
Presenteeism organizational culture	Culture of the organization that encourages *presenteeism* taking work home, inflexible working hours, long working hours, and such expectations
Fear of redundancy	Changes in the organization such as the introduction of new technology systems, political regulation (e.g., nationalization of workforce), and restructuring and downsizing operations with the fear of redundancy
Organizational autocracy	Control over work in which an organization's management layers autocratically decide on work and work patterns and subordinates are subjected to instructions with little consultation and discussion
Nonsupportive to personnel needs	Relationships within the organization in which either a nonsupportive occupational culture exists, or otherwise where the worker is subjected to interpersonal conflicts arising from harassment, bullying, belittling, racism, and family support systems. Interestingly much of these issues are very much important with the transient multicultural workforce in the oil and gas industry in the Middle East
Unclear goals/objectives	Unclear organizational, personnel, and work goals and targets that at times develop further to conflicting demands by supervisors
Overdemanding	Demands on the individual, especially at times when too much is expected of a person in a certain time, the work is too difficult, and the worker is not trained sufficiently, or the task is simply beyond the capability of the worker
Lack of social support to employees	Nonsupportive management to workers' personnel issues and family problems. Lack of social support from colleagues at work
Lack of induction/training	Lack of basic induction and specialist job-related training specifically with certain new work methods
Other individual factors	Individual factors related to level of acceptance and fear from change and the level of motivation inspired by tight deadlines as opposed to fear and panic as a reaction to the same

Source: Adapted from Elfarmawi, M. A. 2008. *American Society of Safety Engineers (ASSE), Middle East Chapter, 8th Professional Development Conference Proceedings*, Kingdom of Bahrain. Ref: ASSE-0208-014, 2008.

Leadership Commitment and Sustainable Safety Culture

Management and leadership skills are very much different. Kotter (1998) explains the significance of both the skills and impact of both leadership and management in organizations. The distinctive yet complementary systems of action in which management is about coping with complexity—the practices and procedures of organizations whereas leadership by contrast is about dealing with change. As such it has become more and more important as a trait in the senior-level employees of a workforce in recent years. More so than before, as business, organizations, and the business world at large have become more changeling and volatile, change management has become a must. It is for this very reason that organizations have found it very difficult to implement effective organizational safety cultural changes. Leading safety and organizational culture change within an organization thus requires the skills of a leader and not only a manager. However, in all organizations it is the managers who are tasked (and expected) with leading the safety cultural development revolutions within organizations. Watson (2007) argues that leadership for safety starts at the board level who must be actively involved.

In their consultative document issued in 2000, the Health and Safety Commission of the United Kingdom (HSE, 2000) listed five action points with respect to the director's liabilities for health and safety at work. Table 18.6 provides a synopsis. The consultative document called on the requirement for the articulation of the health and safety responsibilities of all the board of director (BoD) members in the policy and the organization's statements

TABLE 18.6

Synopsis of the Director's Responsibilities for EH&S

Dimension	Explanation
Collective acceptance of BoD of HSE responsibilities	The BoD of a company needs to accept formally and publicly their collective role in providing health and safety leadership in their organization. This includes the consultative process and communication of the health and safety policy, arrangements, and changes
Individual acceptance of BoD of HSE responsibilities	Each board member needs to accept their individual role in providing health and safety leadership for their organization and here their actions and decisions should reinforce the health and safety policies and statements with no contradiction
Impact of decision making on HSE	The BoD need to ensure that all board decisions reflect their health and safety intentions, as articulated in the health and safety policy statement. As such the belief in and adherence to the spirit of the statements and not only the letter of the statements in the health and safety policy are important. Therefore, the business relationships they have with other companies, organizations, and service providers are equally as important in the sense that they should also have equally if not better policies and health and Safety Management Systems
Recognize role and actively participate	The board needs to recognize their role in the active participation of their staff in improving health and safety, notwithstanding the legal requirements of employee participation, consultation, and engagement in health and safety
Appointment of a "safety champion" at the BoD level	The appointment of one of the BoD members as a "safety champion" or otherwise referred to as the "health and safety director." Thus, reference to a bare minimum of annual health and safety reviews is required. Also, more often reviews are required when changes happen to the structure of a company. The reviews should have outcomes and the BoD must thus influence the business in the application of betterment or corrective actions as and when needed. The champion can be the CEO or the chairman of the Board, but he or she should be assigned

relating to health and safety. This thus confirms that safety must start at the top of the organization.

Mearns et al. (2003) advised that, for example, in the application of health and safety, best practices in the oil and gas business, marking the operations of the BP group, Conoco, and Royal/Dutch Shell group, the prescribed practices included under operation and governance that one managing director has a board-level responsibility for health and safety; a corporate health and safety advisor recommends policy and chairs a committee comprising senior business managers. Furthermore, health and safety are looked at as a core value and part of the company culture in which risks are assessed and targets for safety performance are set and monitored (Zwetsloot & Ashford, 2003).

Most importantly, the moral aspect of preventing harm to people and loss in general is very important indeed and should be part of the guiding principles of a business as confirmed by Hayward (2005). In most legislation it is assumed to be, in fact, the primary reason for managing risk. The employer thus owes all employees a certain duty of reasonable care in order to ensure that he/she is not hurt at work. There is also a societal expectation that each employer shall exercise appropriate care toward each employee, and especially those who are involved in a risky business, providing them with information, training, and appropriate protection measures, including equipment and guidance. This duty of care naturally extends to all those who may be affected by the employer's business operations such as contractors, visitors, and the public. Solnaki (2007) suggested training, establishing clear roles and responsibilities, enforcement, and potential organizational partnership with the regulators when it came to developing line managers to take the lead when it came to safety.

Marsh et al. (1998) discussed in great detail the role of management in the development and sustenance of a safety culture as a key and integrated role in the organization's workings. They argued that the management in organizations has a key role in managing proactively by monitoring performance and applying continuous learning principles within an organization's staff. In fact, by virtue of being managers their KPI depends on their ability to do this successfully and in a sustainable manner in every aspect of the running of the business (Moore, 2008). Thus, their effectiveness in behavioral measurement, goal-setting, motivation of employees, and feedback for improvement is fundamental and key to improving safety behavior at work.

In the aviation industry, management commitment to establishing a thriving and pervasive safety culture will determine, in large part, whether an organization achieves its corporate goals. In such an industry, safety is paramount. The aviation industry is an extreme case in which the operating environment is extreme conditions of unstable environments, and with a twin goal of attaining maximum operational efficiency and preservation of safety and reliability. Van Dyke (2006) argued that whatever the driver, the motivation for management commitment needs to be framed in a solid business case in which the articulation of the best fit and associated returns in the same convincing terms as any business opportunity, including establishing safety as a profit center. This thus includes capital management and finance planning, linked directly to airline objectives and strategy, complete with RAs, gap/sensitivity analysis, as well as timelines and performance measures. This means a fully integrated model of safety in operations with little tolerance to risk taking as such. It is argued with empirical data that the benefits of an effective safety culture are better profitability; efficiency through allowance of better deployment of resources; flexibility as the organizations that have assessed their risks understand how to manage risks when others may not dare because of lack of understanding, in addition to adaptability, continuity, and durability; and finally predictability.

Safety Leadership

Safety leadership, which was discussed earlier in this chapter in light of major disastrous incidents, is also a very vital component of improving safety. Carrillo (2006) explained in their work on safety leadership that successful safety leaders advocate a few basic principles such as: ensure goals are shared and made clear; set an example; create trust by trusting people; view mistakes as learning opportunities rather than create a blame culture which would be counter productive to establishing credibility for gaining support on safety initiatives. Interestingly, safety leadership is argued to be both simple and complex, simple in the sense that it deals with people, so the problem area is easy to identify, but complex because organizational cultural dynamics are complex because again they deal with people. Carrillo (2006) establishes that *trust + credibility × organizational development = results.*

The model presented by Carrillo (2006) spoke of Vision to Results through a five-step process, which is summarized in Table 18.7.

Management commitment can be defined as demonstrated behaviors associated with achieving health and safety goals. Interestingly, although many managers would not admit that they are not committed to safety when asked, their actions and behaviors are the ultimate proof of their commitment. The workforce knows this but Cooper (2006) argues that there is very little empirical work to assess the actual impact of management commitment. Most of the work has been on perceptual questionnaires and/or semi-structured interviews. His field study of a nickel refinery showed that increasing the frequency of management–subordinate interactions positively improved safety performance and that the magnitude of improvement was dependent upon time, ranging from small to large effects in terms of reducing lost time injury rates and minor incident rates.

Barrett (2000) said that safety management is based on a triad of policies and procedures, people, and environment. He explained that if leadership does not make the connection between SMSs and behavior, employees will

TABLE 18.7

The Vision to Results Five-Step Process

Strategy Step	Explanation
Insight	• Set the *insight* with which self-awareness and organizational assessment can be made. This is naturally difficult as it requires that actions are taken on grievances from employees to set that initial level of buy-in and trust from employees
Direction	• Set a *direction* in which the organizational mission, vision, and values are linked with safety and those are aligned with personal mission, vision, and values. Here, leadership rather than simple management needs to be exhibited by the management team
Focus	• *Focus* follows with managers prioritizing and focusing on the top 3–5 issues that need to be addressed and accepted by the management team and workforce and communicates one message to all that is aligned with the business goals
Capability	• The development of capability is then required in which accessible-cultivated personnel relationships are developed through mentoring, coaching, and personnel communications. *Capability* requires investment of time and where needed resource allocation or reallocation to enhance the communication structures
Accountability	• Finally, once the capability and the grounding has been completed, *accountabilities* need to be set getting commitments from all, clear expectations are set and when changes occur at any point in time, the organizational structure requires that explanations are made clearly and again accepted. Here, the integration of performance reviews and incentive rewards should be also considered

Source: Adapted from Carrillo, R. A. 2006. Leadership formula: Trust + Credibility × Competence = Results. A guide to safety excellence through organisational, cultural and personnel change. *American Society of Safety Engineers (ASSE), Middle East Chapter, 6th Professional Development Conference Proceedings*, Kingdom of Bahrain. Ref: ASSE-0306-010.

eventually lose faith in the behavioral process and the process will falter. By working with one another in an open and transparent culture management, better safety results for their organization can be achieved by being promoters and leaders in safety rather than expecting the SMS to drive the behavioral change processes.

Leadership Behaviors for Improving Workplace Safety and Safety Culture

The following are some of the ways in which leadership can help to create a strong safety culture.

1. Leadership support and commitment at all levels of the organization is an absolute requirement at all times. Proactive approaches to

maintaining the safety of workers (employee and contractors alike) sends a powerful message of commitment.

2. Performance management is important for success in safety. There must be strong emphasis on leading indicators as a proactive approach to safety in the workplace. Some organizations have found in team safety meetings that the use of "safety moments" for example have helped reduce incidents. Additionally, when a desired behavior is rewarded, the behavior is sustained. Rewarding performance for focusing on leading indicators can sustain improvements in safety.
3. Leadership must establish safety champions across the organization to proactively sell the message of safety in the workplace. Champions must be continually equipped with new learning based on the experiences and knowledge created by industry peers and other industries. Such champions along with every employee must be continually trained in safety principles and be equipped with safety best practices.
4. Leaders must become role models and mentors for followers. Demonstrated safe leadership behaviors both at home and in personal lives are very influential in the lives and behaviors of followers. Anyone in roles where he/she has the ability to influence others must commit to safe work behaviors.
5. Leaders must demonstrate genuine care for all workers and make safety a part of the organization's work culture. Leadership must create and sustain behaviors that are safety oriented at all times and people must be placed first above all else. Safety training, sharing information such as key learning's from safety orientated at all times and people must be placed first above all else. Therefore safety training, sharing information such as key learning from safety incidents can be very informative to improve the awareness of safety.
6. Safety is everyone's responsibility. Leaders must create a collaborative work environment that involves all stakeholders in finding the right solution to safety concerns. Ownership to solutions is critical for sustained success in safety.
7. Leaders must seek to recognize and treat expenditure in safety and loss management as an investment in the safety culture of the organization. Failure to do so leads to suboptimal safety support from leadership with ultimate failure in developing a strong safety culture.
8. Leadership must seek to create a continuous learning environment. As a consequence, concerted efforts must be placed on continually upgrading the skills and competency of all workers. No employee should ever be asked to do work for which he/she does not possess the right level of training and competency.

9. Inspire the hearts and minds workers to work safely. Workers must be rewarded for behaviors that support safe work. The right to refuse unsafe work must be a core attribute of the organization's safety culture and leadership must support this requirement at all times. When this rule is broken in the interest of production, trust in the leadership's commitment to safety is broken. Trust is earned over time from demonstrated credible and predictable behaviors for doing the right thing at all times.

Developing a Model of Safety Culture

Extensive management research exists in organizational safety cultures through using observation and perception studies, for example, Flin et al. (2006), Clarke (2006), Cooper and Phillips (2004). Through structured observation of *acts* and *conditions* at the workplace we are able to determine if they are safe or unsafe. As the level of *safe operation* of individuals is reflected in behavior, the safety culture can also have its particular *behavioral characteristics*, which become measurable in terms of communication, incidence rates, productivity, and level of proactivity.

In the research of health and safety at work and in particular behavioral safety, less time has been invested in explaining the mechanisms, theories, and models underlying these behavioral characteristics. There is, however, good evidence, for example, Reason (2000), Peters (2008), Farrington-Darby et al. (2005), Duff et al. (1994), and Cooper (2000), which shows that behavioral intervention programs with workers is effective in bringing about positive safety behavioral change. Argyris (1996) explained the importance of balancing effectively using management research in both the theoretical knowledge in the actionable knowledge contexts by managers.

There are various definitions of safety culture as described by Johnson (2003) that relate to the behavior of a person in response to specific conditions. He also notes that behavior can be an act, reaction, or function of a person in response to an activator driven by a motivation, either a goal to be achieved (expectancy) or accessibility of the goal (availability). Furthermore, that behavior is an activity, response with an underlying motivation/meaning and intent. Behavior is an observable action and, therefore, many studies have focused on human behavioral investigations. Cooper (2000) spoke about safety culture as the way things are done in any particular workplace. He defined safety culture as the context for action within an organization that binds together the different parts of an organizational system in pursuit of organizational goals. Organizational culture within an organization in general terms is defined by Buchanan and Huczynski (2004) as organizations that have shared beliefs, customs, and values inherited from prior generations. Harvey et al. (2000) explained that safety culture can be different within organizations, especially between leadership and shop-floor employees. Having different cultures within each group inhibits effective communication and teamwork to the extent of at times inhibiting cooperation.

Training

Organizations generally arrange for training to improve awareness, knowledge, and competency about health and safety. The main purpose of the safety training would be to change (improve) competency and the safety culture. Harvey et al. (2000) assumed that in an organization with an SMS and where safety is of paramount importance, if the safety attitudes changed then safety behavior would also change. In this study, the effectiveness was more profound with the manager levels rather than the shop-floor employee levels. In the context of culture and risk and when risk and safety are put into a cultural context to the employees, it is very much seen as a top-down initiative and if different cultures exist in both groups, the level of acceptance (open-mindedness) of the information may be inhibited. Training programs that are owned and driven by the management and that focus more on loss prevention rather than on the technical aspects of safety were not successful with shop-floor employees becoming less engaged with these initiatives and seeing it as another management "safety scheme."

Al-Kudmani (2008) noted lack of skill and knowledge to being one of the main root causes of critical behaviors leading to the establishment of a safety culture. Peters (2008) argued that fundamentally the failure of training for highway safe driving knowledge and skills is because it has focused on rules and regulations rather than the appreciation of risks. Much of traffic accidents are the outcomes of other factors such as the road conditions and vehicle conditions that account for >85% of accidents and crashes. Effective training and competency assurance is, therefore, essential in reducing accidents, and thus improving safety.

Iskandar (2010) explains that in order to achieve sustainable world-class performance in safety the development of competency in employees is essential and that in fact 87% of injuries are attributed to unsafe or at-risk behaviors as opposed to unsafe conditions. He explains that operator conditioning and observational learning are critical types of learning to lead to an improved safety culture. Brewster (2010) explains the importance of engaging the *imbedded safety coach* within organizations who takes time to review the JSA and observe behaviors and work, and works closely with the work force to improve behaviors through one-to-one coaching and training. Flin (2005) explained that situation awareness, decision-making, and assertiveness of managers during emergencies are critical skills that should be developed in the management and supervisory levels through training and drills.

In some extensive research work done in upstream oil and gas operations, 34% of leaders considered the training and competency levels of their staff to be a high priority for safe work (O'dea and Flin, 2001). This was key to promoting safe behaviors and a sound safety culture. Marsh (2007) argued that more people-focused training was required in order to move toward a more proactive safety culture. He explained that leading organizations provide training to leaders and supervisory personnel on effective coaching, assertion, and ice-breaking to enable greater confidence to lead safety and change cultures.

Awareness and competency development and assurance in the employees of an organization at all levels are important factors for moving toward a strong safety culture. Perhaps not directly, because behaviors are governed by attitudes but effective training should improve *the appreciation of risk* and the *importance of safe behavior* and therefore, should lead to a better safety culture in the workplace. Moreover, as directed effective training is required, employees should therefore be consulted on where they see safety training can help them improve their performance. Training can be effective if there is already a homogeneous safety culture in which both supervisors and managers work effectively together and employees see training as a learning opportunity rather than a necessary evil.

Information Sharing/Reporting Incidents

Groups and teams generally operate effectively and interdependently when they are at the stages of *norming* and *performing* (Buchanan and Huczynski, 2004). One of the reasons is that there is effective communication and information sharing. Safety culture development should move toward having a shared set of values derived from alignment with the SMS and the organizational safety vision. Therefore, it is imperative that individuals share best practice and safety information with one another, both through the formal channels such as safety committees, work groups, briefing meetings, and operator–supervisor discussions. Information sharing through informal groups such as with work-gangs and shift personnel relationships is also important. In fact, near-miss reporting of potential incidents has been argued by many practitioners as being one of the most important and effective ways of reducing incidence rates.

Ajabanoor (2007) suggested that near-miss reporting is the first step to prevent incidents and accidents. From an organizational perspective, all incidents should be prevented. By focusing on where the highest risks are (in terms of consequence and probability), organizations can effectively allocate resources for greatest value in loss prevention. Al Marzoqi (2003) places much emphasis on the importance of near-miss reporting and in-vehicle monitoring systems to study behaviors as tools to prevent incidents. Reason (2000) explained that an informed culture can only be built on the foundations of a reporting culture, which in turn depends upon the establishment of a just culture. He explains that people generally will not readily confess their blunders and mistakes, especially if they fear reprimanding. However, some organizations such as those in the aviation industry have set up effective reporting cultures in which protection is given to employees for reporting, confidentiality, separation between those who report and those who analyze the data, easy reporting systems, and effective feedback mechanisms. In terms of sharing and learning, Fleming (2000) identified communication, participation, learning organization, shared perceptions about safety, and trust for improving safety culture. Organizations with open reporting of both incidents and near misses, which develop strategies for effective analysis and corrective actions through engineering (hard) and training/behavioral change methods (soft), eventually develop a more proactive culture—better safety culture.

Autonomy and Leadership Support

In general, the level of ownership and accountability that workers will take for their safety and the safety performance is linked to the level of empowerment and clarity of role description and expectation. In such organizations, trust is high. Conchie and Donald (2006) explain that the role of trust is predominant and very important in high-risk industries. It acts as a good lubricant for effective communications. In their work, which involved the use of surveys they found that attitudes of distrust were much better predictors of safety performance. Odea and Flin (2001) confirmed that trust and effective communications are fundamental as a facilitator of effective safety leadership, and in particular where managers and supervisors demonstrate commitment, the workers feel an obligation and thus engage in safety behaviors that may even expand formal definitions and they would in turn reduce the risk of incidents. Carroll (2002) explained that a good safety culture would typically rely on strong safety leadership because leadership promotes shared values, positive safety attitudes, and also a commitment to organizations safety policies.

Duff et al. (1994) proposed that safety culture can be developed through key behaviorally based techniques. The study showed that safety performance could be objectively measured and that goal setting and feedback can produce improvements in safety performance and that management commitment to safety can effectively enhance this performance. The commitment of employees empowered to lead with safety (and told that they were responsible and accountable for safety) yields good results but as explained by Reason (2000) does have to be necessarily based on an organization that has a just culture to start with. This would mean that employees feel that they would be treated fairly and supported when they operate or even when they stop an operation.

In the Piper Alpha platform disaster, the Claymore rig continued to pump into the Piper Alpha even with the Piper Alpha on fire and the issuance of the Mayday messages. The manager was not empowered by the head office in Aberdeen to shut down the operation unless he received clearance from them. As such, this led to the fuel being fed to the Piper Alpha rig that was on fire, eliminating any chance of controlling the fire that eventually led to many fatalities (Cullen, 1990).

In the BP Texas refinery explosion, which claimed 15 lives, the investigation committee found serious issues underlying the failures that included failure to implement leadership in the PSM. Baker et al. (2007) suggested the following for a strong PSM culture:

1. A positive work atmosphere with trusting and open environment
2. Clear and effective lines of communication among stakeholders
3. Accountability as a core concept for driving desired conduct

The recommendations of the elaborate investigations included clearly defining expectations and strengthen the accountability for PSM and performance at all levels, executive and operational, as well as ensure that the BP organization continues to support the line organization (CSB, 2007).

With the Buncefield oil terminal explosion, Allars (2007) advised about the need for

1. Effective communication—right tone
2. Commitment—demonstrated by leaders walking the talk
3. Engagement—listen and value others' insights
4. Promote a no-blame work environment
5. Focus on RM and accident prevention
6. Recognize and reward safe behavior and challenge unsafe practices

Among the key drivers for a strong safety culture are learning, reporting, just, flexible, and informed. For a workforce to operate safely in an effective manner they need to be adequately informed and trust that organization's management in terms of their support for decisions made on the basis of safety (Reason, 2000).

Developing a Strong Safety Culture

Many organizations are very interested in understanding and benchmarking its safety culture. The goal is to establish quantitative variables that may be able to provide an insight to worker perceptions about the organizational safety culture. Autonomy and management support help drive employees at all levels take accountability and responsibility for safety and this in turn leads to a workforce more engaged, which forms an effective safety culture. Training builds knowledge and competency develops capability, as the workforce becomes more aware of the safety aspects in terms of risks, risk control, prevention of incidents, safe systems of work, and best practice. Therefore, the training ultimately improves the safety culture by raising awareness in the workforce. This is very similar with knowledge sharing and reporting. If an employee is made aware of the policies, procedures, reasons for incidents that have occurred by information sharing, and works in a culture in which reporting of even minor incident, accidents, and near misses is open and transparent, it raises the awareness of the workforce. So, if there is a raised awareness in the workforce, a better safety culture may result. Therefore, both training and knowledge sharing/reporting help create an effective safety culture through awareness building.

Much of the development in health and safety over recent years has been developed from major accidents like Chernobyl, BP Texas City refinery, and Buncifield. It is also well established that safety management is critical to business continuity and organizational resilience. A great deal has been learnt from all these major incidents, which had become a very important focus for safety practitioners as well as for industrial/organizational physiologists. The concepts of BBS are quite simple and the strategies are not overly complicated. However, it is the implementation of such strategies that is the real challenge.

Most authors have established strong linkages between trust, employee engagement, transparency and openness, shared learning, and leadership commitment as a fundamental element of establishing a sound safety culture. Organizational culture and its impact on safety have been discussed and especially with respect to variations and impacts of various dimensions as per Hofstede (1980) communicating the safety message (Dapo, 2006) and relationship with organizational cultures (Hopkins, 2006). Motivation has also been discussed with respect to theory and with respect to which theories bear particular significance when it comes to BBS programs from practical experiences (Jeneeh, 2005). The impacts of both physical and physiological stress on workers have been discussed and finally the significance of management commitment and visibility especially with leadership and management of safety at work has been addressed from various perspectives.

Ultimately, a strong safety culture requires most significantly the following:

1. Leadership commitment
2. Fairness and organizational transparency and trust
3. A consistent SMS, that is, simple and easy to follow

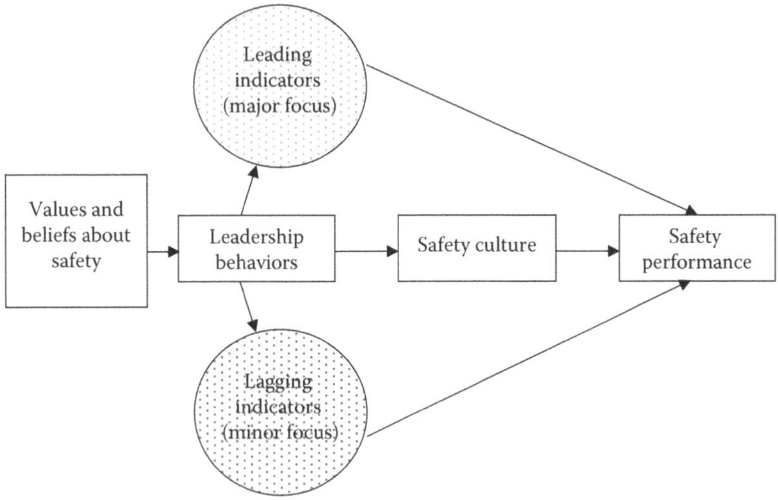

FIGURE 18.1
Safety performance and safety culture.

4. A positive approach to health and safety that rewards the right safety behaviors
5. A motivated workforce
6. A well-communicated, clear, and shared health and safety vision
7. Tactical health and safety deliverables and performance indicators and measures that are properly prioritized and aligned with the health and safety vision
8. A trained, competent, learning and empowered workforce
9. Clearly defined roles and responsibilities as well as accountabilities

Figure 18.1 summarizes how strong safety performance can be derived from the organizational safety culture.

Safety Culture Maturity Assessment

Various methods have been developed to assess and measure the safety culture of an organization. The most commonly used methods in large organizations today are quantitative surveys conducted using tools such as Survey Monkey to collect snapshot data regarding any particular safety culture variable of interest to the organization. As discussed earlier, the following

10 areas of interest are common in determining the safety culture of the organization.

1. Management commitment and visibility
2. Communication
3. Production versus safety
4. Learning organization
5. Availability of safety resources
6. Participation
7. Shared perceptions about safety
8. Organizational and leadership trust
9. Industrial relations and job satisfaction
10. Training

Each area can be assessed using survey questions to determine worker perception at all levels of the organization. Responses to each question can be graphically presented to provide a cultural snapshot at that particular point in time. Surveys are easy to execute and can be very inexpensive compared to other methods of evaluation. Generally, directional information is derived from surveys and observed areas of concerned can be further investigated using interviews and more detailed analysis to understand the underlying concerns.

Other valuable methods for assessing the safety culture of the organization may include qualitative methods that may include performing interviews with randomly selected individuals in the organization and probing with open-ended questions to determine worker perceptions of the organizational performance related to the evaluation criteria/variables listed above. Town hall meetings and small group forums are also very effective in determining the organizational culture. However, these methods are costly and will often provide more specific information regarding the safety culture of the organization such that targeted corrective actions and strategies can be developed.

Establishing a baseline set of data for the safety culture of the organization relative to the criteria defined above provides organizations a benchmark against which progress can be compared and areas for opportunity identified when follow-up safety culture maturity analysis is conducted again. In the early stages of development, safety culture maturity assessment surveys should be conducted annually to determine improvement opportunities and for developing corrective actions and strategies. As the organizational safety culture matures, analysis should be reduced to once every 2–3 years.

19

Implementing an Effective Global Occupational Health Policy and Program: Case Study in the Oil and Gas Industry

Occupational health (OH) continues to receive less focus than safety in the workplace relative to other disciplines of HSEQ programs. Perhaps this is so because of the low impact to the cost associated with workplace health and safety. Unpublished data suggest that only 15% of workplace health and safety cost is associated with occupational health incidents. As a consequence, organizations have failed to place emphasis on OH, thereby failing to capitalize on opportunities to promote further improvements in this area.

One great challenge that many organizations around the world face today is the implementation of effective and meaningful OH strategies and programs that add value to the business. Greatly misunderstood and regrettably at times not fully respected as an independent medical discipline function within many organizations, OH finds it difficult to locate itself within the organizational structures. It requires the independence, yet it is highly connected with human resource policies and enterprise RM and also has strong links with the health and safety functions. The implementation programs require an investment, and more importantly a set of organizational policies and processes that support its existence and working. There is also the issue of funding. The economics of the OH provision within organizations can be easily challenged by the typical organizational financial investment instruments involved, even though economics in itself is very much a social science.

Some aspects within the OH function can be outsourced effectively. However, although the technical expertise and diagnostics can be outsourced, the accountability and responsibilities remain very much within the organization's management. This can pose a significant emotional and ethical pressure on managers who do not quite understand how to go about making OH work for them rather than looking at it as an instrument in the hand of the employee that can be abused. OH requires that organizations understand its value both as employers and employees.

This misunderstanding and the variations in implementation models that naturally must vary significantly from one business to another is what has made it a challenge for the whole organization, from board members to line supervisors. This chapter reviews some of the significant discussions that have taken place, mainly by the Energy Institute and the Institute of Petroleum in various research works and discussion forums by various specialists. The same issues are then discussed and the positioning of this function is critically reviewed. This work presents some explanation to the whole issue of implementation of an effective OH program. The basis of the case study is the ENOC in Dubai. This chapter really focuses on where to best place OH within an organization. The answer to this question is not a *one-size-fits-all*; it truly is not and this chapter discusses this critical issue.

Influenced by much of the legislation and awareness that has come about in the last 20 years with the positive discoveries of work-related illness and diseases, occupational medicine has become increasingly important and a prominent discipline in medicine. In the industry, it has become important to other mainstream departments in companies and functions such as human resource management; environment, health, and safety; legal advisory; as well as productivity and RM. Although the function is usually found within organizations that generally have a formalized primary health function or otherwise, a significant number of employees (3000+) warrants having a medical specialist within the company.

The cost of employee health like many other aspects in organizational behavior and management is very much an important and fundamental critical business parameter or metric. The growth in the power of the trade unions and other such groups that have focused on employee welfare protection at work has all made organizations think long and hard about occupational health like it has with compensation, pensions, and health and safety at work.

In this chapter, we discuss a very important aspect of occupational health, which is the sustainment of the function within an organization. The oil and gas industry is the case study we use here. The case study comes from a diverse mid- and downstream oil and gas company mainly operating in refining, gas processing, retail, marketing, terminal operations, shipping, and fabrication.

Background on Occupational Health Development Perspectives

Looking through much of the occupational health work that has been done in the last two decades, it has developed at different rates in different disciplines. Much of the clinical medicine work has been exploratory in the sense of case management. Physicians have come across conditions that were linked, in most cases, to an increased number of cases showing similar symptoms from similar occupational background patients. Sepai and Blain (2000) in a review of biological monitoring of benzene explain that there is clear evidence of the toxic effects of benzene from studies of human populations exposed to the benzene chemical in their workplace and also from observation studies undertaken on laboratory animals. In the United Kingdom, the HSE EH/40 guidance document gives explanation to the 15 min and 8 h exposure limits and methods of biological monitoring.

The American Group of Industrial Hygienists have developed most of the same short- and long-term exposure limits based on the studies of various effects from epidemiological and toxicological studies. It is a complex discipline in science and exposure studies focus on exposure of workers or direct contact with the different substances. However, in the past decade, the accumulative effects of a substance (e.g., benzene, now confirmed as a carcinogen through impact of small exposures) have now become very important. The synergistic effects of different substances or, for example, exposure to substance and tobacco and/or alcohol use by a worker have also been confirmed.

The above case demonstrates an important compliance aspect because once it became confirmed that a substance (or process) caused ill health, it became important for the employer to look at the protection of the employee. Jean-Phillipe Gennart of ToatlFinaElf explained that the future of health surveillance lies in demonstrating safe workplace as opposed to the identifications of early health effects. Occupational medicine practiced by *company doctors* has been widely scrutinized by employees. This is quite natural as the company doctor is seen as serving the purpose of the organization. This is very true, and as such it must also be true for this function to also align itself with the code of medical ethics and professional conduct to be in the first instance serving the employees.

Roberts (2004) explains that it is necessary to "apply an economic yardstick to medical practice." The occupational health function within an organization can help:

1. Extend the productive years of employees
2. Reduce the costs of medical care
3. Conserve human resources—which can be seen by an organization in business terms as human capital investment

He goes on to explain that the key is to understand this kind of investment as a financial "return on investment (ROI)." He explains that preventative medicine is very much an investment—you eat and live healthier, you live longer and become more productive. Miller (2004) addresses the "economics of OH provision" and argues that traditionally, occupational health strategies could be founded on drivers such as humanitarianism, public health, and social justice perspectives. And although that remains valid to a great extent, the future and sustainability of occupational health are more related to the more internal organizational drivers. Therefore, there must be an economic justification to the provision of resources for occupational health. This is important as OH and HSE practitioners are then positioned more actively to help the decision makers within an organization make decisions based on the mechanisms as discussed by Roberts (2004) above.

Miller et al. (2002) suggested that although difficult and challenging, the demonstration of economic value of occupational health services was possible. They estimated that savings of over 500 Great British Pounds (GBP) per employee could be made on average when considering that about 3.7% of total estimated working time in the United Kingdom is taken in sick leave. They argue that after using three methods of economic evaluation, namely, cost model, contingent evaluation, and empirical approaches, the contingent model is possibly the best as it takes into account a valuation based on a hypothetical premium paid in exchange for OH services rendered. This was based on some extensive work done with the BP organization. Cost–benefit analysis can show the value added by occupational health and safety (OHS) services for each different business unit and in fact this was done on all the business units across BP.

Marson (2001) explained that although the awareness to the cost of ill health and medical injuries compensation in the United Kingdom and the EU is now very high and the payouts are extensive, OH professionals are still struggling to attract the level of resources required or needed to meet the increasing challenge, which is also naturally becoming more regulated even if in terms of strict liability. He offers a four-step process:

1. Strategically choosing a direction through reviewing the current and comparable performances and endorsing the value assessment for each type of business
2. Establishing the issues of most importance to the organization
3. Developing options for successful direction of OH services
4. Implementing, monitoring, and assessing to make improvement

This falls well within the plan-do-check-act cycle that is typical of any quality or SMS.

A better understanding of the losses from illnesses and injuries and loss of productivity is always required. Once we understand and quantify those losses and implement strategies to reduce such losses, the quantification of the financial return becomes more possible, although as Miller (2004) explains, it does require a detailed approach. Traditionally, the impact of lost work days has been the focus for economic justifications. It is difficult to quantify the productivity caused by more fit employees! The future analysis of costs that are related to OH strategy domains includes

1. Direct cost value contribution such as employee *presenteeism* (productivity losses associated with employee work when employees come to work but perform below par due to any kind of illness).
2. Indirect use of values and the impact on morale, staff retention, and corporate image.

3. Option values for employees; as such, it reduces uncertainty for them and they thus derive economic value from it.
4. Nonuse value: existence value; as such, people see it as useful to be there although they may not necessarily use it.

Furthermore, in more recent years OH&S practitioners have been working more closely with a greater understanding and appreciation of the impacts from fatigue at work. As per Gall et al. (2006), fatigue has been recognized around the world as a contributor to many accidents involving means of transportation. This has brought about much research in more recent years involving study of long work hours; night working; working without adequate breaks or adequate rest between shifts; and unfavorable shift patterns, especially in repetitive jobs and even working in comfortable/cozy sitting positions that promote sleep. In fact, today, some discussions and input from OH practitioners are undertaken when formulating work rosters and shift rotations.

Economic theory is derived from welfare economics or more accurately socioeconomics that is founded on the notion of social welfare economics. Therefore, the net economic effect of an activity needs to be defined. Interestingly, Miller (2004) identifies a real challenge in this regard, that is, organizations drive occupational health and therefore consider the benefits from a partial perspective; what is really required is a more holistic societal perspective for the same. What this meant is that consideration is made, which affects the firm or organization. Occupational health supports general employee health and this places thus less of a burden on the national health services and welfare systems. This in turn has a positive impact on both the individual and the society at large. Organizations therefore must not look at the cost–benefit analysis only through using traditional ROI instruments but must also look at the other benefits they give to society that translates to corporate social investment—the net impact from occupational health strategy must therefore be considered in the social perspectives.

Organizations quantify the investment in financial terms, and in return they also look at the return on that investment over a period of time. In occupational health, the investment is combined by the organization and the employees and the return is also combined; the employee is healthier, and so the organization gets more productivity. Society benefits from greater output and the multiplier effect.

Presenteeism can be just as bad in terms of lack of productivity as absenteeism. Many employees suffering from high-risk pregnancy, migraines, depression, musculoskeletal problems like neck and back pain, as well as allergies and even arthritis may come to work but not be very productive at all. Roberts (2004) cites an example in the United States where a survey done by the management indicated that more than 70% of the workforce were overweight [by calculation of body mass index (BMI)] and although the calculation of days lost due to BMI was a very subjective indicator, they did make an effort to change by adding more fruits and vegetables in the plant canteen diet as opposed to the staple fried foods and high-fat-content cuisine offered at the time. The risk of lost time occupational illnesses due to BMI included heart disease, depression, hypertension, musculoskeletal illnesses, diabetes, and asthma.

Knudson (2004) explains that from a business perspective, there are clear correlations between sustainably successful business and health of employees within that business. He commented that it also impacts on the sustainable growth over an extended period of time. Health and safety are at the intersection of economic and social factors. Knudson, who was working with the global human resource operations of one of the largest oil and gas companies in the world, explained that on average the health factors cost the organization anywhere between US$1000 and US$1500 per employee per year due to factors including

1. Sickness absence; reduced performance
2. Occupational illnesses
3. Workers' compensation litigations
4. Environmental incidence

These costs are not easy to measure directly. On the other hand, improved health means improved worker productivity and the benefits of that investment include

1. A healthy workforce
2. Improved worker motivation and performance
3. Reduced incidences of premature retirement and death

Ball (2004) explained that organizations must appreciate that employees do not come to work as part of an economic transaction but more holistically as members of a society they wish to contribute. They and their families see themselves as part of the future in that community, such that they want to remain healthy and active, and to be able to engage in many activities. As such, their health at and off work becomes quite an important factor in their occupational and personal life. It was only until the mid-1990s that issues such as bullying at work, long working hours, and stress, which all contributed to psychosocial causes of ill health, became just as important to deal with as other chemical, biological, and physical (traditional) hazards at work in the United Kingdom. Ashton (2004) raised a very interesting point in his lecture when he argued that many of the illnesses are actually controlled by environmental, socioeconomic, and educational factors. He illustrated this in his discussion on the sharp decline in tuberculosis between 1838 and 1970 in England and Wales to be not based on medical interventions and cures given to patients because he argued that this happened before the specific treatment and prevention were known for it. He attributed this more to

1. Standards of living
2. Environmental conditions (the availability of cheap food in the urban areas, which meant it was more available to low-income people)
3. Developments in hygiene

Further on, Ashton (2004) spoke of the mechanistic solutions to general employee health management that was generally employed since the 1900's. With more knowledge on the subject and with epidemiological studies, we appreciate that the solutions were based on very little understanding of the problem. In saying this, and from a more social and corporate perspective, we must understand and appreciate *who pays, who benefits, and on whom do the costs fall*. Here, the displacement costs of the ill health, for example, a worker diagnosed with an occupational illness or injury, become redundant and although the compensation is paid perhaps by his/her employer, the state and perhaps the family will still have to deal with all the other indirect issues related to prolonged medical treatment, physiological issues such as depression, and many potential social problems. Here, we may argue therefore, that it is in the interest of the state to pursue policies of occupational ill health prevention deterministic goals.

To follow on from this, in the Gulf Corporation Council states in the Middle East, there is a high number of foreign workforce who work in a wide spectrum of industries. In the UAE and Qatar, the percentages are probably the highest among the others and some of the highest in the world. However, with greater awareness in the past decade, many of these states are formulating and promulgating regulations covering occupational health, communicable diseases incident reporting, and prevention as well as compensation. In Bahrain and Kuwait, worker/labor unions are also working closely with both the regulators and employers to address issues related to occupational health.

Bliss (2004) confirms that most general practitioners and OH physicians attribute that 85% of worker ill health was work related. The health and safety executive *securing health together* initiative had goals such as

1. Reduce ill health in workers and the public, where this is caused or aggravated by work.
2. Help people who have been ill to return to work.
3. Improve work opportunities for people currently unemployed due to ill health or disability.
4. Use the work environment to help people maintain their health.

With respect to solutions presented by many commentators on the subject of OH, it follows that effective OH management can help in

- Reduced productivity loss of workers
- Reduced cost associated with sick pay
- Reduced litigation costs
- Reduced medical insurance
- Better reduction management (i.e., prevention of root causes leading to) of medical and social issues such as drug abuse, alcoholism, smoking, domestic violence, and so on.
- Improved motivation of employees and their performance
- Healthier workforce in general with less absenteeism
- Reduced premature retirement and death

The strategies for making improvements through implementation of a successful occupational health strategy require

1. A holistic approach, seeing health in the total sustainable context, balancing economic, social, and environmental needs.
2. Leadership and championship of occupational health programs and initiatives that must be driven from the top down, demonstrating accountability and ownership.
3. A process must be in place that an organization implements effectively and rigorously, which must include
   a. Identification of occupational health issues through research and organizational issues.
   b. Development of a high-level strategy and policy for the occupational health awareness building—throughout the organization, both general and specific programs/campaigns and initiatives.
   c. Development of programs to implement the occupational health strategies.
   d. Implementation of programs, including educational, clinical, observational, and field-based measurements and site visits (i.e., industrial hygiene and ergonomic assessments, etc.).
   e. With the programs in place, these programs should be subject to internal audits and if the facilities are certified, external audits as well.
   f. Feedback and management reporting with strategic development and program refinement and development.

Further information can be obtained in a very comprehensive and practical document for the oil and gas industry, which is the *Guidance on Health Surveillance* (Energy Institute, 2010).

Occupational Health Management

The management of occupational health for an organization must be seen as a *preventative strategy* for the protection of employees from undue harm at work. However, the premise that an employee should go to work and return from work with no health impact is in itself a very difficult matter to ensure. Some of the reasons are that, especially in the high hazard and heavy industrial installation, the inherent hazards that exist in physical, biological, chemical, and physiological impacts are significant and as such those working within the industry will inevitability be exposed to occupational health risks. However, it is based on the *duty of care* principles in common law that organizations have to ensure that they have taken all reasonable care to prevent such impacts from adversely impacting on an employee's health and welfare.

A typical plan for preventing and managing occupational-related diseases includes the following:

1. Identify health-related risks.
2. Establish compliance with legislated requirements.
3. Determine testing requirements and frequencies.
4. Obtain testing equipment and resources.
5. Train employees to test, calibrate, and functionally maintain the testing equipment.
6. Test employees and explain the testing needs to employees. Employee buy-in and support are essential.
7. Test according to required frequency.
8. Share results with employees and teach or provide PPE as required.
9. Report results to regulators as required.
10. Maintain data for due diligence requirements.
11. Provide data to physicians and compensation boards of cases of occupational health disease treatment or for medical claims management.

Once an OH management system is established, the first question is where does occupational health management, organizationally as a function, belong? If we analyze it, then let us consider the following permutations.

Human Recourses Function

As discussed in the previous section, occupational health is very much a social aspect in its holistic perspective. The HR function is an employee compensation and benefit and employee services management function. Again, it has the functional duty as the facilitator for many of the occupational health-related matters such as preemployment medicals, in-employment medicals, suspect and random testing, counseling employees with occupational-related illnesses and injuries, management of sick leave that may have arisen or been aggravated by the job or otherwise, and also compensation packages in case of redundancy due to ill health arisen from occupational and nonoccupational reasons. It usually also has to deal with the most sensitive and challenging job of communicating with the family of those who die at work. It is with particular reference to the latter function that makes the HR department most unsuitable for hosting occupational health management. Generally, it is best for the leader of the business unit or operating area within which the fatality occurred to make deliver the news of the fatality. Furthermore, employee trust is eroded when HR dealing with OH issues. This is so because the HR department is traditionally viewed upon by employees as the judge and executioner from an employee perspective as it relates to worker issues and ultimately protecting the employers' position as their representative.

Risk Management/Productivity Function

Although this function does not exist in many organizations, it does in some that are usually large and progressive in the oil and gas industry and it comes under many different nomenclatures. This function usually looks at implementing policies that are directed toward increasing the company productivity and reducing the risks to the company. Inherently, in its nature it would take the organizational interests in the first instance. This is why at the outset it is probably not a fair function to have occupational health management under. OH ultimately, in the "prevention" paradigm, looks at improving and maintaining good occupational health of the employee to prevent lack of productivity (e.g., partial incapacitation, lowered output, days of work due to illness, and so on—even challenging presenteeism). This function is sometimes called the enterprise risk management (ERM) function.

Health, Safety, and Environment Function

This is probably technically one of the most suitable functions. This is in terms of its closeness to the technical aspects such as JSA, incident investigation, health at work, and so on. The growth in both the science and technology of industrial hygiene also makes this function most suited to manage functionally. However, there remains an issue of independence and also medical confidentiality. HSE departments generally have scientists, technicians, and engineers, and thus ultimately they are not qualified in the first instance to manage in the field of medicine. Finally, being ultimately a loss prevention function, both HSE and occupational health management share the core philosophy of existence. There are synergic effects of having HSE and occupational health work closely. The work of The Institute of Occupational Safety and Health in the United Kingdom and the regulatory body in the U.S. OSHA has developed best practice and standards relating to both occupational health and occupational safety together.

Independent Function

Ideally, the occupational health management should be independent in its policy development, program delivery, and analysis/diagnosis. This, however, is understandably, in many organizations, not a practical alternative. Some organizations that are large enough (over 10–15,000 employees) and have a primary medicine management function have implemented this. Such organizations include Saudi Aramco and Sabic in Saudi Arabia. Ultimately, it is too difficult to manage otherwise.

Outsource the Function

Many organizations, within the oil and gas industry, have outsourced this function. This was done for several reasons:

1. It is easier and cheaper to hire the expertise to conduct this function. However, this is usually limited to conducting medicals and maybe doing some onsite surveys (industrial hygiene).
2. The organization wants to portray both independence and objectivity by engaging a third party to make the evaluations, and so on.
3. With a limited scope (except with diversified operations), and with small companies, that is <3000–5000 employees, it is also cheaper to run through a third party—this and the expertise is as such hired on a "use and pay as you go basis." For more complex analysis, you just pay more and get the analysis you require.

In saying the above, many of the oil majors only engage certified occupational health service centers that execute UKO&G (previously known as UKOOA)-type medicals for fitness, especially for any upstream work.

Generally, the downside here is twofold. First, you are not managing occupational health; you are purely limiting the scope to medicals and hygiene assessments. Second, the focus and emphasis are on reaction more than prevention, although this depends highly on the type of contract with the service provider (e.g., case management). This, as we said at the beginning of this section, does not fully fulfill the primary objective, which is loss prevention. There is also the issue of *continuity* in terms of neither case management nor the building on the knowledge of the occupational health challenges facing the company at large, which becomes important for development of more effective strategies in the future. Figures 19.1 through 19.5 provide strengths/weaknesses/opportunities/threats (SWOT) analyses for locating OH in different organizational structures of the organization.

Functions of Occupational Health Management

For an occupational health function to be effective and to deliver an essentially loss prevention function and have a maximized employee and organizational value-adding proposition, it must have the following components:

1. A risk inventory evaluation system that comes in the form of a JSTA profile for each job. This is a system that uses the following: job

Strengths	Opportunities
o Very close ties with employee services. o HR are in charge of recruitment, renewal of rontracts and dismissal/retirement. o Holistic human resource function.	o Can be used to strengthen employee relations. o Linked directly to HR policies. o Studies can be done to justify manning levels and so on.
Weaknesses	**Threats**
o Not a technical specialist function. o Different type of operation in the function.	o Employees can see it as a function which is monitoring them. o Company Doctors perceived to be only working for the organization's interest.

FIGURE 19.1
SWOT analysis of OH as part of the human resource function.

Strengths	Opportunities
o Very strong technical association with the occupational safety function. o EHS generally works very closely with operational business units and employees. o Training, awareness development is part of the EHS core result areas. o Industrial hygiene function closely linked. o Usually reports within a function closely linked with the CEO and board.	o Can be more effective in running certain studies as a unit—for example, HSEIA studies which are more holistic. o Linked directly to EHS policies and more prevention of harm and loss. o Studies can be done jointly with occupational safety.
Weaknesses	**Threats**
o Many not be seen as totally independent. o May impact on EHS priorities in certain areas of work.	o Employees can see it as a function which is monitoring them. o Company doctors perceived to be only working for the organization's interest.

FIGURE 19.2
SWOT analysis of OH as part of the EHS function.

descriptions, employee and management discussions, field visits, and also industrial hygiene survey data. This would be encapsulated within a controlled and live system, which is updated as the jobs change and new facilities (infrastructure), processes, equipment, and substances handled change. This is well explained in the case of ENOC—see Ghanem and Newson-Smith (2008).

2. The function must have a policy development and program development capability. For this, it requires a qualified and competent person who is conversant with occupational medicine and although may not necessarily be a physician, must have an in-depth understanding of the testing protocols, international standards for medical evaluations, impact of substances on human health, ergonomics,

Strengths	Opportunities
o Good holistic risk management function. o ERM works closely with EHS which generally works very closely with operational business units and employees. o Usually reports within a function closely linked with the CEO and board.	o Can be more effective in running holistic risk management studies as a unit—for example, studies which are more organizationally holistic. o Can help place OH high on the agenda when measured against other risks especially in high risk industries.
Weaknesses	**Threats**
o Many not be seen as totally independent. o May impact on EHS priorities in certain areas of work due to higher other risks within an organization, especially commercial and business continuity risks.	o Employees can see it as a function which is monitoring them. o Company doctors perceived to be only working for the organization's interest.

FIGURE 19.3
SWOT analysis of OH as part of the enterprise risk management.

industrial toxicology, epidemiology, and so on. This function would develop the standards that would be applied either directly within the organization (if there are clinical facilities—either housed within a primary medicine function/facility or a stand-alone occupational health services clinic/center) or otherwise forms the detailed scope of work for evaluation that would be given in the annexure of a contract to a qualified service provider.

3. Industrial hygiene: The industrial hygienists can be either contracted or developed in-house depending on the volume and nature of the work (and also the prevailing local regulations for qualification of industrial hygienists, as their reports may be used as independent evidence-based information at times). In any case, standards for the same have to be developed and in most cases would refer to

Strengths	Opportunities
o Seen as a function which is independent of all other operations and departments within the group. o Can be perceived by the employees to be independent and therefore greater medical confidentiality.	o Can be more effective in running holistic risk management studies as a unit—for example, studies which are more holistic look at the whole organization similar to ERM. o Can help place OH high on the agenda when measured against other risks especially in high-risk industries.
Weaknesses	**Threats**
o Many not be seen as totally independent—directly reporting to top management. o Does not generally have the inter-relationship advantages of being part of a technical EHS function and so on.	o Employees can see it as a function which is monitoring them. o Company doctors perceived to be only working for the organization's interest.

FIGURE 19.4
SWOT analysis of OH as part of an independent function.

particular referenced standards within given limits of occupational exposure standards (OEL/MEL) such as those published by the UK HSE, the EH/40 standard, and the U.S.-based American Conference of Governmental Industrial Hygienists standards and others that are particular reference to certain areas such as the indoor air quality standards American Society of Heating, Refrigerating and Air-Conditioning Engineering, and so on. Even when contracted out, the organization would benefit from training and developing key staff such as the HSE staff to have an understanding on the interpretation of results and so on.

4. Incident investigation: The function should be able to be impartial and pragmatic, and will investigate incidents such as occupational health injuries, illnesses, and fatalities. These would also include communicable illness and diseases that can have an impact on the safety and health in operations, for example, medical fitness of those who are involved in food preparation and handling operations.

Strengths	Opportunities
o Seen as a function which is independent of all other operations and departments within the group. o Can be perceived by the employees to be independent and therefore greater medical confidentiality.	o Can gain greater buy-in from the organization if you are starting the function up. o Can be monitored more closely in terms of the costs.
Weaknesses	**Threats**
o Many not be seen as totally independent—directly reporting to top management. o The function is not a function within the organization and would be limited to clinical activities and may be training—for example, policy will require consultants. o Not many specialized occupational health practitioners and service providers available. o Lost confidentiality.	o No real proper case management. o Employees will always be concerned with how much of their confidential data is shared with the company and other people. o Continuity may be effected if the work is retendered and another hospital or service provider is appointed—must have strong record management. o No organizational learning.

FIGURE 19.5
SWOT analysis of OH as part of an outsourced function.

5. Training and development: For the occupational health program prevention to work and reap rewards, a critical level of knowledge throughout the workforce starting from the BoD right through to the plant operators and technicians is important. Introduction to occupational health programs for managers, supervisors, and employees would help raise awareness and understanding of how to avoid occupational health illness and injuries. These are also very important in demystifying the role of the occupational health policies and programs, gaining in many cases a greater deal of buy-in.

6. Awareness programs: General awareness programs such as seminars, leaflets, e-information, presentations, quizzes, and so on are very important. They can be both function/department specific or otherwise very general in nature. Although these take time to flow through the system, over time, they give all employees a greater appreciation of how they can help themselves (and in fact their families) prevent health problems. If done effectively, they reflect very positively on the organization that has invested in educating its employees and their families.
7. Every organization, regardless of how remotely positioned, operates within a community. Most oil and gas companies need to invest in educating the community and adding value. This can be, and is in fact extensively done by many oil companies, in terms of sponsorships. Organizations that can also help by organizing seminars and field trips for school and university students to help them understand more why occupational health is important are an ethical investment in both the present and the future.
8. Special assignments: Some specific and special assignments that are research related in a similar way to policy development are also important. Such examples would be setting up an effective medical insurance policy, executive health and medicals, medical review boards, presentations of technical research contributions at various conferences and seminars, and so on.

In ENOC, as an example, the above functions and responsibilities have been addressed and dealt with through the policy and procedures development function, clinical OH services center function, training and awareness development function, and the extensive industrial hygiene work. This has also been in close coordination with the environment and fire and safety sections within the group environment, health, safety and quality (EHSQ) compliance directorate.

Challenges of Setting Up an Occupational Health Function

There are many critical aspects that relate to the challenges of occupational health management systems, which include the following:

1. Integrating the function into business operations and functions. As discussed above, questions such as where to place the function, how to run the function, where it should report, limitations of authority,

medical confidentiality, accountability, imitations of accountability, and so on.

2. Producing a business case. This function, in whatever form it is developed, will cost the organization an initial development cost (making a structure, engaging consultants, surveys, discussions groups, seminars, etc.). Where do these costs come from and as discussed above based on which ROI and financial instrument/ model?
3. Occupational health is a philosophy that is borne from the organization's commitment to the health and welfare of employees and also, as explored, their families/community and even the society at large. It is therefore important and critical to be aligned with more than one function such as HSE, HR, enterprise RM, and so on. These functions at times are in themselves not fully aligned, which therefore adds substantially to the challenges that organizations face. Operationally, for the OH function to work effectively, the process management related to employee management has to be integrated with HR employee services, and so on.
4. Application in a globalized organization: Different standards are available in different jurisdictions, and therefore providing a unified strategy and policy needs to be developed in such a way that it sets minimum standards of expectation from the group management. However, it must be flexible enough for the standards to be integrated in such a way that they also meet local authority requirements in where these subsidiaries or operating units work. This requires careful attention from both a technical and legal perspective.

Occupational health management is an important function that is growing in its importance around the world. In the oil and gas industry, it has been recognized and manifests itself in different ways throughout the organization. The function is now recognized as a discipline within the industry. It is particularly important in the oil and gas industry. Different organizations address it differently and this depends not only on many organizational behavior factors but also very much on the nature of the industry itself. There is no doubt that much of the development has been through legal requirements, economic purpose, and sustainable social and employee welfare interest. In unionized states, the influence of unions has also ultimately played a role, like in HSE generally.

There are many challenges as well that are highly dependent on the company structure, legal formation, level of bureaucracy, progressive thinking, and so on. These are organization specific and need to be addressed individually. To overcome and implement any system of any shape or form will depend on the top management commitment and drive toward this and the right amount of time for the organizations to understand, appreciate, and buy-in and then implement the same.

Different organizations will need to implement different strategies for implementation of occupational health management, yet the steps required and stages of development as described in this paper would most probably be very similar for any organization except that only the time frames may differ. Finally, the positioning of the function as an independent department or one that is fostered through a conventional department is really dependent on the size and nature of the operation. In any case, the function must be *preventative* in its approach, must work closely with other sectors of the organization, must be holistic in value addition to both the employee and the organization, and most significantly must be very much independent.

20

Consistent Terminologies and Processes

Health and safety are indeed an area where common language and interpretation of data and information are essential to protect workers. Today, with a highly global and mobile workforce, this need is even greater. More importantly, when contractors as a work group make up 75–80% of the workforce as is the case in the oil and gas industry, it is imperative that organizations seek to develop common languages, terminologies, standards, procedures, and practices so as to minimize the confusion that these highly important groups of workers are exposed to.

Failure to do so results in this critical workgroup being made to learn the same thing over and over again under different names thereby leading to their confusion and frustration. In this chapter, we shall explore the need for a health and safety taxonomy and the areas where common terminologies, standards, procedures, and work practices may be essential. Let us begin by providing a simple definition of taxonomy and its application to health and safety.

In its simplest form, taxonomy is about classification. Where health and safety are concerned, taxonomy is essential for classifying the following in standardizing knowledge, to name a few:

1. Documentation hierarchy
2. Types of incidents
3. Role descriptions
4. Standards and procedures

Let us take a few moments to discuss the need for taxonomy in these areas.

Document Hierarchy

Often, confusion among workers is with regard to which takes precedence over the other. Is a standard more important than an SOP or vice versa? Figure 20.1 shows the hierarchy of documents in health and safety.

As shown in the figure, the law is the highest level of authority in documentation, and organizations will seek to comply with the law at all times. Conversely, check sheets and work tools are the lowest level of authority and

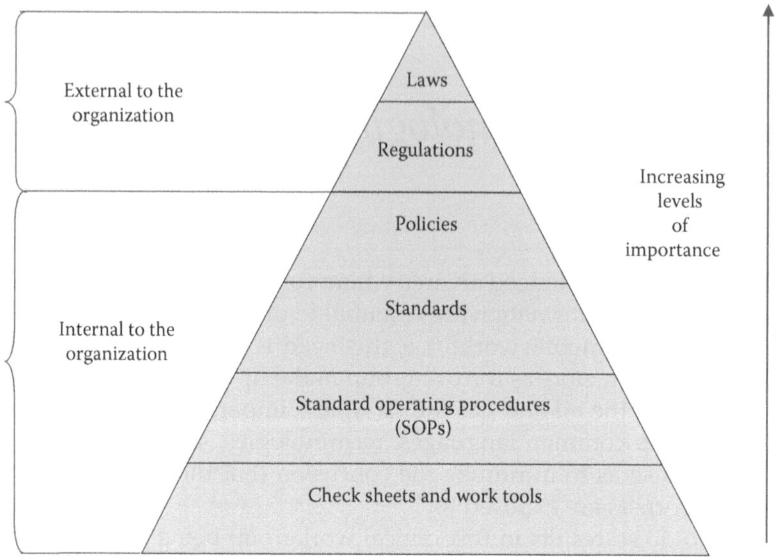

FIGURE 20.1
Hierarchy of documents in health and safety.

provide opportunities for organizations to be creative in its quest for optimization and efficiency in stewardship of processes.

Figure 20.1 also shows those documents that are internal to the organization as well as those external to the organization to which the organization must comply. Check sheets and work tools, SOPs, standards, and policies are intended to support an organization's desire to meet regulatory requirements and to comply with the law. This, to some extent, has limited or has hindered the effective implementation of health and safety in organizations. Many organizations still see that compliance with health and safety is a compliance issue only and not an effective way of doing business. More progressive organizations that have their SMS or PSM system as part of their organization's workings and performance would thus have better performance. As discussed in the previous chapters, this is why some have gone to integrating health and safety into operational excellence models rather than stand-alone systems that are audited separately and not in the context of the overall organizational workings.

Check Sheets and Work Tools

Check sheets and work tools are simple prompts to remind workers of all the activities that must be completed when doing any particular task. They are not mandatory; however, they provide a documented trail that the task was done properly and completely. Businesses must encourage the use of check sheets when performing task with multiple requirements. Check sheets are very effective in equipment inspection activities since they remind the worker of all the different activities that must be done during the equipment inspection.

Standard Operating Procedures

SOPs provide a step-by-step process for performing no-routine work and are designed to help personnel perform assigned work safely and effectively. SOPs provide a sequential set of steps to personnel such that when followed properly a worker who is not familiar with the equipment or process will be able to safely and effectively perform an assigned task. SOPs must be followed in all instances.

When difficulties in following the procedure are encountered, or a step is no longer valid, an appropriate level or authority is required to make the required changes to the procedure. Where errors are identified in the SOP, the use of the procedure must be stopped immediately and corrective actions taken. This is a very complex process and should not be taken lightly. Changing an SOP is not simply making the changes and moving on. Changes require that a change management process is followed to ensure all new hazards introduced in the process are identified and addressed. Additionally, all changes must be communicated to all users of the SOP; they must be trained and their competency assessed so that full due diligence is completed when changes to the SOP are made. The same applies when improvement changes are made to the SOP.

According to Lutchman (2010), an SOP will typically provide the user the following details:

1. A reference number and title for identifying and searching for the procedure
2. Date the SOP was written
3. Author of the SOP
4. Person(s) who reviewed the SOP
5. Revision number of the SOP
6. Page numbers of the SOP—normally numbered in the format page *x* of *y*
7. Steps involved in starting up, shutting down, or operating the equipment or machinery

8. Hazards identified
9. Precautionary measures and PPE required

Occasionally, an SOP may also include a hazard analysis assessment and recommendations for addressing each hazard identified. All steps in SOPs must be followed as outlined in the procedure.

Standards

Essentially, the standard identifies *what* is required by the business. *How* this is achieved may be dependent upon the business unit/area leadership. Standards do not require a standardized approach for achieving the *what* across the organization for compliance. Nevertheless, the outcome is important. The *what* must be consistently achieved across the entire organization. How this is achieved may be entirely up to the business unit leadership. Organizations often seek to standardize how results are achieved when effective processes are developed within a business unit that can be shared.

Standards are the absolute minimum that organizations require from their workers regarding various business activities in the organization. Standards stand above SOPs and reflect the way work is done for the particular business activity in the organization. When we consider PSM in Chapter 10, each element of PSM is a business process and a standard is required for each PSM element. As shown in Appendix 2, the standard covers several areas of business requirements and against which functional units of the organization must demonstrate compliance.

Requirements of the standard are often translated into SOP in order to guide personnel in fulfilling the intent of the standard. Standards are also designed to meet or surpass regulatory requirements and will often define the success of the business. When organizations develop standards that exceed regulatory requirements, they are sending a clear message to workers that the health and safety of its workers are paramount to organizational success and this cannot be compromised.

Policy

A corporate policy is a governing document that requires all workers to comply with. Policies advise workers, service providers, and contractors the compliance required for working with the organization. Examples of policies include drug and alcohol, cell phone usage during work, incident reporting, and SMSs requirements, to name a few. Policies are governing principles of the organization and must be interpreted and applied consistently across the entire organization.

Types of Incidents

The taxonomy around classifying incidents varies from across companies and between companies and contractors for obvious reasons. As discussed earlier, the health and safety performance of an organization speaks volumes about the organization and its corporate social responsibility image. Moreover, where contractors and service providers are concerned, a poor safety performance will determine the amount of work the organization will receive from owner organizations. As a consequence, there is unintended support for classifying incidents based on the impact of the classification on your business.

Classifying workplace injuries is an area for tremendous manipulation of data to the advantage of organizations. Table 20.1 shows some personal injuries taxonomy challenges that the oil and gas industry must contend with.

Resolving these challenges requires a united effort by industry leaders. However, many organizations are not prepared to even consider a standardized approach to classify these incidents since inconsistent interpretation provides opportunities for some organizational safety performance to appear better than they really are. Incentives for a standardized taxonomy

TABLE 20.1

Personal Injuries Taxonomy and Challenges

Type of Personal Incident	Taxonomy Concerns
Near miss	• There is no clear and consistent industry definition • Allows for interpretation and will often not be reported because of potential loss of work if severe outcomes were possible
First aid	• There is no clear and consistent industry definition • No consistent criteria for determining whether or not a first aid is considered a recordable injury—value judgment required
Recordable injury	• There is no clear and consistent industry definition • No consistent criteria for determining whether or not a first aid is considered a recordable injury—value judgment required
Loss time injury	• There is no clear and consistent industry definition • Can an injured worker be asked to perform another job and not be classified as an LTI?
Modified work	• What percentage of work performance is required to be classified as modified work?
Disabling injury	• Depends on expert medical and health care advice to be classified

are weak and may not gain support from business, given the shelter provided from the ability to interpret in an inconsistent manner.

Similar challenges exist when attempting to classify injuries. For example, when we attempt to classify hand injuries, what do we consider to be included as hand injuries? Does it include all injuries up to the shoulder (i.e., inclusive of the arm) or is it limited to the fingers and the hand area from the wrist to the tips of each finger? In most instances, this is left for interpretation by the organization.

Role Descriptions

An area of inconsistency for many industries is in the way roles are defined for similar responsibilities within industries. Consistency in role description is very important for many industries, particularly where a large mobile workforce (such as a contracting workforce) is used within an industry. For regions such as the oil sands of Alberta, where multiple oil and gas producers are involved in macrooperating and construction projects and are all dependent upon the same contracting workforce, consistency in language helps to eliminate confusion among contract workers.

More importantly, consistency in work practices helps organizations and contractors better address safety concerns around work. An opportunity for industry enhancement lies in the interface management between contractors and owners at the frontline where hands on work are performed and where most confusion and risk potential exist. Managing the interface requires the role of a contract coordinator with consistent responsibilities within the industry. Consistent responsibilities for the contract coordinator will provide opportunities for contractors to perform work in a safer way and with reduced confusion.

When defined consistently, the role of the contract coordinator can be standardized to reflect:

1. A single point of contact for the contractor at the frontline where hands on work is performed.
2. Addressing risks related to contractor activities at the field locations.
3. The application of consistent practices and direction and field performance while monitoring the contractor's obligations as defined in the contractual agreements.

Specific duties of the contract coordinator may include the following:

- Acts or serves as a single point of contact within the owner's organization at the frontline that interfaces with and monitors contractor safety performance on a daily basis.

- Develops strong working relationships with contractors and service providers at the field levels and provides field supervision to ensure that continuous work is performed safely at all times.
- Participates in delivering key messages to contractors that ensure alignment in owner expectations with the contractor.
- Serves as the primary interface between field operating leadership, contractor supervision, and health and safety at the frontline.
- Works jointly with health and safety experts to advise and inform senior site leadership on safety concerns relating to contractor performance.
- Participates in and leads, when necessary, safety meeting beneficial to the contractor at the frontline.
- Supports and responds to all internal and external contract inquiries, issues, and risks.
- Identifies and supports the contract change request process when triggered.
- Ensures contractors are viewed as valued partners.
- Monitors the performance of owner and of suppliers related to contract terms.
- Reports all contract compliance discrepancies, highlighting repeat offences, to the owner leadership.
- Maintains a visible performance scorecard that addresses safety, quality, and schedule performance.
- Where necessary, provides access to information such as owner policies, SOPs, and other relevant information required by contractors to perform assigned tasks safely.
- Coordinates the contractor performance evaluation and closeout activities upon completion of a contract.
- Acts as the point of collection for shared learning in contractor safety management.

In the aforementioned example, when roles and responsibilities are standardized within industries, tremendous improvements in health and safety and efficiency in performance can be derived.

Standards and Standard Operating Procedures

Standards and SOPs provide excellent opportunities for organizations to improve health and safety within industries. This is particularly important

TABLE 20.2

Benefits and Challenges of Consistently Applied Standards and SOPs

Benefits	Concerns
• Standardized and consistent training; training can be outsourced for the industry • Consistent safety behaviors within a mobile workforce	• Not every organization aspires for the same standard. Meeting regulatory requirements may be adequate • Meeting higher standards can be costly. Different owners may not be prepared to accept the same cost associated with higher standards
• Removes confusion among contractor workforce	
• All workers required to meet the same standards regardless of the place of work	• Differences in standards and SOPs may lead to competitive advantages that organizations may seek to exploit themselves
• Industry improvements in health and safety performance	

for industries such as the oil and gas industry. Consistency on standards and SOPs drive consistent safe behaviors within the industry since all workers will be required to meet the same standard and follow the same/similar SOP when similar work is to be done.

Consistent standards and SOPs have substantial benefits for efficiency improvements from a training and development perspective. In addition, with consistent standards and SOPs, real differences between organizations within an industry begin to appear thereby providing opportunities for continuous improvements and gap closure strategies. More importantly, a mobile contractor workforce will not be required to relearn things differently from one site to another. Table 20.2 provides the benefits and challenges of a consistent set of industry standards and SOPs.

Addressing these challenges may require a consortium approach to resolution and may be best handled by an independent organization intimately aware of the benefits and challenges associated with similar industry standards. A huge effort will be required to gain support from organizations to undertake such an effort. Until this is viewed by organizations as an industry gap, consistent standards and SOPs within an industry will continue to be an unexploited opportunity.

21

Conclusion

Workplace health and safety are of tremendous importance to every organization and industry. Health and safety provide one of the greatest opportunities for enhancing value creation in the workplace through direct and indirect cost reduction, reduced numbers of incidents, and the unquantifiable benefits of a motivated and safe workforce. Providing a healthy and safer workplace is not only a legal requirement but also a moral obligation of all leaders, and deliberate failure to do so should not go unpunished.

Our global business environment is continually evolving and changing. As we have seen from the 2010 BP Gulf of Mexico spill, with changing business environments come changing risks, hazards, and consequences. Leaders are challenged to accept more risks and to make faster decisions. Failure to do so generally results in reduced competitive advantage, which can ultimately affect the bottom line performance of the organization. However, all it takes is one major incident to dramatically impact on the fortunes of an organization as we saw with BP.

To support organizational growth and venture into unchartered territories, therefore, there is greater pressure on health and safety professionals to ensure that leaders are not only aware of the associated risks but are also educated on the most effective solutions so that accidents and incidents can be prevented. More importantly, safety professionals and leaders must belong to that group of workers who must be classified as continuous learners in order to function properly and effectively in their roles.

Undoubtedly, there is tremendous room for improvement in workplace health and safety. However, such improvements start with leadership commitment to do something different and stand out against industry peers. To be different is often a very challenging and difficult decision for any leader, particularly when shareholder interests tend to focus primarily on the profits and net earnings generated by the organization. However, when senior leaders and executives recognize the long-term benefits of a healthy and safe workplace, there is little resistance to climbing onboard and doing it right.

In this book, we discussed health and safety in the workplace from a practical perspective and attempted to focus on areas of opportunities where real improvements can be made to improve the health and safety of our workforce. From our perspectives, we saw real opportunities for improving the health and safety of our workforce in the following key areas:

1. Select leaders with a strong commitment to health and safety or better educate leaders about the benefits of health and safety in the workplace—creating genuine leadership commitment to health and safety as opposed to doing so from a convenience standpoint.
2. Promote and upgrade the way learnings are shared in organizations bearing in mind >90% of incidents are repeated—we can do a better job in the way we share learning within organizations, within industries, and across industries.
3. Improving understanding of PSM and enhancing the ways PSM is integrated into business processes and operations.
4. Improving the way contractors are managed in organizations—the way contractor safety is managed in organizations. The need to recognize contractors as our greatest business partners; they want to do a good job and require collaboration, support and access to information, and knowledge to perform their work safely.
5. Shifting the mindset of leaders around health and safety audits—the opportunity to shift mindsets regarding audits from policing and punitive to collaborative, proactive gap identification and closure in a nonfault-finding way.
6. Improving the skills and competencies of frontline supervisors and leaders—frontline supervisors and leaders provide one of the best opportunities for improving health and safety in any business. However, when we examine the way we develop the leadership skills and competencies of our frontline leaders there are gaping holes that must be addressed.
7. Transition focus in the way we measure and reward improvement in health and safety from lagging indicator—rearview mirror focus to forward-looking leading indicator focus.
8. Creating a work environment whereby a no-blame culture is created—the need to create a no-blame safety culture such that near misses and incidents are brought forward early such that constructive solutions can be developed and shared across the organization.
9. Developing a common language and taxonomy for health and safety—collusion for profits is unacceptable; however, collusion for improvements of the health and safety of workers must regarded as one of the most sacred rules in business. Knowledge in health and safety in many organizations is said to be nonproprietary and organizations are willing to work together for genuine improvements. An organized approach to a common taxonomy for health and safety will bring tremendous value to all industries.

As we all know, there is no single solution for improving health and safety in the workplace. As leaders, we must continually work to find better and safer ways to do work. It is not acceptable to say incidents are a part of the cost of doing business. Instead, and in the words of a highly successful business leader Rick George, CEO of Suncor Energy Inc., our mantra should be: *Nothing is more important than protecting ourselves and others from harm. That means safety is never compromised. If we can't do it safely, we don't do it.*

We sincerely hope that you have found this information useful and valuable and that you may be able to take away one bit of knowledge that will eventually impact on your behavior and that of others you may influence. Your health and safety and that of your peers and coworkers are important and should never be compromised. Always seek ways to improve workplace health and safety. We wish you and your coworkers a healthy and safe journey throughout your careers and in any personal work you may ever undertake. Remember health and safety must be practiced both at home and in the workplace and at all times.

*Appendix 1: Contractor or Service Provider Prequalification Questionnaire**

Company Information 1

★1.1 Legal Company Name:

2	410578	37	Edit	

★1.2 Name of person who completed this questionnaire?

2	410578	635	Edit	

★1.3 Phone number of person who completed this questionnaire?

2	410578	603	Edit	

★1.4 Email of person who completed this questionnaire?

2	410578	604	Edit	

★1.5 What year was your company established?

2	410578	53	Edit	

★1.6 Select your form of business:

2	410578	63	Edit

- ○ Corporation
- ○ LLC (Limited Liability Corporation)
- ○ LLP (Limited Liability Partnership)
- ○ Partnership
- ○ Sole Owner
- ○ AG
- ○ GmbH
- ○ Plc
- ○ Ltd
- ○ Ultd
- ○ ULC
- ○ Other

★1.7 What is your parent company's name?

2	410578	48	Edit	

★1.8 What is your subsidiary company's name(s)?

2	410578	49	Edit

* Approval from Pacific Industrial Contractor Screening (PICS) required to reproduce this data.

★1.9 Has your company, or the owners of your company, operated under a different name in the last three years?

2	410578	52	Edit

○ Yes

○ No

★1.10 If you answered YES to the previous question, what was the name & location of the company?

2	410578	54	Edit

★1.12 Has your company, or any officer of your company, ever been charged, indicted, convicted, or fined for any kind of offense?

2	410578	1462	Edit

○ Yes

○ No

★1.13 If yes, please explain:

2	410578	1463	Edit

Company Information 1.1

Physical Address:

★1.1.1 Street

1141	410578	38	Edit	

★1.1.2 City:

1141	410578	39	Edit	

★1.1.3 State/Province:

1141	410578	40	Edit

★1.1.4 Zip/Postal Code:

1141	410578	41	Edit	

★1.1.5 Country:

1141	410578	42	Edit

Mailing Address:

★1.1.6 Mailing address is same as physical address

1141	410578	871	Edit

○ Yes

○ No

★1.1.7 Street/P.O. Box

1141	410578	646	Edit

★1.1.8 City

1141	410578	647	Edit

★1.1.9 State/Province:

1141	410578	648	Edit

★1.1.10 Zip/Postal Code

1141	410578	822	Edit

★1.1.11 Country

1141	410578	649	Edit

★1.1.12 Telephone:

1141	410578	43	Edit

★1.1.13 Fax:

1141	410578	44	Edit

★1.1.14 Website:

1141	410578	47	Edit

Company Contacts 1.2

★1.2.1 President:

376	410578	72	Edit

★1.2.2 Vice President:

376	410578	73	Edit

★1.2.3 Operations Manager:

376	410578	74	Edit

★1.2.4 Safety Manager:

376	410578	75	Edit

Contact info for the highest ranking safety professional in the company:

★1.2.5 Name:

376	410578	76	Edit	

★1.2.6 Title:

376	410578	1452	Edit	

★1.2.7 Phone number:

376	410578	622	Edit	

★1.2.8 Address:

376	410578	623	Edit

★1.2.9 Email:

376	410578	624	Edit	

Contact info for the highest ranking environmental professional in the company (if different from above):

★1.2.10 Name:

376	410578	1436	Edit	

★1.2.11 Title:

376	410578	1453	Edit	

★1.2.12 Phone:

376	410578	1438	Edit	

★1.2.13 Address:

376	410578	1439	Edit

★1.2.14 Email:

376	410578	1437	Edit	

Contact information of individual to speak to regarding job bids:

★1.2.15 Name:

376	410578	61	Edit	

★1.2.16 Phone:

376	410578	629	Edit	

★1.2.17 Email:

376	410578	630	Edit	

NAICS 1.3

★1.3.1 What is your company's primary NAICS Code?

1139	410578	57	Edit	

★1.3.2 What is your company's secondary NAICS Code?

1139	410578	58	Edit	

US Company Information 1.4

★1.4.2 State of Incorporation:

1135	410578	96	Edit

★1.4.5 When did your company begin operation?

1135	410578	95	Edit	

example: 12/31/1999

★1.4.6 EEO (Equal Employment Opportunity) Category?

1135	410578	70	Edit	

★1.4.7 What percentage of your services are performed for the.U S. Government?

1135	410578	65	Edit	

Other Information 1.7

Which of the following personnel do you normally employ?

★1.7.1 Union Personnel

375	410578	66	Edit	☐	X

★1.7.3 Non-union personnel

375	410578	77	Edit	☐	X

★1.7.5 If you answered Union to the previous question, please list collective bargaining agreements and expiration dates:

375	410578	68	Edit

★1.7.7 List all Provinces/territories/states/regions licensed to work in:

375	410578	3247	Edit

★1.7.8 Is your company ISO certified?

375	410578	93	Edit

○ Yes

○ No

★1.7.9 If your company is ISO certified, please list level of certification and any specific details (including expiration date of certification):

375	410578	94	Edit

★1.7.10 Is your company OHSAS 18001 certified?

375	410578	2909	Edit

○ Yes

○ No

Company Work History 2

★2.1 Amount of largest job during the last three (3) years?

6	410578	84	Edit

★2.2 What is your firm's desired minimum project size?

6	410578	91	Edit

★2.3 What is your firm's desired maximum project size?

6	410578	92	Edit

★2.4 Are there any judgments, liens, claims of any nature, or lawsuits pending or outstanding against your company?

6	410578	86	Edit

○ Yes

○ No

★2.5 If you answered 'yes' to the above question please describe in detail:

6	410578	87	Edit

★2.6 Are you currently involved in any bankruptcy proceedings?

6	410578	88	Edit	○ Yes

○ No

★2.7 Are there any labor issues that could interfere with your work?

6	410578	1440	Edit	○ Yes

○ No

★2.8 If yes, please explain. Be sure to identify which of the PICS consortium sites may be affected by this labor issue.

6	410578	1441	Edit

★2.9 Have you ever been involved in any bankruptcy proceedings?

6	410578	89	Edit	○ Yes

○ No

★2.10 Are you currently involved in any reorganization proceedings?

6	410578	90	Edit	○ Yes

○ No

Major jobs in progress or completed in the last 3 years

★2.11 Customer/Location

6	410578	664	Edit	

★2.12 Type of Work You are Performing:

6	410578	665	Edit	

★2.13 Project Size ($)

6	410578	666	Edit	

★2.14 Customer Contact

6	410578	667	Edit	

★2.15 Telephone

6	410578	668	Edit	

★2.16 Email

6	410578	814	Edit	

★2.17 Customer/Location

6	410578	669	Edit	

★2.18 Type of Work You are Performing:

6	410578	670	Edit	

★2.19 Project Size ($)

6	410578	671	Edit	

★2.20 Customer Contact

6	410578	672	Edit	

★2.21 Telephone

6	410578	673	Edit	

★2.22 Email

6	410578	815	Edit	

General Safety Information 4

★4.1 List any safety awards your organization received in the last three (3) years that you would like noted:

8	410578	1460	Edit

★4.2 Does your company have a policy pertaining to criminal background checks on employees?

○ Yes

○ No

★4.3 What types of background checks does your company perform?

★4.4 Does your company use formal safety committees?

○ Yes

○ No

★4.5 Does your company have a written safety incentive program?

○ Yes

○ No

★4.6 Does your company provide paid safety/health/environmental training?

○ Yes

○ No

★4.7 Does your company conduct on-site and equipment inspections?

○ Yes

○ No

○ NA

★4.8 Do you have a preventive maintenance program for tools and equipment?

○ Yes

○ No

○ NA

★4.9 Does your company have a policy stating that no weapons or firearms of any type are allowed on the worksite?

○ Yes

○ No

US Safety Information 4.1

★4.1.1 Does your company actively participate in the Transportation Worker Identification Credential (TWIC) program?

○ Yes

○ No

Citations 4.3

★4.3.1 Has your company received any inspections from a regulatory agency during the last three (3) years?

○ Yes

○ No

★4.3.2 Has your company experienced any fatalities during the last three (3) years?

○ Yes

○ No

★4.3.3 If you answered YES to the previous question, please provide details. Was a citation issued? By what agency? If there was a lawsuit filed, provide the name, court, number, and status of the lawsuit.

★4.3.4 Are there any lawsuits filed relating to the recorded fatalities?

○ Yes

○ No

★4.3.5 If yes, please provide the name of the lawsuit, the court, the number, and the status.

Medical 4.4

★4.4.1 Does your company conduct medical examinations for pre-employment?

○ Yes

○ No

★4.4.2 Does your company conduct medical examinations for re-employment?

○ Yes

○ No

★4.4.3 Does your company conduct medical examinations for hearing function (audiograms)?

○ Yes

○ No

★4.4.4 Does your company conduct medical examinations for vision?

○ Yes

○ No

★4.4.5 Does your company conduct medical examinations for Pulmonary Function (PFT)?

○ Yes

○ No

★4.4.6 Do you have personnel trained to perform first aid and CPR?

○ Yes

○ No

★4.4.7 Describe how you will provide first aid and other medical services for your employees while on site, and please specify who will provide this service:

★4.4.8 On large projects do you employ a paramedic, nurse, or physician of occupational medicine? Experience at the worksite?

○ Yes

○ No

○ NA

Risk Assessment 4.5

★4.5.1 Does your company perform mechanical services that require the use of heavy machinery?

○ Yes

○ No

★4.5.2 Does your company perform mechanical services that require the use of hand/power tools?

○ Yes

○ No

★4.5.3 Can all the services you provide be conducted from an office setting?

○ Yes

○ No

★4.5.4 Do you perform services that would require your employees to wear fall protection or provide for fall protection safety measures?

○ Yes

○ No

★4.5.5 What risk level do you believe your company should be rated? See examples below:

- Low—Delivery, janitorial, off site engineering, security, computer services, etc.

- Medium—On site engineering, safety services, landscaping, inspection services, etc.
- High—Mechanical contractor, remediation, industrial cleaning, general construction, etc.

○ Low

○ Medium

○ High

Work Orders and Practices 4.6

Please identify type of stop work orders and/or fines (check all that apply):

4.6.1 Administrative Fines

☐

4.6.2 Convictions

☐

4.6.3 Stop Work Orders

☐

★4.6.4 Are there any HSE-related judgments, claims or suits pending or outstanding against your company?

○ Yes

○ No

★4.6.5 Are workers informed of the job/task specific hazards?

○ Yes

○ No

Do you have specialized rules/practices in place for the following: (check all that apply):

4.6.6 All Terrain Vehicles

☐

4.6.7 Anti-virus

☐

4.6.8 Barricading Flagging

☐

★4.6.13 Has your company been involved in any reportable spills or releases in the past three years?

○ Yes

○ No

★4.6.14 If yes, please provide additional information.

Health, Safety, and Environmental Policies/Procedures 5

HSE Programs 5.1

★5.1.1 Does your company have a written Health, Safety, & Environmental Program / Policies (safety manual)?

○ Yes

○ No

★5.1.2 If you answered yes to the previous question, what is the documented date of your last revision to the safety manual? Your manual should indicate either a header or footer with the specified date and/or a management of change page which indicates the last revision date.

example: 12/31/1999

★5.1.3 Does your program address management commitment and expectations?

○ Yes

○ No

★5.1.4 Does your program address employee participation?

○ Yes

○ No

★5.1.5 Does your program address accountability and responsibility for managers, supervisors, and employees?

○ Yes

○ No

★5.1.6 Does your program address resources for meeting health and safety requirements?

○ Yes

○ No

★5.1.7 Does your program address periodic health and safety performance appraisals for all employees?

○ Yes

○ No

★5.1.8 Does your program address hazard recognition and control?

○ Yes

○ No

★5.1.9 Does your company have a written Abrasive Blasting/Sandblasting program? (Ref: NIOSH Pub. 92.102)

○ Yes

○ No

○ NA

★5.1.10 Does your company have a written Accident Reporting, Investigation, and Root Cause Analysis program?

○ Yes

○ No

★5.1.11 Does your company have a written Asbestos Awareness program?

○ Yes

○ No

○ NA

★5.1.12 Does your company have a written Behavioral Based Safety Program? If YES, what program are you using?

○ Yes

○ No

Comments:

★5.1.13 Does your company have a written Chemical Exposure Awareness program?

○ Yes

○ No

○ NA

★5.1.14 Does your company have a written Bloodborne Pathogens program?

○ Yes

○ No

○ NA

★5.1.15 Does your company have a written Compressed Gas Cylinder program?

○ Yes

○ No

○ NA

★5.1.16 Does your company have a written Concrete Construction program?

○ Yes

○ No

○ NA

★5.1.17 Does your company have a written Confined Space program?

○ Yes

○ No

○ NA

★5.1.18 Does your company have a written Crane Operator Program to all required updates per Fed OSHA 1926.1427?

○ Yes

○ No

○ NA

★5.1.19 Does your Crane Program address qualifications for assembly/dis-assembly personnel and assessment of Ground conditions?

○ Yes

○ No

○ NA

★5.1.20 Does your Crane Program address inspections procedures for applicable equipment?

○ Yes

○ No

○ NA

★5.1.21 Does your Crane Program address power line safety?
○ Yes
○ No
○ NA

★5.1.22 Does your Crane Program address signals and signal persons qualifications?
○ Yes
○ No
○ NA

★5.1.23 Does your Crane Program address Fall Protection requirements?
○ Yes
○ No
○ NA

★5.1.24 Does your Crane Program address Operator Qualifications and Certification Requirements?
○ Yes
○ No
○ NA

★5.1.25 Does your company have a written Demolition program?
○ Yes
○ No
○ NA

★5.1.26 Does your company have a written Disciplinary program?
○ Yes
○ No

★5.1.27 Does your company have a written Diving program?
○ Yes
○ No
○ NA

★5.1.28 Does your company have a written Electrical Equipment Grounding Assurance program?

○ Yes

○ No

○ NA

★5.1.29 Does your company have a written Electrical Safety program?

○ Yes

○ No

○ NA

★5.1.30 Does your company have a written Elevated Work Platform (Aerial Lift) program?

○ Yes

○ No

○ NA

★5.1.31 Does your company have a written Emergency Prepardness program?

○ Yes

○ No

★5.1.32 Does your company have a written Fall Protection program?

○ Yes

○ No

○ NA

★5.1.33 Does your company have a written Fire Protection program, including portable fire extinguishers?

○ Yes

○ No

○ NA

★5.1.34 Does your company have a written First Aid/CPR program?

○ Yes

○ No

○ NA

★5.1.35 Does your company have a written Forklift program?

○ Yes

○ No

○ NA

★5.1.36 Does your company have a written Ground Disturbance Level 1 or 2 program?

○ Yes

○ No

○ NA

★5.1.37 Does your company have a written Harassment Policy?

○ Yes

○ No

★5.1.38 Does your company have a written Hazardous Waste Operations and Emergency Response program?

○ Yes

○ No

○ NA

★5.1.39 Does your company have a written Hearing Conservation program?

○ Yes

○ No

○ NA

★5.1.40 Does your company have a written Heavy Equipment program?

○ Yes

○ No

○ NA

★5.1.41 Does your company have a written Housekeeping program?

○ Yes

○ No

★5.1.42 Does your company have a written program requiring your employees to conduct a Hazard Analysis before every task? For example a Job Safety Analysis (JSA) /Job Hazard Analysis (JHA)/Task Hazard Analysis (THA)

- ○ Yes
- ○ No
- ○ NA

★5.1.43 Does your company have a written Lead, Cadmium, Metals Exposure program?

- ○ Yes
- ○ No
- ○ NA

★5.1.44 Does your company have a written Lifting/Mobile Equipment/Crane program?

- ○ Yes
- ○ No
- ○ NA

★5.1.45 Does your company have a written Lockout/Tagout program?

- ○ Yes
- ○ No
- ○ NA

★5.1.46 Does your company have a written PPE program?

- ○ Yes
- ○ No

★5.1.47 Does your company have a written Portable Electrical Power Tools program?

- ○ Yes
- ○ No
- ○ NA

★5.1.49 Does your company have a written Process Safety Management (PSM) program?

- ○ Yes
- ○ No
- ○ NA

★5.1.50 Does your company have a written Respiratory program?

- ○ Yes
- ○ No

○ NA

★5.1.51 Does your company have a written Rigging program?

○ Yes

○ No

○ NA

★5.1.52 What classifications or job descriptions perform rigging operations within your company? e.g., Boilermaker/Rigger, etc.

★5.1.53 Does your company have a written Scaffolding & Ladder User program?

○ Yes

○ No

○ NA

★5.1.54 Does your company have a written Steel Erection program?

○ Yes

○ No

○ NA

★5.1.55 Does your company have a written Trenching/Shoring/Excavation program?

○ Yes

○ No

○ NA

★5.1.56 Does your company have a written Vehicle Safety/Defensive Driving program?

○ Yes

○ No

★5.1.57 Does your company have an inspection process for any company owned vessels/totes?

○ Yes

○ No

○ NA

★5.1.58 Are the inspection records for the vessels/totes available on-site for review upon request?

○ Yes

○ No

○ NA

★5.1.59 If NO, where are the records located?

★5.1.60 Does your company have a written Water Survival/Working Over Water program?

○ Yes

○ No

○ NA

★5.1.61 Does your company have a written Welding, Cutting & Hot Work program?

○ Yes

○ No

○ NA

US HSE Programs 5.1.1

★5.1.1.1 If your company works in the State of California, do you have an Injury, Illness, Prevention Program (IIPP)?
Cal/OSHA Injury & Illness Prevention Program

○ Yes

○ No

○ NA

★5.1.1.2 Does your program satisfy your responsibility under the law for ensuring your employees follow the safety rules of the facility?

○ Yes

○ No

★5.1.1.3 Does your program satisfy your responsibility under the law for advising the owner/operator of any unique hazards presented by the contractor's work and any hazards found by the contractor?

○ Yes

○ No

★5.1.1.4 Does your company's Electrical program include NFPA 70E?
NFPA 70E

- ○ Yes
- ○ No
- ○ NA

★5.1.1.5 Does your company have a written H2S Awareness Specific program?

- ○ Yes
- ○ No
- ○ NA

★5.1.1.6 Does your company have a written HAZCOM program?

- ○ Yes
- ○ No
- ○ NA

★5.1.1.7 Does your company have a written HAZWOPER/Emergency Response program?

- ○ Yes
- ○ No
- ○ NA

★5.1.1.8 Does your company have a written NORM (Naturally Occurring Radioactive Material) program?

- ○ Yes
- ○ No
- ○ NA

★5.1.1.9 Does your company have a written PSM (Process Safety Management) program?

- ○ Yes
- ○ No
- ○ NA

Insurance 6

General Insurance Information 6.1

★6.1.1 In the last five years has any insurance carrier, for any form of insurance, refused to renew the insurance policy for your firm?

○ Yes

○ No

★6.1.2 If YES, list the name the insurance carrier, the form of insurance, and the year of the refusal.

★6.1.3 Within the last five years has there ever been a period when your firm had employees but was without workers' compensation insurance or self-insurance?

○ Yes

○ No

★6.1.4 If YES, please explain the reason for the absence of workers' compensation insurance.

★6.1.5 Are you self insured for Worker's Compensation Insurance?

○ Yes

○ No

Your company contact person for insurance information:

★6.1.6 Name:

★6.1.7 Phone number:

★6.1.8 Email:

Insurance Carrier(s)

★6.1.9 Name:

★6.1.10 Type of Coverage:

★6.1.11 Telephone:

★6.1.12 Name:

★6.1.13 Type of Coverage:

★6.1.14 Telephone:

★6.1.15 Name:

★6.1.16 Type of Coverage:

★6.1.17 Telephone:

US General Insurance Information 6.1.1

★6.1.1.1 What are your general liability limits?

★6.1.1.2 Does your company have a modified duty/return-to-work program?

○ Yes

○ No

★6.1.1.3 Is your company self-insured?

Training 7

★7.1 Are employees' job skills certified where required by regulatory or industry concensus standards?

○ Yes

○ No

★7.2 Do you have a Health, Safety & Environmental orientation program for new hires?

○ Yes

○ No

★7.3 Do you have training records for your employees?

○ Yes

○ No

★7.6 Does your company have quantifiable training requirements (job specific) for new employees or experienced employees new to their position?

○ Yes

○ No

US Training 7.1

★7.1.1 Do you have training/certification programs for certifying Craftsmen?

○ Yes

○ No

○ NA

★7.1.2 Do you have practical (hands-on) testing programs for certifying Craftsmen?

○ Yes

○ No

○ NA

★7.1.3 Is each employee instructed in the known potential of fire, explosion, or toxic release hazards related to his/her job, the process, and the applicable provisions of the emergency action plan?

○ Yes

○ No

○ NA

★7.1.4 Do you have a specific health, safety, and environmental training program for supervisors?

○ Yes

○ No

○ NA

★7.1.5 Do you have written Workforce Development Policies and Procedures?

○ Yes

○ No

○ NA

★7.1.6 Does your company have a Short Service Employee (SSE) policy that identifies new employees or experienced employees new to your company or new in their position?

○ Yes

○ No

★7.1.7 If YES, does your SSE policy include a means to visually identify a SSE?

○ Yes

○ No

○ NA

★7.1.8 If YES, does your SSE policy include a mentor being assigned to the SSE?

○ Yes

○ No

○ NA

★7.1.9 If YES, does your SSE policy define the roles and responsibility of the SSE mentor?

○ Yes

○ No

○ NA

Communications 11

★11.1 Do your employees read, write, and understand English such that they can complete on-site safety training and perform their job tasks safely, without an interpreter?

○ Yes

○ No

★11.2 If you answered NO to the previous question, please provide a description of your plan to assure that they can safely perform their jobs:

★11.3 If language barriers are an issue, are they addressed in your HSE programs?

○ Yes

○ No

○ NA

★11.4 If YES, what is the common language?

★11.5 In what language(s) is/are your HSE policies and procedures written?

★11.6 Has your company had an incident(s) where a causal factor was a language barrier/issue?

○ Yes

○ No

★11.7 If YES, please describe what was done to address language barriers/issues as a causal factor:

Personal Protective Equipment 12

★12.1 Does your company perform equipment checks on all PPE?

○ Yes

○ No

★12.2 Does your company PROVIDE Fire Retardant Clothing (FRC)?

○ Yes

○ No

○ NA

★12.3 When applicable, does your company REQUIRE Fire Retardant Clothing (FRC)?

○ Yes

○ No

★12.4 Does your company PROVIDE Chemical Protective Equipment?

○ Yes

○ No

○ NA

★12.5 When applicable, does your company REQUIRE Chemical Protective Equipment?
○ Yes

○ No

★12.6 Does your company PROVIDE Welding Garments?
○ Yes

○ No

○ NA

★12.7 When applicable, does your company REQUIRE Welding Garments?
○ Yes

○ No

★12.8 Do you provide required PPE at no charge to the employee?
○ Yes

○ No

US Personal Protective Equipment 12.1

★12.1.1 Does your company PROVIDE eye protection (ANSI-Z87.1)(29 CFR 1910.133 & 1926.102)?
○ Yes

○ No

○ NA

★12.1.2 When applicable, does your company REQUIRE eye protection (ANSI-Z87.1)(29 CFR 1910.133 & 1926.102)?
○ Yes

○ No

★12.1.3 Does your company PROVIDE fall protection (29 CFR 1926.502, 104 & 105)?
○ Yes

○ No

○ NA

★12.1.4 When applicable, does your company REQUIRE fall protection (29 CFR 1926.502, 104 & 105)?

○ Yes

○ No

★12.1.5 Does your company PROVIDE hard hats (ANSI-Z89.1)(29 CFR 1910.135 & 1926.100)?

○ Yes

○ No

○ NA

★12.1.6 When applicable, does your company REQUIRE hard hats (ANSI-Z89.1)(29 CFR 1910.135 & 1926.100)?

○ Yes

○ No

★12.1.7 Does your company PROVIDE hand protection (29 CFR 1910.138)?

○ Yes

○ No

○ NA

★12.1.8 When applicable, does your company REQUIRE hand protection (29 CFR 1910.138)?

○ Yes

○ No

★12.1.9 Does your company PROVIDE hearing protection (29 CFR 1910.95 & 1926.101)?

○ Yes

○ No

○ NA

★12.1.10 When applicable, does your company REQUIRE hearing protection (29 CFR 1910.95 & 1926.101)?

○ Yes

○ No

★12.1.11 Does your company PROVIDE Personal Flotation Devices (33 CFR 142.45 & 29 CFR 1926.106)?

○ Yes

○ No

○ NA

★12.1.12 When applicable, does your company REQUIRE Personal Flotation Devices (33 CFR 142.45 & 29 CFR 1926.106)?

○ Yes

○ No

★12.1.13 Does your company PROVIDE respiratory protection (29 CFR 1910.134 & 1926.103)?

○ Yes

○ No

○ NA

★12.1.14 When applicable, does your company REQUIRE respiratory protection (29 CFR 1910.134 & 1926.103)?

○ Yes

○ No

★12.1.15 Do you have employees who wear respirators? (Canister or SCBA)

○ Yes

○ No

★12.1.16 Does your company PROVIDE safety shoes (ANSI-Z41.1)(29 CFR 1910.136 & 1926.96)?

○ Yes

○ No

○ NA

★12.1.17 When applicable, does your company REQUIRE safety shoes (ANSI-Z41.1)(29 CFR 1910.136 & 1926.96)?

○ Yes

○ No

★12.1.18 Does your company provide PPE for flash protection as defined in NFPA 70E?
NFPA 70E

○ Yes

○ No

○ NA

Accident and Incident Procedures 13

★13.1 Does your company have a written policy that describes roles and responsibilities that will be initiated in the event of an accident?

○ Yes

○ No

★13.2 Is your accident & incident policy communicated so all employees understand your company's position?

○ Yes

○ No

★13.3 Does your company require an authorized individual to accompany injured employees to the medical provider for initial treatment?

○ Yes

○ No

★13.4 Does your company have a policy requiring written accident/incident reports (spills, injuries, property damage, near misses, fires, explosions, etc.)?

○ Yes

○ No

★13.5 Does your company conduct accident/incident investigations?

○ Yes

○ No

★13.6 Are accident/incident reports reviewed by managers/supervisors?

○ Yes

○ No

★13.7 Does your company investigate and document near-miss accidents?

○ Yes

○ No

★13.8 Does your company have a written restricted duty/light duty policy?

- Yes
- No

★13.9 Does your company utilize a specific medical provider that understands your company's restricted duty/light duty policy?

- Yes
- No

★13.10 Comments for the previous question, if any:

★13.11 Does your company have a written process in place to share the lessons learned from accidents with the entire workforce?

- Yes
- No

★13.12 Does your company conduct routine emergency drills?

- Yes
- No

★13.14 Have personnel who perform incident investigations received formal investigative training?

- Yes
- No
- NA

★13.15 If you answered YES to the previous question, list the formal training programs attended.

★13.16 Are root cause investigation techniques/protocols used?

- Yes
- No

★13.17 Do you have a system in place to track incident investigation corrective action findings to closure?

○ Yes

○ No

Drug and Alcohol Program 16

★16.1 Does your company have a written policy regarding drug & alcohol screening or testing of your employees?

12	410578	318	Edit	○ Yes

○ No

★16.2 Does your drug & alcohol program include drug & alcohol awareness training for supervisors?

12	410578	1458	Edit	○ Yes

○ No

○ NA

★16.3 Are your company's employees subject to Post Accident drug screening?

12	410578	328	Edit	○ Yes

○ No

★16.4 Are your company's employees subject to Probable Cause drug screening?

12	410578	329	Edit	○ Yes

○ No

★16.5 Are your company's employees subject to Random drug screening?

12	410578	330	Edit	○ Yes

○ No

★16.6 Are your company's employees subject to drug screening at Return to Duty?

12	410578	331	Edit	○ Yes

○ No

★16.7 Are your company's employees subject to Follow-Up drug screening?

12	410578	332	Edit	○ Yes

○ No

★16.8 Are your company's employees subject to drug screening for other reasons? Please explain:

12	410578	333	Edit

US Drug and Alcohol Program 16.1

★16.1.1 Does your drug & alcohol program conform to DOT requirements?

1203	410578	319	Edit

○ Yes

○ No

○ NA

★16.1.2 Does your drug & alcohol testing program satisfy DOT regulation: Federal Aviation Administration 14 CFR, Part 91.17

1203	410578	320	Edit

○ Yes

○ No

○ NA

★16.1.3 Does your drug & alcohol testing program satisfy DOT regulation: Federal Motor Carrier Safety Administration 49 CFR, Part 382

1203	410578	321	Edit

○ Yes

○ No

○ NA

★16.1.4 Does your drug & alcohol testing program satisfy DOT regulation: Federal Railroad Administration 49 CFR, Part 219

1203	410578	322	Edit

○ Yes

○ No

○ NA

★16.1.5 Does your drug & alcohol testing program satisfy DOT regulation: Pipeline and Hazardous Materials Safety Administration (PHMSA) 49 CFR, Part 199

1203	410578	323	Edit

○ Yes

○ No

○ NA

★16.1.6 Does your drug & alcohol testing program satisfy DOT regulation: United States Coast Guard 33 CFR, Part 95

1203	410578	324	Edit	○ Yes

○ No

○ NA

★16.1.7 Are your company's employees subject to Pre-Employment drug screening?

1203	410578	325	Edit	○ Yes

○ No

★16.1.8 Does you company provide an EAP (Employee Assistance Program)?

1203	410578	1459	Edit	○ Yes

○ No

○ NA

What Drug and Alcohol testing services/consortiums does your company use?

★16.1.9 DISA

1203	410578	334	Edit	○ Yes

○ No

★16.1.10 ASAP

1203	410578	833	Edit	○ Yes

○ No

★16.1.11 PTC (Pipeline Testing Consortium)

1203	410578	834	Edit	○ Yes

○ No

★16.1.12 OTHER

1203	410578	835	Edit	

Industrial Hygiene/Occupational Health 17

★17.1 Are your company employees exposed to substances that would require IH monitoring?

17	410578	892	Edit	○ Yes

○ No

★17.2 Do you perform IH monitoring on your employees?

17	410578	378	Edit	○ Yes

○ No

○ NA

★17.3 Where are IH monitoring records kept?

17	410578	383	Edit

Please indicate what substances your company performs IH testing on?

★17.4 Asbestos

17	410578	379	Edit	☐	X

★17.5 Benzene

17	410578	837	Edit	☐	X

★17.6 Hydrogen Sulfide (H2S)

17	410578	838	Edit	☐	X

★17.7 Lead

17	410578	839	Edit	☐	X

★17.8 Radiation

17	410578	840	Edit	☐	X

★17.9 Silica

17	410578	841	Edit	☐	X

★17.10 Total Hydrocarbons

17	410578	842	Edit	☐	X

★17.11 Welding Fumes

17	410578	843	Edit	☐	X

★17.12 Other

17	410578	844	Edit	

★17.13 Have your applicable employees been trained on how to properly wear a respirator?

17	410578	628	Edit	○ Yes

○ No

★17.14 Are your employees who wear respirators medically cleared?

17	410578	385	Edit	○ Yes

○ No

★17.15 Are your employees who wear respirators annually fit tested?

17	410578	386	Edit	○ Yes

○ No

Environmental Issues 18

Environmental Programs and Policies 18.1

★18.1.1 Is your company required to have any Federal, state, or local environmental licenses or permits to perform their service(s) (for example, NORM, asbestos, DOT, lead, explosives, etc.)?

393	410578	254	Edit	○ Yes

○ No

★18.1.2 If YES, list types of environmental licenses/permits and state of issue:

393	410578	255	Edit

★18.1.3 Does your company have a written environmental program?

393	410578	256	Edit	○ Yes

○ No

○ NA

★18.1.4 Does your EMS (Environmental Management System) address management commitment and expectations?

393	410578	567	Edit	○ Yes

○ No

○ NA

★18.1.5 Does your EMS (Environmental Management System) address accountability and responsibilities for managers, supervisors, and employees?

393	410578	568	Edit	○ Yes

○ No

○ NA

★18.1.6 Does your EMS (Environmental Management System) address resources for meeting environmental requirements?

393	410578	569	Edit	○ Yes

○ No

○ NA

★18.1.7 Does your EMS (Environmental Management System) address periodic environmental performance appraisals for all employees?

393	410578	570	Edit	○ Yes

○ No

○ NA

★18.1.8 Does your EMS (Environmental Management System) address hazard recognition and operational environmental control?

393	410578	571	Edit	○ Yes

○ No

○ NA

★18.1.9 Does your EMS (Environmental Management System) identify anticipated significant environmental risks?

393	410578	572	Edit	○ Yes

○ No

○ NA

★18.1.10 Does your EMS (Environmental Management System) have objectives and targets to minimize significant environmental risks?

393	410578	573	Edit	○ Yes

○ No

○ NA

★18.1.11 Does your EMS (Environmental Management System) address written procedures for minimizing significant environmental risks?

393	410578	574	Edit	○ Yes

○ No

○ NA

★18.1.12 Does your EMS (Environmental Management System) address a method for management of required certifications, documentation, calibrations, and permits?

393	410578	575	Edit	○ Yes

○ No

○ NA

★18.1.13 Does the program satisfy your responsibility under the law for ensuring your employees follow the environmental requirements of the facility?

393	410578	576	Edit	○ Yes

○ No

○ NA

★18.1.14 Does the program satisfy your responsibility under the law for advising owner facilities of any unique hazards or issues presented by the contractor's work and any hazards or issues found by the contractor?

393	410578	577	Edit	○ Yes

○ No

○ NA

★18.1.15 Does your company have written programs (which include training documentation when applicable) for stormwater pollution prevention when performing any digging or excavating?

393	410578	578	Edit	○ Yes

○ No

○ NA

★18.1.16 Does your company have written programs (which include training documentation when applicable) for waste management, such as RCRA, Universal Waste Handling and DOT hazardous material training?

393	410578	579	Edit	○ Yes

○ No

○ NA

★18.1.17 Does your company have written programs (which include training documentation when applicable) for permitting equipment with internal combustion engines (including hydroblasters, generators, welders, light towers, filtering systems, hydrolazers, Baker Tanks, and Vacuum Trucks)?

393	410578	581	Edit	○ Yes

○ No

○ NA

★18.1.18 Does your company have written programs (which include training documentation when applicable) for tank seal inspections?

393	410578	582	Edit	○ Yes

○ No

○ NA

★18.1.19 Does your company have written programs (which include training documentation when applicable) for painting?

393	410578	584	Edit	○ Yes

○ No

○ NA

★18.1.20 Does your company have written programs (which include training documentation when applicable) for solvent usage?

393	410578	585	Edit	○ Yes

○ No

○ NA

★18.1.21 Does your company have written programs (which include training documentation when applicable) for container hazard communication labeling?

393	410578	586	Edit	○ Yes

○ No

○ NA

★18.1.22 Does your company have written programs (which include training documentation when applicable) for fugitive hydrocarbon or particulate emissions?

393	410578	587	Edit	○ Yes

○ No

○ NA

★18.1.23 Does your company have written programs (which include training documentation when applicable) for drains?

393	410578	588	Edit	○ Yes

○ No

○ NA

★18.1.24 Does your company have written programs (which include training documentation when applicable) for visible emissions?

393	410578	589	Edit	○ Yes

○ No

○ NA

★18.1.25 Does your company have written programs (which include training documentation when applicable) for odors?

393	410578	590	Edit	○ Yes

○ No

○ NA

★18.1.26 Does your company have written programs (which include training documentation when applicable) for spill and release prevention?

393	410578	591	Edit	○ Yes

○ No

○ NA

★18.1.27 Does your company have written programs (which include training documentation when applicable) for source testing?

393	410578	592	Edit	○ Yes

○ No

○ NA

★18.1.28 Does your company have written programs (which include training documentation when applicable) for pollution prevention?

393	410578	593	Edit	○ Yes

○ No

○ NA

Refrigerant Management 18.2

★18.2.1 Does your company provide services that involve refrigerant/freon environmental issues?

394 | 410578 | 2067 | Edit ○ Yes

○ No

★18.2.2 Do your technicians have the appropriate certification?

394 | 410578 | 2068 | Edit ○ Yes

○ No

○ NA

★18.2.3 Do you have documentation of certification for refrigerant recovery and recycle equipment used by your personnel and your subcontractors, including loaned and rented equipment?

394 | 410578 | 2069 | Edit ○ Yes

○ No

○ NA

★18.2.4 Do your records include the certification information for reclaimers that receive materials from you for reclamation?

394 | 410578 | 2070 | Edit ○ Yes

○ No

○ NA

Contractor's Licensing 19

★19.1 Is your company required to have a state contractor's license?

21 | 410578 | 1240 | Edit ○ Yes

○ No

★19.2 If your company is not required to have a State contractor's license, please explain why:

21 | 410578 | 1241 | Edit

★19.3 Has your contractor's license been revoked at any time in the last five years?

21 | 410578 | 257 | Edit ○ Yes

○ No

○ NA

★ 19.4 If YES, please explain:

21	410578	258	Edit

★ 19.5 If any of your firm's license(s) are held in the name of a corporation or partnership, list the names of the qualifying individual(s) listed on the records who meet(s) the experience and examination requirements for each license:

21	410578	267	Edit

★ 19.6 Has any contractor's license held by your firm or its Responsible Managing Employee (RME) or Responsible Managing Officer (RMO) been suspended within the last five years?

21	410578	268	Edit

○ Yes

○ No

○ NA

★ 19.7 If YES, please explain:

21	410578	269	Edit

Quality Assurance/Quality Control 21

★21.1 Does your company have a Quality Assurance/Quality Control Department?

19	410578	244	Edit

○ Yes

○ No

○ NA

★21.2 Does your company have a Quality Assurance/Quality Control manual?

19	410578	245	Edit

○ Yes

○ No

○ NA

★21.3 Does your company have a management of change process/program?

19	410578	246	Edit

○ Yes

○ No

○ NA

★21.4 Does your company have a document control policy?

19	410578	247	Edit

○ Yes

○ No

○ NA

★21.5 Does your company keep preventative maintenance and equipment inspection records?

19	410578	250	Edit

○ Yes

○ No

○ NA

★21.6 Do you maintain the applicable inspection and maintenance certification records for operating equipment?

19	410578	252	Edit

○ Yes

○ No

○ NA

★21.7 Does your company perform equipment checks on all equipment (cranes, forklifts, JLG's etc.)?

19	410578	253	Edit

○ Yes

○ No

○ NA

PSM (Process Safety Management) 22

★22.1 Does your company perform work activities at facilities regulated by the Process Safety Management Standards?

32	410578	828	Edit

○ Yes

○ No

○ NA

★22.2 Does your company train employees in the work practices necessary to perform their jobs safely?

32	410578	829	Edit

○ Yes

○ No

○ NA

★22.3 Does your company have documentation that shows employees have received and understood the training provided?

32	410578	830	Edit

○ Yes

○ No

○ NA

★22.4 Describe how your company verifies that its employees are instructed in the known potential fire, explosion, or toxic release hazards and the applicable provisions of the emergency action plan.

32	410578	831	Edit

★22.5 Describe how your company informs the host facility of any unique hazards associated with the company's work activities.

32	410578	832	Edit

Upload Files 27

Safety Manual 27.2

NOTE: Only electronic copies of your safety manual/IIPP will be accepted. If a hard copy is provided, a fee will be assessed in order to convert your manual into a PDF document.

★27.2.1

Please upload your manual as a single pdf or word document. NOTE: (Manual audits are performed every 3 years. This section will require an updated manual to be uploaded if you are scheduled for an audit this year) If you work in California you will need to upload your IIPP in the IIPP category if it is not included in your safety manual (we need your entire safety manual in this section, not your IIPP). Depending on the size of the document and your connection speed, this may take a long time.

Business Continuity Plan 27.3

★27.3.1
Please upload your Business Continuity Plan.

Anti Drug and Alcohol Misuse Programs 27.4

★27.4.1 Please upload your Drug & Alcohol program in this section as it needs to be audited for DOT OQ (operator qualification) purposes. This must meet PHMSA requirements.

449	410578	1532	Edit

★27.4.2 Please upload a document that indicates that at least 25 percent of your OQ covered employees have undergone random drug testing. This should be in the form of a printout from a drug consortium or company that manages your random drug screening program

449	410578	1581	Edit

IIPP (Injury and Illness Prevention Program) 27.5

★27.5.1 Please upload your IIPP (Injury and Illness Prevention Program) if you work in California and if it is not included in your safety manual.

448	410578	1435	Edit

QA/QC (Quality Assurance/Quality Control) Manual 27.6

★27.6.1 If you have a QA/QC (Quality Assurance/ Quality Control) Manual that your clients need to view, please upload it here.

445	410578	1431	Edit

Security 29

Security 29.1

★29.1.1 Do you perform background investigations and screenings on your employees?

420	410578	1385	Edit	○ Yes

○ No

★29.1.2 If so, how far back do you review?

420	410578	1410	Edit	

★29.1.3 What Company does your organization use to perform background investigations?

420	410578	1386	Edit	

★29.1.4 Federal Criminal Record Check (7 years, all residential addresses)?

420	410578	1389	Edit	○ Yes

○ No

★29.1.5 Are you checking for a time period other than 7 years? If so, what time frame?

420	410578	1390	Edit	

★29.1.6 County / Parish Criminal Record check?

420	410578	1391	Edit	○ Yes

○ No

★29.1.7 Employment verification of last employer?

420	410578	1392	Edit	○ Yes

○ No

★29.1.8 Education verification of the highest level attained?

420	410578	1393	Edit	○ Yes

○ No

★29.1.9 Do you 'grandfather' or exclude any employees from the Background Investigations?

420	410578	1394	Edit	○ Yes

○ No

★29.1.10 If so, how?

420	410578	1395	Edit

Has any employee been denied admission to the United States or removed from the United States who:

★29.1.11 Is a representative of a political, social, or other similar group whose public endorsement of acts of terrorist activity the Secretary of State has determined undermines United States efforts to reduce or eliminate terrorist activities?

420	410578	1397	Edit	○ Yes

○ No

★29.1.12 Has used the alien's position of prominence within any country to endorse or espouse terrorist activity, or to persuade others to support terrorist activity or a terrorist organization, in a way that The Secretary of State has determined undermines United States efforts to reduce or eliminate terrorist activities?

420	410578	1398	Edit

○ Yes

○ No

★29.1.13 Is state issued identification required of all potential employees?

420	410578	1400	Edit

○ Yes

○ No

○ NA

★29.1.14 Are any other forms of identification used to verify the identity of the applicant?

420	410578	1401	Edit

○ Yes

○ No

○ NA

★29.1.15 If so, what form of ID do you accept?

420	410578	1402	Edit

★29.1.16 Is each employee checked to determine citizenship status?

420	410578	1403	Edit

○ Yes

○ No

○ NA

★29.1.17 Are motor vehicle records reviewed for each employee who drives company vehicles and who operates machinery?

420	410578	1404	Edit

○ Yes

○ No

○ NA

★29.1.18 Do you verify employee address and contact information?

420	410578	1405	Edit

○ Yes

○ No

○ NA

★29.1.19 If so, how?

420	410578	1406	Edit

★29.1.20 Do you maintain a personnel file on each employee?

420	410578	1407	Edit

○ Yes

○ No

○ NA

★29.1.21 Do you verify previous employment?

420	410578	1408	Edit

○ Yes

○ No

○ NA

★29.1.22 If so, how far back do you review?

420	410578	1409	Edit	

★29.1.23 Does your company have badges that readily identify your employees?

420	410578	1411	Edit

○ Yes

○ No

○ NA

★29.1.24 Are company owned movable items marked by a company logo/name (i.e., vehicles, high dollar Tools, trailers, etc.?

420	410578	1412	Edit

○ Yes

○ No

○ NA

★29.1.25 Does your company have an incident reporting system (i.e., suspicious, illegal, unsafe, unethical activities, etc.)?

420	410578	1413	Edit

○ Yes

○ No

○ NA

★29.1.26 Are your employees aware of the system and how to use it?

420	410578	1414	Edit	○ Yes

○ No

○ NA

★29.1.27 Does your company make formal reports of known incidents?

420	410578	1415	Edit	○ Yes

○ No

○ NA

★29.1.28 Does your company have a written disciplinary policy that addresses serious security/safety-related incidents?

420	410578	1416	Edit	○ Yes

○ No

○ NA

US Security 29.1.1

Does the background investigation include the following:

★29.1.1.1 Verification of SSN and address?

1212	410578	1388	Edit	○ Yes

○ No

★29.1.1.2 Do you restrict employees from working on Owner premises if in the seven (7) year period ending on the date of the background investigation an employee has been convicted of the following? A crime referred to in Section 46306, 46308, 46312, 46314, or 46315 of Title 49 of the United States Code (U.S.C.) or chapter 465 of Title 49 or Section 32 of Title 18?

1212	410578	1396	Edit	○ Yes

○ No

○ NA

Transportation Worker Identification Credential (TWIC) 29.2

★29.2.1 Do any of your employees work at facilities considered 'Secure Areas' under the maritime Transportation Security Act? (e.g., ports)

421	410578	1807	Edit	○ Yes

○ No

★29.2.2 Do all applicable employees have TWIC cards?

421	410578	1808	Edit	○ Yes

○ No

○ NA

★29.2.3 If No, please explain any special circumstances.

421	410578	1809	Edit

Supplier Diversity 30

US Supplier Diversity 30.1

★30.1.1 Does your company hold any Supplier Diversity Certifications (Small Business, Women Owned, Minority Owned, etc.)

428	410578	2478	Edit	○ Yes

○ No

Small Business Certifications

★30.1.2 Is your business considered a small business based upon the criteria and size standards defined by the Small Business Act (13CFR Part 121.1) and Public Law (95-507)?

428	410578	2340	Edit	○ Yes

○ No

★30.1.3 Is your business registered with the SBA via the Central Contractor Registration (CCR) database (formerly Pro-Net)?

428	410578	2341	Edit	○ Yes

○ No

★30.1.4 Small Disadvantaged Business (SDB)?

428	410578	2342	Edit	☐	X

★30.1.5 8(a) Business Development Program?

428	410578	2345	Edit	☐	X

★30.1.6 Service Disabled Veteran Owned?

428	410578	3543	Edit	☐	X

★30.1.7 HUBZone?

428	410578	2348	Edit	☐	X

★30.1.8 DOT Disadvantaged Business Enterprise?

428	410578	2351	Edit	☐	X

★30.1.9 Upload your Small Business certification.

428	410578	2349	Edit

★30.1.10 Expiration Date

428	410578	2350	Edit	

example: 12/31/1999

Minority Owned Business Certifications

★30.1.11 Is your business currently certified as a Minority Owned Business?

428	410578	2354	Edit

○ Yes

○ No

★30.1.12 Hispanic American

428	410578	2355	Edit	☐	X

★30.1.13 Native American

428	410578	2358	Edit	☐	X

★30.1.14 Asian-Pacific American

428	410578	2361	Edit	☐	X

★30.1.15 African American

428	410578	2364	Edit	☐	X

★30.1.16 Asian-Indian American

428	410578	2367	Edit	☐	X

★30.1.17 Other

428	410578	2370	Edit	☐	X

Comments:

★30.1.18 Upload your Minority-Owned Certification.

428	410578	2371	Edit

★30.1.19 Expiration Date

428	410578	2372	Edit	

example: 12/31/1999

Women-Owned Business Certifications

★30.1.20 Is your business currently certified as a Women-Owned Business?

428	410578	2373	Edit

○ Yes

○ No

★30.1.21 Upload your Women-Owned Certification.

428	410578	2374	Edit

★30.1.22 Expiration Date

428	410578	2375	Edit	

example: 12/31/1999

Appendix 2: Contractor Safety Standard*

Process Safety Management System

Contractor Safety Standard

Document No

OWNER CORPORATION

Senior Executive, e.g., CEO/COO	Senior Leader, e.g., VP EH&S or Director of Process Safety	09/06/01		
Document Owner	**Document Contact**	**Date Developed**	**Revision Date (YY/MM/DD)**	**Last Reviewed On:**

Standard		
Department:	**Sustainable Development/Environment, Health and Safety**	Document Number:
Subject:	**Contractor Safety**	Revision: **June 1/09**

Rev	**Revision History**
1	N/A

APPROVED BY

____

CEO/Chief Operating Officer

Standard		
Department:	**Sustainable Development/Environment, Health and Safety**	Document Number:
Subject:	**Contractor Safety**	Revision: **June 1/09**

TABLE OF CONTENTS

Standard		
Department:	**Sustainable Development/Environment, Health and Safety**	Document Number:
Subject:	**Contractor Safety**	Revision: **June 1/09**

Purpose and Scope

1.1 It is essential that contract employees complete all tasks safely and in accordance with established Owner's procedures and/or safe work practices, and be consistent with the principles and essential features of process safety management (PSM). Contractor environment, health, and safety (EHS) work practices need not be identical to The Owner's procedures, but they shall afford an equivalent level of protection.

The purpose of this Standard is to define The Owner's requirements to assist management in developing and implementing procedures and practices to control or minimize the risks associated with on-site contractor activity including injuries and illnesses, environmental harm, property damage and business losses.

1.2 This Standard applies to the Owner's and subsidiaries over which the Owner has operational control or majority ownership worldwide (collectively "Owner" or "Company"). References in this document to "Owner Personnel" include directors, officers, employees, contract workers, consultants and agents of the Owner.

1.3 This Standard is applicable to all on-site contractors and subcontractors that perform work on, or in areas where hazardous processes are present.

1.4 All aspects of this Standard shall be in effect unless otherwise noted. Where regulations impose requirements not reflected in this Standard, the most stringent requirements shall apply. Reference the Owner's Legal and Other Requirements Register (database).

Responsibilities

This section presents the responsibilities of the Owner's leaders and organizations to implement this Standard and supplements Section 2 of the

Standard		
Department:	**Sustainable Development/Environment, Health and Safety**	Document Number:
Subject:	**Contractor Safety**	Revision: **June 1/09**

Owner's Process Safety Management Core Standard. A Responsible-Accountable-Consulted-Informed (RACI) chart for each of these roles is in Appendix A.

Business Area and Functional Leaders

Business Area and Functional Leaders that have contractors working in their areas of responsibility shall implement and conform to this Standard.

Business Area and Functional Leaders shall demonstrate their leadership and commitment to process safety by:

a. Providing sufficient competent resources to implement this Standard including:
   - Identifying a safety resource qualified to address contractor safety.
   - Designating contract coordinators to oversee the contracting process and work done under contracts.

b. Establishing written procedures that define the roles and responsibilities of personnel involved in contract administration and coordination.

c. Setting goals, objectives, and expectations for contractor performance.

d. Establishing accountability for performance against goals and objectives.

e. Establishing a process for evaluating contractor safety performance.

f. Periodically evaluating the contractor environment, health and safety management process to determine if changes should be made.

g. Auditing and monitoring contractor activities and performance.

Contract Coordinators

The contract coordinator role is responsible for assuring the implementation of an established process so that contract work is administered to meet the Owner's requirements. The role is dependent on the size and complexity of the operation or the project scope.

Standard		
Department:	**Sustainable Development/Environment, Health and Safety**	Document Number:
Subject:	**Contractor Safety**	Revision: **June 1/09**

Contract coordinators shall:

a. Confirm that the contractor's proposal meets the bid package's mandatory requirements.
b. Confirm that applicable environment, health and safety hazard information is communicated to contractor employees and contractor's environment, health and safety documents are reviewed and documented, including qualification and certifications, safety orientation and communications, and training required in the contract or by law.
c. Assure that contractor orientation and training are completed and documented as outlined in this Standard.
d. Assure when additional work is required by contractor employees, that the specific training required for that work is conducted before the work is performed.
e. Assure that work is done in accordance with the contract terms by means of audits, reports, and reviews.

Supply Chain Leaders

Supply chain leaders shall:

a. Assure requests for quotation include:
   - Safety requirements and limitations.
   - Descriptions of where work is to be done in relation to hazardous process operations.
   - Boundaries of any areas to be given over to control of prime contractors or constructors.
b. Set prequalification criteria and assure compliance to them.

Contractors

Contractors shall assure:

a. Each contract employee has the appropriate job skill training and is qualified to perform the contracted work safely.
b. Each contract employee is instructed in the known potential fire, explosion, or toxic release hazards related to his or her job and the process.

Standard		
Department:	**Sustainable Development/Environment, Health and Safety**	Document Number:
Subject:	**Contractor Safety**	Revision: **June 1/09**

c. Each contract employee receives and understands training regarding the following:
   - Safety rules of the facility
   - Applicable provisions of the emergency response and control plan
   - Applicable safe work practices of the facility

d. Records are maintained to demonstrate that each contract employee has received and understood the instruction and training described above.

e. Each contract employee follows the safety rules and applicable safe work practices of the facility.

f. A program is implemented to assure that contract employees who work in hazardous process facilities are fit for duty.

g. Advise the Owner's business area management of any unique hazards presented by the contractor's work, or any hazards found by the contractor's work.

Sustainable Development/Environment, Health and Safety Group Leaders

The Sustainable Development/Environment, Health and Safety (SD/EHS) Group Leaders shall provide interpretation and maintenance of this Standard.

Referenced Standards

References are provided in the Owner's Process Safety Management Core Standard document.

Definitions

Pertinent terms are defined in a separate Definitions document. Process Safety Management Definitions Document

Standard		
Department:	**Sustainable Development/Environment, Health and Safety**	Document Number:
Subject:	**Contractor Safety**	Revision: **June 1/09**

Standard

This section covers the requirements of this Standard. Appendix B provides recommended entry-level safety awareness information that is aligned with the Owner's Environment, Health and Safety Policy that should be conveyed to contractors.

Contractor Prequalification Criteria

Contractors shall not be used for jobs the Owner's employees would not be assigned, unless the contractor's special expertise and training allow these jobs to be performed safely.

Businesses shall use contractors with demonstrated commitment to safety. When prequalifying a contractor, information on safety programs and performance shall be obtained and evaluated for indications of satisfactory safety performance. Each organization utilizing contractors shall develop a process for issuing variances for each contractor that does not meet prequalification criteria.

Contract Coordination

Contract activity shall be overseen to assure:

a. Work is done in accordance with contract terms.
b. Safety of the owner's and contract employees is not compromised by the activity of the other.

Contractor Orientation and Training

Each business area shall document procedures and practices for training and orienting contractors to minimize risk of an incident as a result of contracted activity.

Contractor orientation and training shall include as a minimum:

a. Establishing clear lines of communication between the owner and the contractor.

Standard		
Department:	**Sustainable Development/Environment, Health and Safety**	Document Number:
Subject:	**Contractor Safety**	Revision: **June 1/09**

b. Informing the contractor of the known potential fire, explosion, reactive chemicals, or toxic release hazards related to the contractor's work and any nearby processes.

c. Informing the contractor of applicable safety rules and procedures of the facility, including safe work practices for the control of hazards, entry into process areas, and mandatory PPE requirements. All safety orientations/communications shall be documented.

d. Explaining to the contractor the applicable provisions of the emergency action plan.

e. Establishing a process for reporting process incidents to the Owner's management.

f. Reviewing the contractor's performance on safety and other responsibilities periodically.

g. Explaining activities which will take place during their time on a site such as:
   - Field audits of contractor work activities; and
   - Post-job evaluations.

Special Considerations for Staged Contractors

Where business areas have developed specific procedures for the Owner's-owned assets/equipment and provided these procedures to maintenance and/or operation contractors the following shall be developed and implemented:

a. The staged contractor company shall develop and maintain systems to provide training to contract employees and help assure that they are qualified to conduct assigned tasks in accordance with the applicable Owner's procedures.

b. A system to help assure that any revisions or updates to specific procedures are coordinated between Owner's and the applicable contractor company.

c. Owner's evaluation and auditing of the contractor training and qualification process as it applies to use of Owner's maintenance or operations procedures.

Standard		
Department:	**Sustainable Development/Environment, Health and Safety**	Document Number:
Subject:	**Contractor Safety**	Revision: **June 1/09**

d. A management of personnel change process administered by the contractor for newly assigned contractor supervisory personnel who oversee PSM-critical maintenance or operations tasks, training and qualifications of contract employees, and related functions.

Contractor Work Coordination and Auditing

Contractor work shall be administered and audited to meet the Owner's process safety and environment, health and safety requirements.

Contract Evaluation

Contractor performance shall be evaluated on a continual and/or post project basis.

Metrics

Metrics shall be reported in accordance with the requirements outlined in the Owner's Process Safety Management Core Standard.

Implementation

The contact for this Standard is the Owner's Director of Process Safety in the Sustainable Development/Environment, Health, and Safety Group.

Interpretation and Updating

Management Records

Documentation shall be maintained as specified in this Standard.
Records shall be retained in compliance with the Owner's document control and records requirements.

Standard		
Department:	**Sustainable Development/Environment, Health and Safety**	Document Number:
Subject:	**Contractor Safety**	Revision: **June 1/09**

Audits

Audits shall be carried out as specified in this Standard and in accordance with the Owner's Operations Integrity Audit Standard.

Standard Renewal Process

This Standard shall be reviewed and revised as necessary and, at a minimum, not later than five years from the date of the last review.

Deviation Process

Deviations from this Standard must be authorized by the Owner's Vice President Sustainable Development and authorized by the Chief Operating Officer using the management of change process. Deviations must be documented, and documentation must include the relevant facts supporting the deviation decision. Deviations shall have a stated expiry date not to exceed five years. Any extensions must be re-authorized.

Training and Communications

Training in and communication of process safety management to employees and contractors shall be carried out in accordance with this Standard and the Owner's Training and Performance Standard.

List of Appendices

Appendix A—RACI Chart

Appendix B—Guidance

Addenda

N/A

Standard		
Department:	**Sustainable Development/Environment, Health and Safety**	Document Number:
Subject:	**Contractor Safety**	Revision: **June 1/09**

Appendix A: Raci Chart—Contractor Safety

RACI Symbols

R = Responsible for making it happen.
A = Accountable for acceptability of the results (products, services, etc.).
C = Consulted by "R" before plans are final or decisions taken. "C" must be consulted and given the opportunity to influence "R's" plans/decisions.
I = Informed after the fact.

Key Activities / Position	BAL	CC	SCL	Contractors	SD/EHS
Provide sufficient, qualified resources to implement this Standard	A,R		R		
Audit contractors for compliance to all environment, health and safety requirements in the contract	A,R		R		I
Identify a safety resource qualified to assess contractor safety	A,R				
Designate contract coordinators to oversee contracting process	A,R		R		
Establish a process to evaluate contractor safety performance	A,R		R		
Periodically evaluate the contractor environment, health and safety management process to determine if changes are needed	A,R		R		C
Set objectives and expectations for contractor performance and establish accountability for objectives and expectations	A,R		R		
Set prequalification criteria for contractor safety performance	C		I		A,R
Confirm that contractor's proposal meets the bid package mandatory requirements			A,R		
Confirm that applicable environment, health and safety hazard information is communicated to contractor	R		A		
Document when all safety orientations or communications are provided to contractors	A,R		R		
Complete and document any work-site specific training for contractors has been completed before work begins	A,R				

Standard		
Department:	**Sustainable Development/Environment, Health and Safety**	Document Number:
Subject:	**Contractor Safety**	Revision: **June 1/09**

Key Activities / **Position**	**BAL**	**CC**	**SCL**	**Contractors**	**SD/EHS**
Assure that requests to employ contractors includes all safety requirements and limitations, descriptions of work location in reference to hazardous processes, boundaries for any areas where control is given over to contractors or constructors and assure compliance to pre-qualification process	R		A,R		
Assure contract employees have the appropriate job skill training and are qualified to perform the contracted work safely		A,R	R	R	
Assure contract employees are instructed on the known potential fire, explosion and toxic release hazards related to their job and the process where they are assigned	A	R			R
Assure that contractors receive and understand training regarding site safety rules, applicable provisions of the emergency response and control plan and applicable safe work practices	A	R		R	
Training records are maintained to show that each contract employee has received and understood the training described above	A	R			
Contractors follow the safety rules and applicable safe work practices of the facility	A	R		R	
Assure that contractors that work in hazardous facilities are fit for duty	A	R		R	
Advise Owner's area business management of any hazards presented by the contractor's work or hazards discovered during contractor's work	I	I		A,R	
Provide interpretation and maintenance of this Standard					A,R
Legend BAL Business Area and Functional Leaders CC Contractors Contractors SCL Supply Chain Leaders	Contractors Contractors SD/EHS Sustainable Development/ Environmental, Health and Safety Group Leaders Health and Safety Group Leaders				

Standard		
Department:	**Sustainable Development/Environment, Health and Safety**	Document Number:
Subject:	**Contractor Safety**	Revision: **June 1/09**

Appendix B: Guidance

It is an Owner's practice to contract with companies that embrace the Owner's Environment, Health and Safety (EHS) Policy. This Standard includes provisions to assist managers to develop and implement procedures and practices to control or minimize the risks associated with on-site contracted activity. Where a contractor is utilizing special expertise to perform unusual tasks (e.g., high stack repair, hot tapping, and steel erection), Owner's may rely on the contractor's special safety expertise subject to the prequalification criteria mentioned in Section 5 of this Standard.

Support Resources

Support resources for managing contractor safety include the following:

- Business unit and area environment, health and safety and contract administration personnel
- Sustainable Development/Environment, Health and Safety Group
- Employee Safety Leadership Network
- Supply Chain Management
- Owner's Contractor Safety and Health Network

The Six-Step Process

The Owner's six-step contractor safety management process is designed to provide a methodology for managing the risks associated with contractor activities at Owner's facilities. All Owner's facilities using contractor services on-site should observe the provisions of the six-step process, which are as follows:

1. Contractor prequalification
2. Contract preparation
3. Contract award
4. Contractor orientation and training
5. Contractor work coordination and auditing
6. Contract evaluation

Standard		
Department:	**Sustainable Development/Environment, Health and Safety**	Document Number:
Subject:	**Contractor Safety**	Revision: **June 1/09**

Contractor Prequalification Criteria

Safety criteria should include the following:

- Injury/illness performance data
- Special skills and training (e.g., certifications, registrations, licenses, and experience)
- Environment, health and safety program
- Work history with Owner's

Contract Preparation

The contract bid package should be developed to include safety performance expectations and conditions for execution of the work. A safety resource should be involved in preparing the bid package to identify environment, health and safety hazards and include appropriate safety requirements.
The following elements should be included in the bid package:

- Scope of work
- Environment, health and safety requirements
- Provisions for contract termination for failure to comply with Owner's Environment, Health and Safety Policy requirements

"Site Conditions" and "Special Conditions" should be included in the bid package. An addendum should be issued when additional safety information and instructions are requested, or if additions, modifications, corrections, and clarifications to bid documents occur.

Contract Award for Bid Jobs

Confirmation of contractors' proposals should be accomplished before the contract is awarded. Other activities that should be performed during the contract bid and award process as applicable are:

- Bidders' meeting
- Bid clarification meeting
- Pre-mobilization (i.e., pre-job meeting)

Standard		
Department:	**Sustainable Development/Environment, Health and Safety**	Document Number:
Subject:	**Contractor Safety**	Revision: **June 1/09**

Contractor Orientation and Training

Orientation

The purpose of a general safety orientation is to provide entry-level safety awareness information that outlines safety expectations and general emergency procedures and conveys Owner's position on safe work practices.
Before contracted work begins, it is important for Owner's to provide an environment, health and safety orientation to familiarize contractor employees with essential environment, health and safety information such as hazards posed by Owner's operations. Environment, health and safety information to be shared with contractor employees can be divided into three broad categories:

- Owner's Environment, Health and Safety expectations
- General worksite information, including rules and procedures
- Area-specific information (where applicable)

Contract coordinators should evaluate the effectiveness of the contractor's safety orientations, communications, and training.

Owner's Environment, Health and Safety Policy and Expectations

Owner's Environment, Health and Safety Policy and expectations should be conveyed to all contractors, including the following:

- All injuries and occupational illnesses, as well as environment, health and safety incidents, are preventable, and our goal for all of them is zero.
- The company builds, operates, and maintains all our facilities so they are safe and protect the environment.
- Compliance with the Owner's rules and procedures, applicable laws, and safe work practices is the responsibility of every employee and contractor acting on our behalf and is a condition of employment or contract terms.

Standard		
Department:	**Sustainable Development/Environment, Health and Safety**	Document Number:
Subject:	**Contractor Safety**	Revision: **June 1/09**

General Business Area Information

The business area should provide a general safety orientation on the following items:

- Security measures (e.g., access/egress, badging, and contraband restrictions).
- Emergency procedures, alarm signals, assembly areas, and reporting expectations.
- Basic safety pertaining to the type of work performed.
- Personal responsibility for safety, with emphasis on absolute adherence to business area rules and local regulations.
- Procedures for injury and incident reporting.
- Medical treatment and first-aid access, procedures, and policies.
- Business area smoking, eating, dress, and personal hygiene policies and procedures.
- Site rules for vehicular and pedestrian traffic, including transporting personnel.
- Use of appropriate special procedures, highlighting the information specific to the work being performed by the contractor.
- Permits (e.g., work, flame, confined space, crane, and excavation).
- The lock, tag, tests, and try procedure (with emphasis on its importance as well as related personal actions).
- Site chemical hazards communications procedures.
- Co-occupancy procedures when working in operating areas (with emphasis on the importance of proper communications regarding the use of hazardous chemicals and equipment).

Area-Specific Information

The business area should provide a specific environment, health and safety orientation on the following items:

- Emergency evacuation procedures.
- Chemical and physical hazards and associated concerns inherent in the area.

Standard		
Department:	**Sustainable Development/Environment, Health and Safety**	Document Number:
Subject:	**Contractor Safety**	Revision: **June 1/09**

- Process hazards inherent in the local operations.
- Permit authorization and coordination.
- The lock, tag, tests, and try procedure (with emphasis on area communication/coordination and location of lockout boxes).
- Co-occupancy procedures when working in operating areas (with emphasis on coordination between different work groups and the status of operating processes).

Training

Each function, business unit and area should train appropriate personnel to carry out their responsibilities as outlined in this Standard. Unless an imminent hazard exists, communications should typically be conducted between the Owner's contract coordinator and the contract supervision.

Examples of additional training include use of special PPE because of unusual facility/process hazards and special procedures and expectations for control of hazardous energy, line breaks, or confined space entry.

Special Considerations for Resident Maintenance and Operations Contractors

Owner's sites may use a permanent or "resident" contractor maintenance group to perform periodic tasks on hazardous process equipment. This may include testing and inspection, routine maintenance on valves, piping, pumps, safety instrumented systems, and other equipment-specific procedures on PSM-critical equipment or components. In certain circumstances, written maintenance procedures developed by the Owner's may be provided to contractors conducting these equipment-specific tasks.

Similarly, the Owner's sites may engage contractor personnel to perform specific operations tasks on hazardous processes (e.g., raw material unloading, product packaging, and shipping) where written operating procedures developed by Owner's are provided to operations contractors.

Contractor Work Coordination and Auditing

Engineers and field contract coordinators may be used to assist in the contracting process. The contract coordinator should hold the prime contractor

Standard		
Department:	**Sustainable Development/Environment, Health and Safety**	Document Number:
Subject:	**Contractor Safety**	Revision: **June 1/09**

accountable for subcontractors' (i.e., any subcontractors used to fulfill the contract) compliance with all environment, health and safety expectations. Field contract coordinators should:

- Coordinate contractor work.
- Specify that contractors notify the Owner's before hazardous materials are brought on site.
- Conduct audits of contractor work, communicate the findings with the contractor, and require timely correction.
- Validate that the contractor reports and investigates all incidents occurring on-site.
- Hold the contractor accountable for compliance with all applicable laws and regulations.

Contract Evaluation

The documented contracting process should identify which contracted work activities should be evaluated and at what frequency. The contract coordinator should confirm that the contractor's safety performance is evaluated according to that process.
The evaluation of contractor's environment, health and safety performance should include:

- Injury/illness performance
- Environment, health and safety incidents
- Audit observations and corrective actions
- Overall commitment to Owner's environment, health and safety expectations
- Quality of the safety program
- Behavior and effectiveness of supervision

Appendix 3: Ground Disturbance Attachment and Sample Work Agreement

GROUND DISTURBANCE ATTACHMENT GD# 2205

Part I - Pre-Job planning Checklist	
Location (work unit, LSD, GPS):	Project Start Date:

Mandatory Attendance

Project Leader / Job Coordinator:	• (The Person Responsible to clarify scope of Project. This could be a planner, engineer or other. This person should provide drawings, material specification and other details of the excavation)
Team Leader, Shift Lead or designated Alternate:	• Responsible; Verify planning integrity, approve all variances to Critical Procedures or Ground Disturbance Attachment Protocol.
Representative for Operating Authority:	• Responsible; Observe, Commend, Correct, Quality Assurance, Compliance, Documentation, Safe Work Practices. **As defined in the Safe Work Agreement Code of Practice.**
Performing Authority:	• Responsible; Regulatory and Critical Procedure Compliance, Safe Work Practices. **As defined in the Safe Work Agreement Code of Practice.**
Ground Disturbance Representative:	• Optional attendee at this planning phase of the excavation, but must attend the pre-excavation meeting, and is ultimately responsible for safe compliant excavation practices. **As defined in the Ground Disturbance Code of Practice.**

(YES - if task has been completed / NO - if further action required)	**YES**	**NO**	**ACTION / OWNER BY WHEN**
1) P&ID's plot plans (reference material) • Review available drawings; process piping / electrical, / Cathodic protection /anode bed / communication / P&ID's / pipeline / operating maintenance manual, plot plans, utilities / Others			
2) Visual Worksite Inspection • Review location/equipment access and egress requirement / Other surface hazards / Existing Equipment / Obstruction / P/L markers / Soil Conditions / Overhead Hazards / Land Owner Boundaries / Fencing / Environmental Concerns / Topography / Spoil Pile and Top Soil staging locations / Pipeline right of ways / Consider referencing facility turn over agreement / Document			
3) Landowner Agreement • Ensure landowner land use agreement is completed. (Surface Land Representative may be required) • Consider First Nations Consultation - Archeological and project monitoring issues. (Consult Surface Land Rep) * All applications should be made to Owners *at least* 48 hours prior to commencing excavation / daylighting * All Third Parties should be informed *at least* 24 hour prior to covering exposed equipment to accommodate inspections where required			
4) Owner Inspection Program P/L Forms / Documentation (These forms can be found on Leaders Guide Element 2.2.19) • Right of way crossing information sheet • Pipeline Inspections Report • Creek and Water Course Crossing Guidelines • Road Crossing Construction Guidelines • Heavy Equipment Crossing Construction Guidelines • Pipeline Crossing Pipeline Guidelines			
5) Regulatory & Other Notification / Approval Considerations Federal Regulatory Authorities - National Energy Board - ABSA • Provincial Regulatory Authorities - OH&S / Alberta First Call / Alberta Environment / AEUB / Forestry / Pipelines / Lands and Parks / Fish & Wildlife / Archeological. • Surface Lands Group • Municipal District of Wood Buffalo - Road Usage / Traffic controls / • Public Utilities - Water / Sewer / Communications / Hydro Gas transmission / Electrical *above and below ground* / • Private Utilities - Communications / Hydro / Gas transmission / Electrical *above and below ground* / Irrigation / Other Local Producers. (Brainstorm List of Potentials for Follow up) • Others; Asset Team / Partners / Gas Control / Marketing / Gas Transmission Utilities / OIP & Maintenance coordinators should be informed as to optimize potential maintenance synergy.			
6) Critical Procedure Review • Ensure adequate instruction, Supervision, of workers(Ground Disturbance Representative to attend all Daylighting activities). • Step-by-step review of Critical Procedures for location of Underground Hazards, and Excavation procedure with appropriate personnel. • Consider other procedures; Zero Energy / Lockout Electrical & Mechanical; blinding; breaking integrity of pressure system > 1" and other applicable site specific procedures			

Printing Unlimited L.P. 22159

7) Sub-Surface equipment detection required (Location & Depth) • Consider optimum equipment practical to locale or daylight sub-surface equipment, such as Hydrovac, line locating survey, Radarscan, radio frequency surveys, radar, hand exposure (Verify availability of your desired option) • Consider control options to minimize risk during excavation or Daylighting; such as pipeline blow down and other utility de-energizing • Record, stake, identify ALL Underground Lines & Equipment, type & depth of U/G equipment within 30 M of proposed excavation. • Cross reference Sub-Surface Equipment survey results with P&ID's or other available materials (Assess)			
8) Confined Space Review • Review CSE decision flow Diagram to determine if CSE is applicable			
9) Update P&ID, plot maps, documentation • Identify personnel responsible to update P&ID, pipeline manuals, OIP inf., plot plans, etc... and date of delivery after /ground disturbance task completed. • Consider computer data base, photos, file updates, Q/C, etc... if applicable]			
10) Environmental Issues Addressed • Provide follow up plan for all applicable environmental considerations indentified within this checklist • Assess potential environmental hazards (i.e.: pipeline liquids / Drainage / Siltation / Forestry Codes / Fire / Erosion Controls) Provide necessary safe guards to prevent further environment degradation. • Consider methods of containment, removal & disposal of hazardous wastes (soil, liquids, etc...).			
11) Personnel / Contractor Procurement • Assure approved / preferred excavation contractors. • Assure competent First Aid, Rescue, Quality Assurance personnel are available • Assure Orientations are planned prior to job day or use previously prepared workers.			
12) Equipment Procurement • Identify equipment such as; Fire Extinguishers, Gas & NORM's Monitors, Rescue & Respiratory equipment, barricades, fall protection, excavation & construction equipment is properly scoped for the project.(equipment may require Positive Air Shut off depending on the circumstances of the excavation or ground disturbance)			
13) Emergency Response Plan Review • Review ERP 3 minute AID for specific work unit for all workers to ensure understanding. • Emergency procedures in event of live line strike, worker injury or emergency situation (such as; fire, cave-in, and emergency evacuation). • Identify muster sites.			
14) Post Disturbance Activities • Update P&ID / Plot Plans / Pipeline Manuals / OIP Records • Re-energize Cathodic Protection • Follow up with land owners, surface lands and other third parties • Provide turn-on or start up notifications to stakeholders • Sign off Appropriate Safe Work Agreement			

Part II - Ground Disturbance Pre-Job Meeting

<u>Mandatory Attendance</u> **Date if different than Part 1:** ____________________

Operating Authority / Perfoming Authority / Ground Disturbance Representative / All Workers Directly involved in the Excavation

YES - if task has been completed / NO - if further action required		NO	YES	Action Required	ACTION OWNER
1.	Have all action items on Part 1 Pre-Job Meeting been completed?				
2.	Review Critical Operating Procedures indentified during the Pre-job Planning (Section 6 of part 1)				
3.	Safework Agreement completed and attached (Job Scope)				
4.	ERP review completed (Job/Site Specific) (Section 13 part 1)				
5.	Confined Space Entry review completed (Section 8 part 1)				
6.	Site Specific Safety Orientation completed (Section 11 part 1)				
7.	Environmental Controls review completed (Section 10 part 1)				
8.	Review initial Exposure Technique (Daylighting) - (Section 7 part 1)				
9.	Operating Authority identified and on location				
10.	Performing Authority Identified				

I have reviewed and understand the requirements for this Ground Disturbance Activity.

Print	Sign	Print	Sign
____________ /	____________	____________ /	____________
____________ /	____________	____________ /	____________
____________ /	____________	____________ /	____________
____________ /	____________	____________ /	____________
____________ /	____________	____________ /	____________

File this U/G Pre-Job attachment with all the respective Safework Agreements (2 years minimum)

AREA LOCATION:

SAFE WORK AGREEMENT

PART 1	JOB SCOPE IDENTIFICATION						/10
Date:	HOT	COLD	Confined Space	#	Ground Disturbance	#	SWA#

Performing Authority: ☐	☐ Contractor Company Name:			Contact:		
Number of Workers in Crew:	Contractor Orientation Complete:	Y	N	New/Inexperienced Worker:	Y	N
Vehicle Entry Required: Y N	If 'Yes' Positive Air Shutoff Devices are mandatory for all diesel engines					
Equipment # / Description:				SAP WO#		
Scope of Work:						

Additional Attachments / Documentation		
☐ Critical Procedure #	☐ Rescue Plan	☐
☐ P&IDs	☐ MSDS	☐

PART 2			ATMOSPHERIC TESTING			/10
INITIAL GAS TEST REQUIRED	Y	N	Test Results:	%LEL	%Oxygen	H2S ppm
Test Performed by: ☐ Operating Authority			☐ Other (Contract Worker's Name)			Time:
RETESTING REQUIRED	Y	N	Mandatory for hot work & confined space entry - fill in information on PART 9 pink copy			

PART 3				EQUIPMENT ISOLATION IDENTIFICATION			/10
Equipment Isolation Required:		Y	N	Safety Systems Bypassed	Y	N	Describe System
Type of Isolation Required:	Group Lock Out (GLOP)	Y	N	For those with 3 or more isolation points			GLOP Box#
	Simple Isolation	Y	N	For those with 2 or less isolation points (must be identified below)			A separate Job Safety Assessment is required if Zero Energy cannot be verified

SIMPLE ISOLATION IDENTIFICATION LIST - LOCATION / DESCRIPTION	
Location # 1	
Location # 2	

We have verified that all Personal Locks are installed & the System in a state of Zero Energy

Operating Authority Isolation Verified by:	SIGN	Performing Authority Isolation Verified by:	SIGN

PART 4	HAZARD IDENTIFICATION / CONTROLS		/10
Minimum PPE Required	**Supplementary PPE**	**Hot Work**	**Respiratory**
Fire Resistant Clothing	Face shield / goggles	Fire Watch identified	Respirator w/ correct cartridge
Hard hat (CSA approved)	Chemical resistant boots	Fire Eyes on bypass	Self Contained Breathing (SCBA)
Steel Toed boots (CSA)	Chem. Resist. Gloves / Suit	Fire extinguisher available	Supplied Air Breathing (SABA)
Safety glasses (CSA)	Safety harness / lifeline	Intrinsically safe tools req'd	Dust Mask
Hearing protection	PPE identified as per WHMIS	Fire blanket	Continuous Ventilation
Gloves (for task)		Equipment grounded	Man made mineral fibers
Personal 3-gas monitor			
Radio			
Emergency Response	**Environmental**	**Other Protective Measures**	**Other Protective Measures**
Muster area identified	Waste Disposal identified	Falling Objects	Hot/cold surfaces identified
Safety Shower Location/Test	Spill potential reviewed	Barrier / signs / flagging	Heat/cold environment hzrds.
Rescue personnel notified	2nd containment / drip trays	Lighting	Overhead hazards
Working alone plan	Spill kits available	Slips, trips, fall hazards	Fall protection Requirements
Rescue plan reviewed	Open Excavations	Spotter required / identified	Ground Fault Interrupter
	Sewer / drains covered	Critical lift	WHMIS / MSDS reviewed
			Sharps/Pinch points/clearance
Other: (Specify)			

PART 5	AUTHORIZATION / AGREEMENTS - WORK INITIATION	/10
PART 5		
We have reviewed all attachments and documentation associated with the identified tasks, and accept all terms and conditions associated with the job and the agreement. We understand the nature and the extent of the work and the precautions/procedures to be followed in completing the work.		
Shift Lead: I authorize work to proceed for any **Hot Work, Ground Disturbance** or **Confined Space** activities.	PRINT	SIGN
Operating Authority: I authorize work to proceed according to conditions specified on this Safe Work Agreement and attachments.	PRINT	SIGN
Performing Authority: I agree to comply with this Safe Work Agreement and attachments, and I will share the condition of this agreement and attachments with all applicable workers under my direction and control. I understand that work must stop if emergency alarms sound, conditions in the area change, or I reach the expiry time reached in the agreement below.	PRINT	SIGN
Time of SWA issue:	Time of SWA Expiry:	
SWA TRANSFER EXTENTION	**New Issue Time**	**New Expiry Time**
Operating Authority: SWA transfer extension accepted by:	PRINT	SIGN
Performing Authority: SWA transfer extension accepted by:	PRINT	SIGN

PART 6	AUTHORIZATION / AGREEMENTS - WORK CLOSURE						/10
	PERFORMING AUTHORITY			OPERATING AUTHORITY			COMMENTS
Work is Complete	Y	N		Y	N		
Workspace is clean & orderly	Y	N		Y	N		
Bypassed Safety Systems are restored	Y	N	N/A	Y	N	N/A	
Guards & barriers restored	Y	N	N/A	Y	N	N/A	
Personal Locks removed	Y	N	N/A	Y	N	N/A	
Performing Authority:	TIME	SIGN OFF					
Operating Authority:	TIME	SIGN OFF					

Distribution: WHITE (Control Room) / YELLOW (Operating Authority) / PINK (Performing Authority)

13587(09/10)
Rev 05

SAFE WORK AGREEMENT PREJOB MEETING FORM

Operations Guiding Principles:

Zero Harm to People
Zero Harm to the Environment
Nothing is so Urgent or Routine that it cannot be done Safely!

PART 7	PRE-JOB TAILGATE SAFETY MEETING - ALL WORKERS MUST ATTEND /10
Part 1 Review	Performing Authority Will: 1. Review Part 1 of your Safe Work Agreement - Job Scope Identification. 2. Review all identified attachments as applicable / identified within Part 1. 3. Ensure Inexperienced workers are identified, and effective supervision or mentors provided to compensate for the lack of experience. Identify the inexperienced worker, mentor or supervisor below in Part 8. 4. Ensure Inexperienced workers are identified with "Inexperienced Worker Orientation Sticker" on their hard hat. 5. Clarify Leadership Responsibilities with Supervisors, Foreman. Lead Hands or Delegates. 6. Communicate any changes in Part 1 back to the operating Authority.
Part 2-4 Review	Performing Authority Will: 1. Conduct FILED LEVEL RISK ASSESSMENT at worksite to supplement SWA process. 2. Review Part 10 below with regard to Emergency Response. 3. Review Atmospheric testing requirements as per Part 2. Assign responsible person for re-testing if required under Part 9. 4. Review Equipment Isolation Identification as per Part 3. confirm with workers that al hazards have been identified. 5. Review Hazard Identification / Controls as per Part 4. Confirm with workers that all hazards have been identified. 6. Review States of Mind: a) Fatigue b) Complacency c) Frustration d) Rushing 7. Review Common Errors: a) Eyes not on task b) Mind not on task c) Line of Fire d) Balance, traction or grip 8. Review and Verify the additional Protective Measures required for this activity 9. Ensure affected workers are trained and competent in use of the described protective measures. 10. Ensure all workers understand their responsibilities if emergency alarm sounds or they experience an emergency. 11. Ensure workers are aware of their "Rights to Refuse" 12. Communicate any changes in Parts 2-4 back to the Operating Authority.

PART 8	RECORD PRE-JOB MEETING ATTENDANCE		/10

INEXPERCIENCED WORKER IDENTIFICATION SECTION

Definition: Workers that have not worked in the Oil and Gas sector or their current Trade greater than One year.
Examples: Carpenter becoming a Pipefitter, Summer Students, "New hire" fresh out of school.

	VERIFY GREEN ORIENTATION STICKER	Names (print)	Signatures	Current Trade
	NEW / INEXPERIENCED			
	MENTOR / SUPERVISOR			
	NEW / INEXPERIENCED			
	MENTOR / SUPERVISOR			
	NEW / INEXPERIENCED			
	MENTOR / SUPERVISOR			

PART 9 — ATMOSPHERE RE-TESTING /10

The site is for continuous atmospheric monitors. Documented re-testing is required when the workers leave and then return to the work area. Examples of include 1) lunch breaks, 2) emergency response & muster, and 3) unexpected stoppages of work. The Responsible Person will document the Atmospheric conditions before work can re-commence.

RESPONSIBLE PERSON: (Worker's Name)

Time	%LEL	O2	H2S	Other (specify)	Worker's Signature

PART 10	EMERGENCY RESPONSE INFORMATION /10
COMMUNICATION: Use Channel # or ###-#### / State "Emergency, Emergency, Emergency" to initiate response	
REVIEW EMERGENCY ALARMS	
IDENTIFY MUSTER AREA:	

Zero Harm - Promoting Safety and Health - Preventing Injury and Illness

Return this Form to the Operating Authority / Control Room when work is complete and your Agreement is signed off

Glossary of Terms

Audits: Refers to the collaborative process of an external assessment of the business unit or business area's compliance to the policies, standards, process, and practices established by the organization. Audits must be presented as a proactive gap-assessment process to improve compliance to the organizational requirements as opposed to a fault-finding process.

Competency: The effective performance and integration of knowledge, skills, attitudes, and cognitive abilities that enable a worker to perform work of a given standard under a given set of operating conditions.

Competency Assurance: A verification of the application of knowledge and skills gained from training to do work in a safe and efficient manner. Competency assurance is generally done by an experienced leader or subject matter expert who observes the worker while performing the work and in collaboration with the worker, signs off on a verification check sheet that becomes a permanent record of the worker's competency.

Core Values: Refers to the way an organization does business and the behaviors that define the organization. Core values determine the ethics of the organization and are fundamental to the organizational culture. If health and safety are regarded as a core value, then the organization is likely to have a strong safety culture.

Critical Practices: Work practices that can result in severe injuries or death to workers and or damage to the environment or plant property and equipment if not followed as defined in the practice. A critical practice is very much like an SOP and must be followed precisely as defined in the practice. Critical practices like SOPs must also be updated at an approved frequency.

Due Diligence: This refers to an organization doing all that is reasonably expected of it to protect the health and safety of workers, asset, and the environment and to protect stakeholders from loss.

EH&S, HSE, HSEQ, SHE: All these represent the functional services of environmental, health, safety, and quality management systems in an organization. These terms have been used interchangeably throughout this book to mean the same thing.

Frontline Leaders and Supervisors: Refers to leadership personnel who generally interface with workers at the first level of the organization in a leadership or supervisory role. The term can be broadened to include all leaders and supervisors who are required to influence the behaviors of workers at the frontline to get work done. This group of leaders and supervisors are very influential in the way (safely, timely) physical work is done at the frontline and the amount of work (welding, cleaning, installing, etc.) that is completed on a daily basis. Frontline leaders and supervisors may extend two to three levels up from the frontline workers.

Gap Analysis: Refers to the difference between the current state of an organization regarding compliance to a policy, standard, or procedure and the desired future state of the organization. A gap analysis is often accompanied by a gap closure strategy for taking the organization to the desired state of compliance.

Incident: An unplanned event that has resulted in injury or fatality to people or damage to or loss of equipment, material, quality, product, process, or the environment.

Management of Change (MOC): A procedure required for managing risks associated with changes made in an industrial process environment. MOC procedures require the right levels of expertise to evaluate the impact of a proposed change on an operation before a change is approved. The procedure requires the right level of authority for approval of the proposed change and a mechanism for ensuring that newly introduced risks are properly assigned and mitigated once the proposed change is made.

Near Miss: An event that would have led to an injury had it occurred, fatality, damage to equipment, assets, environment, or corporate image. A loss to the organization had it occurred.

Presenteeism: Productivity losses associated with employee work when employees come to work but perform below par due to any kind of illness.

Risk: Defined as a combination of the probability of a hazard or potential failure occurring and the impact of that hazard or potential failure on the business. Risk is determined differently by organizations; however, where health and safety are concerned, a common language is required so that risks are comparable across organizations and within industries.

Safety Management System (SMS): An SMS is a set of processes, behaviors, and tools developed by organizations to meet the health and safety requirements for protecting people, the environment, and assets in the same order at the workplace. SMS varies in complexity and scope based on the type of work or business engaged in. Generally, an SMS is built upon organizational policies, standards, procedures, work practices, and work tools all designed to ensure the health and safety of all workers and to prevent loss to the organization and damage to the environment during its operations.

Shared Learning: Refers to the process of sharing knowledge within an organization and across all business units or business areas or functions of the organization. Shared learning extends to the organized sharing of knowledge within an industry as well as across industries.

Shared learning should be done in a consistent simple format that caters to the learning need of all stakeholders in the workplace.

Situational Leadership: A model for leadership that requires leaders to individually consider the work maturity level of the worker such that appropriate leadership styles can be applied for the specific maturity level of the worker. Situational leadership requires leaders to accurately identify the maturity level of the worker and to use leadership behaviors consistent with the maturity level of the worker. The situational leadership model was developed by Paul Hershey and Ken Blanchard (Bass, 1990) and has been one of the more successful models for developing the workforce.

Standard Operating Procedures (SOPs): Procedures developed for guiding inexperienced workers with complex work activities in a process environment or facility. SOPs also guide experienced workers on jobs with which they may be unfamiliar. SOPs are generally derived from the operating manuals provided by vendors or may be developed based on experiences of operating personnel. SOPs must be updated at a facility-approved frequency so as to ensure that improvements are incorporated into the procedure over time.

Training: Any process that is designed to improve knowledge, skills, attitudes, and/or behaviors of workers. Training is designed to improve the competency of the worker in performing work efficiently and safely. Training can take the form of instructor-led or self-led and can be offered in terms of field demonstration, university or school courses, and seminars and e-courses. Training is generally focused on business needs and is driven by time-critical business skills and knowledge, and its goal is often to improve worker performance.

Training Matrix: A training matrix is a simple matrix that represents the training required by personnel on one axis and the personnel to be trained on the other. The intersection point between specific training course and an individual may be color coded to visibly show whether an individual does not require training in that area or is untrained, trained, or competent.

References

Abdallah, S. and Burnham, G. (n.d.). *Public Health Guide for Emergencies*. The Johns Hopkins and Red Cross/Red Crescent. Retrieved January 16, 2011 from http://pdf.usaid.gov/pdf_docs/PNACU086.pdf.

Agrusa, J. and Lema, J. D. 2007. An examination of Mississippi gulf coast casino management styles with implications for employee turnover. *UNLV Gaming Research & Review Journal, 11*(1), 13–26. Retrieved April 11, 2007 from EBSCOhost database.

Ahmed, I. M. 2008. BAPCO's experience in implementing behavioral based safety (BBS) process. *American Society of Safety Engineers (ASSE), Middle East Chapter, 8th Professional Development Conference Proceedings*, Kingdom of Bahrain. Ref: ASSE-0208-027.

Ajabnoor, A. M. 2008. Near-miss analysis—Accidents prevention. *American Society of Safety Engineers (ASSE), Middle East Chapter, 8th Professional Development Conference Proceedings*, Kingdom of Bahrain. Ref: ASSE-0208-013.

Al-Kudmani, A. 2008. Building a safety culture-our experience. *American Society of Safety Engineers (ASSE), Middle East Chapter, 8th Professional Development Conference Proceedings*, Kingdom of Bahrain. Ref: ASSE-0208-021.

Al Hajiri, M. Q. 2008. Behavior based safety leading to safety excellence. *American Society of Safety Engineers (ASSE), Middle East Chapter, 8th Professional Development Conference Proceedings*, Kingdom of Bahrain. Ref: ASSE-0208-038.

Al Marzooqi, A. i. 2006. Improving road safety standards. *American Society of Safety Engineers (ASSE), Middle East Chapter, 6th Professional Development Conference Proceedings*, Kingdom of Bahrain. Ref: ASSE-0306015, 2003.

Allars, K. 2007. *Bunciefiled, The Story So Far—Case Study Presentation Delivered at the Platts: Creating Value in Oil Storage*. Budapest, Hungary: UK-HSE.

Allen, W. R., Drevs, R. A., and Rube, J. A. 1999. Reasons why college-educated women change employment. *Journal of Business & Psychology, 14*(1), 77–93. Retrieved April 17, 2007 from EBSCOhost database.

Al-Sulby, I. S. 2008. A challenge during implementation Hadeed safety management plan: Truck drivers illiteracy. *American Society of Safety Engineers (ASSE), Middle East Chapter, 8th Professional Development Conference Proceedings*, Kingdom of Bahrain. Ref: ASSE-0208-032.

American History. 2003. The Panama Canal. *American History, 38*(4), 18. Retrieved September 25, 2010 from EBSCOHost database.

Anderson, G. 2008. Creating a culture of safety. *American Society of Safety Engineers (ASSE), Middle East Chapter, 8th Professional Development Conference Proceedings*, Kingdom of Bahrain. Ref: ASSE-0208-019.

Argyris, C. 1996. Actionable knowledge: Design, causality in the service of consequential theory. *Journal of Applied Behavioral Safety Culture Sciences, 32*, 390–406.

Ashton, J. 2004. *Environment and Health—Health as a Business Management Issue in the 21st Century*. Institute of Petroleum, Energy Institute. ISBN 0-85293-403-3.

Aspen Publishers Inc. 2009. ExxonMobil won't appeal verdict. *Oil Spill Intelligence Report*, *32*(28), 1. Retrieved February 6, 2010 from EBSCOHost database.

Asseri, A. H. 2008. Behavioral-based safety (BBS), program implementation pitfalls. *American Society of Safety Engineers (ASSE), Middle East Chapter, 8th Professional Development Conference Proceedings*, Kingdom of Bahrain. Ref: ASSE-0208-026.

Associated Press. 2010. *Poll: BP's Image Improving after Oil Spill; Fears of Seafood Persist*. Associated Press. Retrieved November 22, 2011 from http://blog.al.com/live/2010/08/poll_bps_image_improving_after.html.

Austin, S. 2005. Targeting employees under 30: Effective safety methods for a new generation of workers. *Occupational Hazards*, 27–29. Retrieved October 10, 2009 from EBSCOhost.

Ayers, D. 2007. Contractor safety: Building trust and communication. *Occupational Hazards*, *69*(10), 64–74. Retrieved March 11, 2011 from EBSCOhost database.

Baker, J. (III), Bowman, F. L., Glen, E., Gorton, S., Hundershot, D., Levison, N., Priest, S., Rosentel, T. P., Wiegmann, D., and Wilson, D. 2007. *The Report of the B.P. U.S. Refineries Independent Safety Review Panel*. Retrieved March 10, 2011 from http:///www.csb.gov.

Baldwin, D. 2001. How to win the blame game. *Harvard Business Review*, *79*(7), 55–62. Retrieved March 11 from EBSCOhost database. *Communications and Conflict*, *13*(1): 9–19. Retrieved March 11, 2011 from EBSCOhost database.

Ball, C. 2004. *Workforce View of Health Interventions in the Workplace. Health as a Business Management Issue in the 21st Century*. Institute of Petroleum, Energy Institute, ISBN 0-85293-403-3.

Barrett, G. A. 2000. Management's impact of behavioral safety, safety management. *Journal of Professional Safety, American Society of Safety Engineers, March Edition*.

Bass, B. M. 1990. *Handbook of Leadership Theory, Research, and Managerial Applications* (3rd ed.). New York: The Free Press.

BBC Video Case Study. 1991. Spiral to disaster: Piper Alpha Rig Disaster. *BBC Education & Training Production, Stone City Films*.

Bell, A. 2007. Using vision to shape the future. *Leader to Leader*, (45), 17–21. Retrieved July 2, 2010 from EBSCOhost database.

Bell, K. J., O'Connell, M. S., Reeder, M., and Nigel, R. 2008. Predicting and improving safety performance. *Industrial Management*, *50*(2), 12–16.

Bird, E. F., Germain, L. D., and Clark, D. M. 2003. *DNV, Loss Control Management. Practical Loss Control Leadership* (3rd ed.). Georgia: Det Norske Veritas.

Blanchard, K. 2004. Leadership and the bottom line. *Executive Excellence*, *21*(9), 18. Retrieved July 26, 2008 from EBSCOhost database.

Blanchard, K. 2008. Situational leadership. Adapt your style to their development level. *Leadership Excellence*, *25*(5), 19. Retrieved July 25, 2008 from EBSCOHost database.

Bliss, A. 2004. *The Government Perspective of Occupational Health Provision. Health as a Business Management Issue in the 21st Century*. Institute of Petroleum, Energy Institute, 2004, ISBN 0-85293-403-3.

Boardman, J. and Lyon, A. 2006. Defining best practice in corporate occupational health and safety governance. In *Health and Safety Executive (HSE), Research Report #506*. London: Acona Ltd. Retrieved December 30, 2010 from www.acona.co.uk\reports\rr506.pdf.

Boyle, T. 2000. *Health and Safety: Risk Management*. Wigston, LE/IOSH Services Limited London: IOSH.

Brahmasrene, T. and Sanders, S. S. 2009. The influence of training, safety audits, and disciplinary action on safety management. *Journal of Organizational Culture, Production in Industrial Firms. Safety Science-Pergamon*, (41). Retrieved March 10, 2011 from www.elsevier.com.

Brewster, B. 2010. The embedded safety coach: Global solution to localized challenges. *Paper No. ASSE-MEC-2010-36, 9th ASSE-MEC Professional Development Conference*. Kingdom of Bahrain, February 20th–24th 2010.

British Safety Council (BSC). 2005. International diploma in occupational safety and Health, module A, sub-element British Safety Council, Issue 2. January 2006, British Safety Council Awards.

British Safety Council (BSC). 2006. International diploma in occupational safety and health, module a sub-element. *British Safety Council*, (2). British Safety Council Awards.

Broadribb, M. 2010. Lessons from Texas. *TCE: The Chemical Engineer*, 824, 20–21. Retrieved July 3, 2010 from EBSCOhost database.

BSI (British Standards), 2007. *Occupational Health and Safety Management System—Requirements*. OHSAS 18000:2007. ISBN 978-0-580-59404-5. Copyright 2007.

Buchanan, D. and Huczynski, A. 2004. *Organizational Behaviour* (5th ed.). Prentice Hall-Financial Times.

Bucklow, A. 2001. *A Safety Culture Is For Life—An Independent Paper and Resources Book for Railway Safety, Facilit8*. Communications Ltd. Retrieved March 10, 2010 from www.facilit8.com.

Bufton, M. W. and Melling, J. 2005. Coming up for air: Experts, employers, and workers in campaigns to compensate silicosis sufferers in Britain, 1918–1939. *The Journal of the Society for the Social History of Medicine/SSHM, 0951-631X, 18*(1), 63–86. Retrieved April 22, 2010 from EBSCOhost database.

Burgherr, P. and Hirschberg, S. 2008. A comparative analysis of accident risks in fossil, hydro, and nuclear energy chains. *Human & Ecological Risk Assessment, 14*(5), 947–973. Retrieved February 3, 2010 from EBSCOHost database.

Burns, C., Mearns, K., and McGeoroge, P. 2006. Explicit and implicit trust within safety culture. *Journal of Risk Analysis (Society of Risk Analysis), 26*(5).

Burtch, S. D. 2008. *Beyond a Safety Culture: Protecting People, Processes and Operations*. Copyright © 2008 DuPont. The miracles of science™.

Burton, W. M., McCallester, K. T., Chen, C., and Edington, D. W. 2005. The association of health status, worksite fitness center participation, and two measures of productivity. *Journal of Occupational and Environmental Medicine, The American College of Occupational and Environmental Medicine, 47*(4).

Cable, J. 2005. Generation Y safety: The challenges of reaching the under-30 worker. *Occupational Hazards*, (54), 20–21. Retrieved October 10, 2009 from EBSCOhost database.

Canadian Broadcasting Centre News. 2009. *No Signals from Locator Beacons in Crashed Helicopter: Officials*. The Canadian Press. Retrieved November 10, 2010 from http://www.cbc.ca/canada/newfoundland-labrador/story/2009/03/12/offshore-helicopter.html.

Canadian Broadcasting Centre News. 2010. *BP Gulf Spill Cost Near $40 B US*. The Canadian Press. Retrieved November 10, 2010 from http://www.cbc.ca/money/story/2010/11/02/bp-profit-112.html.

Canadian Centre for Occupational Health and Safety. 2005. Workplace health and safety best practices inventory. *File # WHSPL-0501. Report prepared for Alberta*

Human Resource and Employment (AHRE) Work Safe Alberta by CCOHS. Retrieved December 30, 2010 from http://employment.alberta.ca/documents/WHS-PUB_bp-inventory.pdf.

Canadian Centre for Occupational Health and Safety. 2011. *OH&S Legislation in Canada—Due Diligence*. Retrieved February 23, 2011 from http://www.ccohs.ca/oshanswers/legisl/diligence.html.

Capital Confidential. 2010. *OSHA: BP Less Safe than Other Oil Companies*. Retrieved November 21, 2010 from http://biggovernment.com/capitolconfidential/2010/06/25/osha-bp-less-safe-than-other-oil-companies/.

Carrillo, R. A. 2006. Leadership formula: Trust + Credibility × Competence = Results a guide to safety excellence through organizational, cultural and personnel change. *American Society of Safety Engineers (ASSE), Middle East Chapter, 6th Professional Development Conference Proceedings*, Kingdom of Bahrain. Ref: ASSE-0306-010.

Carroll, I. S. 2002. *Leadership and Safety in Nuclear Power and Health Care—Presentation at the Managers, and Safety in High Reliability Organizations Seminar*, Aberdeen University.

Chacos, B. 2010. *Company Plant Safety Team Audit Checklist*. Retrieved March 13, 2011 from http://www.ehow.com/list_7298475_company-safety-team-audit-check list.html#ixzz1E0UeGJuB.

Chamat, W. 2008. The psychology of behavioral safety. *American Society of Safety Engineers (ASSE), Middle East Chapter, 8th Professional Development Conference Proceedings*, Kingdom of Bahrain. Ref: ASSE-0208-037.

Chemical Safety Board (CSB). 2007. Investigation of the March 23rd 2005 Explosion and Fire at the BP Texas City Refinery. Presentation by the Chemical Safety and Hazard Investigation Board, March 20th, 2007.

Chemical Safety Board (CSB). 2010. Safety deficiencies at all levels at BP. *TCE: The Chemical Engineer*, 790, 3. Retrieved February 7, 2010 from EBSCOHost database.

Chemical Safety Board (CSB), 2010a. Completed Investigations. Retrieved May 15, 2010 from http://www.csb.gov/investigations/investigations.aspx?Type=2&F_All=y

Clarke, S. 2006. The relationship between safety climate and safety performance, a meta-analytic review. *Journal of Occupational Health Psychology, 11*(4), 315–327.

Codjia, M. 2010. *Compliance Audit Checklist*. e.How.com. Retrieved March 11, 2011 from http://www.ehow.com/info_7735520_compliance-audit-checklist.html.

Coles, S. 2009. Canada's crunch. *Employee Benefits*, 28–28. Retrieved February 15, 2010 from EBSCOhost database.

Conchie, S. M. and Donald, I. J. 2006. The role of distrust in offshore safety performance. *Risk Analysis, 26*(5).

Cooling, R. 2007. A systems approach to managing worker involvement in safety and health: Behavior, culture and management of change session. *Institution of Occupational Safety and Health Conference*, Telford, UK.Cooper, L. 2007. Proactive pressure management in a changing world: Managing pressure and preventing workplace stress session. *Institution of Occupational Safety and Health Conference 07*, Telford, UK.

Cooper, M. D. 2000. Towards a model of safety culture. *Safety Culture Science*, 362000, 111–136.

Cooper, M. D. 2001. Improving safety culture—A practical guide. *Applied Behavioral Sciences*. Retrieved March 10, 2011 from http://www.bsafe.co.uk.

Cooper, M. D. 2003. The impact of management's commitment on employee behavior: A field study. *American Society of Safety Engineers (ASSE), Middle East Chapter, 6th Professional Development Conference Proceedings*, Kingdom of Bahrain. Ref: ASSE-0306-013.

Cooper, M. D. 2004. Examining your safety culture to identify where best to focus your improvement efforts, Session No. 625. *The Worldwide Program of Research on Safety Culture*, Indiana University, Bloomington, USA.

Cooper, M. D. 2005. Explanatory analysis of managerial commitment and feedback consequences on behavioral safety maintenance. *Journal of Organizational Behavior Management*.

Cooper, M. D. 2006. The impact of management's commitment on employee behavior: A field study. *American Society of Safety Engineers (ASSE), Middle East Chapter, 6th Professional Development Conference Proceedings*, Kingdom of Bahrain. Ref: ASSE-0306-013.

Cooper M. D. and Phillips, R. A. 2004. Exploratory analysis of the safety climate and safety behavior relationship, paper accepted. *Journal of Safety Research, 35*, National Safety Council—USA. Retrieved March 10, 2011 from www.nsc.org.

Cooper-Hakim, A. and Viswesvaran, C. 2005. The construct of work commitment: Testing an integrative framework. *Psychological Bulletin, 131*, 241–259. Retrieved October 28, 2006 from EBSCOhost database.

Copper, M. D. and Phillips, R. A. 2004. Explanatory analysis of the safety climate and safety behavior relationship. *Journal of Safety Research*, 352004, 497–512.

Cox, S., Johns, B., and Collinson, D. 2006. Trust relations in high-reliability organizations, *Journal of Risk Analysis (Society of Risk Analysis), 26*(5).

Cullen, W. D. 1990. *The Public Enquiry into the Piper-Alpha Disaster*. HMSO, ISBN 0101-1102X.

Dapo, O. 2006. Achieving world-class safety performance in a multi-cultural environment: A PDO case study. *American Society of Safety Engineers (ASSE), Middle East Chapter, 6th Professional Development Conference Proceedings*, Kingdom of Bahrain. Ref: ASSE-0306-004.

Darby, M. 2010. The 9-Vector view of human performance. *T + D, 64*(4), 38–40. Retrieved August 5, 2010 from EBSCOhost database.

Davis, J. H., Schoorman, F. D., Mayer, R. C., and Hwee Hoon, T. 2000. The trusted general manager and business unit performance: Empirical evidence of a competitive advantage. *Strategic Management Journal, 21*(5), 563–576. Retrieved November 6, 2006 from EBSCOhost database.

Desselle, S. P. 2005. Job turnover intentions among certified pharmacy technicians. *Journal of the American Pharmacists Association*, 45(6), 676–683. Retrieved May 28, 2007 from http://www.medscape.com/viewarticle/518371.

Dols, J., Landrum, P., and Wieck, K. L. 2010. Leading and managing an intergenerational workforce. *Creative Nursing, 16*(2), 68–74. Retrieved November 17, 2010 from EBSCOhost database.

Duff, A. R., Roberson, I. T., Phillips, R. A., and Cooper, M. D. 1994. Improving safety by the modification of behavior. *Journal of Construction Management and Economics*, 12, 67–78.

Dunleavy, K. N., Chory, R. M., and Goodboy, A. K. 2010. Responses to deception in the workplace: Perceptions of credibility, power, and trustworthiness.

Communication Studies, 61(2), 239–255. Retrieved December 26, 2010 from EBSCOhost database.

DuPont. 2008. Leading indicators—Key to injury prevention. *The Miracles of Science.* Wilmington, DE.

Edic, J. 2009. A different way of thinking about safety. *Professional Contractor, 3rd Quarter*, 26–27. Retrieved July 4, 2010 from EBSCOhost database.

Elfarmawi, M. A. 2008. Stress is a root cause for accidents. *American Society of Safety Engineers (ASSE), Middle East Chapter, 8th Professional Development Conference Proceedings*, Kingdom of Bahrain. Ref: ASSE-0208-014, 2008.

Energy Institute (EI). 2010. *Guidance on Health Surveillance* (1st ed.). London: Energy Institute Publishing, 2010.

ENFORM Canada. 2010. Contractor management systems guidelines. *The Safety Association for the Oil and Gas Industry.* Retrieved December 11, 2010 from http://enform.ca/media/86272/contractor_management_may2010.pdf.

Esposito, P. 2009. Safety audits comparing three types of assessments. *Professional Safety, 54*(12), 42–43. Retrieved March 11, 2011 from EBSCOhost database.

European Agency for Safety and Health at Work. 2009. *Safe Maintenance—For Employers Safe Workers-Save Money.* Retrieved January 30, 2011 from http://osha.europa.eu/en/publications/factsheets/89.

Facione, P. 2006. Critical thinking: What it is and why it counts. Retrieved January 19, 2009 from http://nsu.edu/iea/image/critical_thinking.pdf.

Farrington-Darby, T., Pickup, L., and Wilson, J. R. 2005. Safety culture in railway maintenance. Safety. *Safety Science*, 43, 39–60. Retrieved August 14, 2011 from http://www.elsevier.com.

Fast, N. 2010. How to stop the blame game. *Harvard Business Review*, May 14. Retrieved on January 28, 2011. http://www.businessweek.com/managing/content/may2010/ca20100514_703477.htm.

Fearing, J. 2008. Prevention, control, sustainability, plan and conduct effective safety audits. *Industrial Engineer: IE, 40*(5), 38–43. Retrieved March 11, 2011 from EBSCOhost database.

Finegold, D., Mohrman, S., and Spreitzer, G. M. 2002. Age effects on the predictors of technical workers' commitment and willingness to turnover. *Journal of Organizational Behavior, 23*(5), 655–675. Retrieved October 21, 2006 from EBSCOhost database.

Fleming, M. 2001. Safety culture maturity model. *Health and Safety Executive Report Series*, HSE-UK.

Flin, R. 2005. *Manager's Resilience*. Industrial Psychology Research Centre, University of Aberdeen, Safety Management Conference, Dubai.

Flin, R., Burns, C., Mearns, K., Yule, S., and Robertson, E. M. 2006. Measuring safety climate in healthcare. *Quality Safety Health Care*, 15, 109–115.

Flint-Taylor, H. 2007. Work-life\balance and improving employee well-being: Occupational health & workplace wellbeing session. *Institution of Occupational Safety and Health Conference 07*, Telford, UK.

Francesco, A. M. and Gold, B. A. 2005. International organizational behavior. *Text, Cases, and Exercises* (2nd ed.). New Jersey: Pearson Education.

Friend, G. and Zehle, S. 2004. Guide to business planning. *Copyright Economist Intelligence Unit*. Retrieved March 2010 from EBSCOhost database.

Gall, B., Bryden, R., Hoad, K., Jeffries, P., Miles, R., Reeves, G., Scanlon, M., Symonds, J., and Wilkinson, J. 2006. *Improving Alertness through Effective Fatigue Management*. Energy Institute/Institute of Petroleum (IP), ISBN 978-0-85293-460-9.

Geller, E. S. 2006. From good to great in safety: What does it take to be world class? *Professional Safety*, *51*(6), 35–40. Retrieved January 16, 2010 from EBSCOhost database.

Geller, E. S. 2008. People -based leadership: Enriching a work culture for world class safety. *Professional Safety*, *53*(3), 29–36. Retrieved July 1, 2010 from EBSCOHost database.

Gentry, W. A. 2006. Human resource officers' opinions of their own companies and "the big three": A qualitative study profiling businesses in a southeastern city. *Organization Development Journal*, *24*(2), 33–42. Retrieved April 26, 2007 from EBSCOhost database.

Ghanem, W. S. 2009. Enhancing company operational performance through effectively engaging safety culture within the Emirates National Oil Companies Distribution Operations, United Arab Emirates. *Down Stream Safety Management (ACI - Conference), 17–18th June 2009, London, UK.*

Ghanem, W. and Newson-Smith, M. 2008. Application of job safety (task) analysis for the purpose of effective occupational health at work evaluations. *American Society of Safety Engineers—Middle East Chapter (161), 8th Professional Development Conference & Exhibition*, Kingdom of Bahrain. www.asse-mec.org.

Gill, C. 2009. Union impact on the effective adoption of high performance work practices. *Human Resources Management Review*, *19*(1), 39–50. Retrieved April 24, 2010 from EBSCOHost database.

Gillis, J. 2010. Size of oil spill underestimated, scientists say. *The New York Times*. Retrieved February 5, 2011 from http://www.nytimes.com/2010/05/14/us/14oil.html.

Glisson, C. and James, L. R. 2002. The cross-level effects of culture and climate in human service teams. *Journal of Organizational Behavior*, 23(6), 767–794. Retrieved November 12, 2006 from EBSCOhost database.

Government of Alberta. 2006. *Building an Effective Health and Safety Management System*. Retrieved July 3, 2010 from http://employment.alberta.ca/documents/WHS/WHS-PS-building.pdf.

Graham, B., Reilly, K. W., Beinecke, F., Boesch, D. F., Garcia, T. D., Murray, C.A., and Ulmer, F. 2011. National Commission on the BP deepwater horizon oil spill and offshore drilling. Deep water: *The Gulf Oil Disaster and the Future of Offshore Drilling*. Retrieved January 16, 2011 from https://s3.amazonaws.com/pdf_final/DEEPWATER_ReporttothePresident_FINAL.pdf.

Guthrie, J. P. 2001. High-involvement work practices, turnover, and productivity: Evidence from New Zealand. *Academy of Management Journal*, 44(1), 180–190. Retrieved November 9, 2007 from EBSCOHost database.

Halama, L. M., Kelly, D., Rideout, S., and Robinson, P. 2004. Leading indicators best practices presentation. Construction Owners Association of Alberta. Available March 21 from http://www.coaa.ab.ca/LinkClick.aspx?link=pdfs/Leading%20Indicators/COAA%20Leadiing%20Indicators%20%20Best%20Practices%20Conference%20Presentation.pdf&tabid=154.

Harrald, J. R., Marcus, H. S., and Wallace, W. A. 1990. The Exxon Valdez: An assessment of crisis prevention and management systems. *Interfaces, 20*(5), 14–30. Retrieved February 6, 2010 from EBSCOHost database.

Harris, K. J., Kacmar, K. M., and Witt, L. A. 2005. An examination of the curvilinear relationship between leader–member exchange and intent to turnover. *Journal of Organizational Behavior, 26*(4), 363–378. Retrieved April 11, 2007 from EBSCOHost database.

Harris, R. 1998. *Introduction to Creative Thinking*. Retrieved January 20, 2009 from http://www.virtualsalt.com/crebook1.htm.

Harter, J. K., Schmidt, F. L., Killham, E. A., and Asplund, J. W. 2006. Q12 Meta-Analysis. *The Gallop Organization*.

Harvey, J., Bolam, H., Gregory, D., and Erdos, G. 2000. The effectiveness of training to change safety culture and attitudes within a highly regulated environment. *Safety & Health Practitioner, Personal Review, 30*(6), 615–636. MCB University Press 00-48-3496.

Harvey, J., Erdos, G., Bolam, H., and Gregory, D. 2001. *An Examination of Different Safety Cultures in an Organization Where Safety Is Paramount*. University of Newcastle Upon Tyne, School of Management, England.

Haukeild, K. 2005. Safety Culture–What Works. University of Oslo, A research study funded by the Research Council of Norway, presented at the Health & Safety Culture Management for Oil and Gas Industry, Amsterdam, Netherlands, May 2005.

Hayward, T. 2005. Working safely—A continuous journey, special topic-safety. *First Break Magazine, 23, EAGE*.

Health & Safety Authority (HAS). 2004. *Improving Safety Behavior at Work*. Guidance for Employers, Managers and Self-Employed, HAS, Dublin, Ireland.

Health & Safety Commission (HSC). 2001. Consultative document: Health and safety responsibilities of directors. *Health and Safety Executive*. Retrieved March 10, 2001 from www.hse.gov.uk/condocs/.

Health & Safety Executive (HSE). 1996. *Cost of Accidents at Work. Health & Safety Guideline (HSG-96)*. HSE Books.

Health & Safety Executive (HSE). 1998. Working alone in safety—Controlling the risks of solitary work. *INDG73 (Rev 3)*, Retrieved March 10, 20910 from www.hse.gov.uk.

Health & Safety Executive (HSE). 2000. Successful health & safety management. *Health & Safety Guideline (HSG-65)*, HSE Books.

Health & Safety Executive (HSE). 2005a. A review of safety culture and safety climate literature for the development of the safety culture inspection toolkit. *Human Engineering*, Research Report 367: HSE Books.

Health & Safety Executive (HSE). 2005b. Reducing error & influencing behavior. *Health & Safety Guideline (HSG-48)*, HSE Books.

Hein, M. 2003. The blame game. In: *Heinsights*. Retrieved March 13, 2011 from http://www.westfallteam.com/Papers/The_Blame_Game.pdf.

Helman, C. 2010. *The BP Oil Spill Blame Game*. Retrieved March 13, 2011 from http://www.forbes.com/2010/05/04/bp-oil-transocean-business-energy-gulf-spill.html.

Hemdi, M. A. and Nasurdin, A. M. 2006. Predicting turnover intentions of hotel employees: The influence of employee development human resource management practices and trust in organization. *Gadjah Mada International Journal of Business, 8*(1), 42–64. Retrieved November 9, 2007 from EBSCOhost database.

Hernan, P. 2009. Hoover dam: History is not for the faint of heart. *Pit & Quarry, 88*(9). Retrieved September 25, 2010 from EBSCOHost database.

Herzburg, F. 2003. One more time: How do you motivate employees. *Harvard Business Review on Motivating People*, HBS Press.

Hibbert, L. 2008. Averting disaster. *Professional Engineering, 21*(11), 20–23. Retrieved February 3, 2010 from EBSCOHost database.

Higgins, L. and Hack, B. 2004. Measurement in the 21st century white paper. *American Productivity & Quality Center (APQC)*. Houston, TX.

Hill, C. 2000. Benchmarking and best practices. *American Society for Quality (ASQ). ASQ's 54th Annual Quality Congress Proceedings*, pp. 715–717.

Hoffman, M. 2007. Employee retention begins with effective leadership. *Hotel & Motel Management, 222*(1), 58–58. Retrieved April 16, 2007 from EBSCOHost database.

Hofsteade, G. 1980. Motivation, leadership, and organization: Do American theories apply abroad? *Organizational Dynamics, Summer 1980, AMACOM*, a division of American Management Associates.

Hopkins, A. 2006. Studying organizational cultures and their effects on safety. *International Conference on Occupational Risk Prevention Seville*, Safety Culture Sciences, 0925-7535.

Hopkins, A. 2006. Studying organizational cultures and their effects on safety, Safety Science. *Volume X-Article in Press. International Conference on Occupational Risk Prevention, Seville, 3rd International Conference, Working on Safety*, The Netherlands.

Hopkins, P. 2008. The skills crisis in the pipeline sector of the oil and gas business. *Journal of Pipeline Engineering, 7*(3), 149–172. Retrieved February 15, 2010 from EBSCOhost database.

Hopkins, S. M. and Weathington, B. L. 2006. The relationships between justice perceptions, trust and employee attitudes in a downsized organization. *Journal of Psychology, 140*(5), 477–498. Retrieved November 24, 2007 from EBSCOHost database.

Horizon Terminals Ltd (HTL). 2007. *EHSSQ Handbook for Horizon Terminals Ltd Group*. Internal Company Guideline Document.

Horseman, S. 2008. Workplace wellness programs: Healthy employees = safe employees? *American Society of Safety Engineers (ASSE), Middle East Chapter, 8th Professional Development Conference Proceedings*, Kingdom of Bahrain. Ref: ASSE-0208-039.

Hudson, P., Verchuur, W. L. G., Paker, D., and Lawton, R. 1996. *Bending the Rules: Managing Violation in the Workplace*. Centre for Safety Science, Leiden University/Department of Physiology, Manchester University, Gerard van der Graaf, SIEF.

Hughes, P. and Ferrett, E. 2009. Promoting a positive health and safety culture. In: *Introduction to Health and Safety at Work*. Elsevier, Butterworth-Heinemann.

Human Resources. 2009. The corporate manslaughter and corporate homicide act. *Hrmagazine co.uk*. 28–28. Retrieved July 4, 2010 from EBSCOhost database.

ILO, 2008. http://www.ilo.org; http://www.ilo.org/global/About_the_ILO/lang--en/index.htm

Industrial Safety & Hygiene News. 2010. BP creates new safety division with "sweeping powers." *ISHN*, *44*(11), 18. Retrieved March 13, 2011 from www.ishn.com.

International Association of Oil and Gas Producers (OGP). 2009. Safety performance indicators 2008 data. *Report # 419*. Retrieved December 11, 2010 from http://www.ogp.org.uk/pubs/419.

Iskandar, M. 2010. Behavior and human factors in safety. Paper No. ASSE-MEC-2010-30, *9th ASSE-MEC Professional Development Conference*, Kingdom of Bahrain.

Janeeh, S. 2005. Effective motivational strategies. *2nd Behavioral Based Safety Conference*, Manama, Kingdom of Bahrain, Marcus Evans Conferences, 2005.

Johnson, S. E. 2003. Behavioral safety theory. *Professional Safety*, *48*(10), 39.

Johnson & Johnson. 2009. Our responsibility. *2009 Sustainability Report*, p. 40. Retrieved December 30, 2010 from http://www.investor.jnj.com/2009sustainability.

Johnstone, J. E. 2001. How EHS auditing programs will evolve for the new millennium. Paper presented at the *SPE/EPA/DOE Exploration and Production Environmental Conference*. SPE 66579, Society of Petroleum Engineers, San Antonio, TX, February 2001.

Kletz, T. 2009. *What Went Wrong*? Gulf Professional Publishing. ISBN-13: 97818561753192009.

Knudson, T. 2004. *Management View of Health Interventions in the Workplace. Health as a Business Management Issue in the 21st Century*—Institute of Petroleum, Published by the Energy Institute, 2004, ISBN 0-85293-403-3.

Korkmaz, M. 2007. The effects of leadership styles on organizational health. *Educational Research Quarterly*, *30*(3), 22–54. Retrieved April 16, 2007 from EBSCOHost database.

Kotter, J. P. 1998. *What Leaders Really Do*. Originally published in May/June 1990, Reprint 90309, Harvard Business Review on Leadership, HBS Press, 1998.

Krishnan, V. 2009. Victims fight on 25 years later. *Chemistry & Industry*, *23*, 7. Retrieved February 3, 2010 from EBSCOHost database.

Leemann, J. E. 2007. Puncturing the myth of world class safety. *Industrial Safety & Hygiene News*, *41*(6), 1–37. Retrieved January 16, 2010 from EBSCOhost.

Leigh, J. P., Markowitz, S., Fahs, M., and Landrigan, P. 2000. The cost of occupational injuries and illnesses. *Frontline*. Retrieved November 16, 2010 from http://www.pbs.org/wgbh/pages/frontline/shows/workplace/etc/cost.html.

Leman, J. 2010. OSHA is on the prowl. *Ward's Dealer Business*, *44*(12), 6–7. Retrieved March 11, 2011 from EBSCOhost database.

Levinson, H. 2003. One more time: How do you motivate employees. *R0301H, Harvard Business Review on Motivating People*, HBS Press.

Lewis, R. 2007. *When Cultures Collide, Leading across Cultures* (3rd ed.). Nicholas Brealey International.

Liberty Mutual 2001. Safety yields high returns. *Professional Safety*, *46*(11), 12. Retrieved September 25, 2010 from EBSCOhost database.

Liberty Mutual. 2009. *The Most Disabling Workplace Injuries Cost Industry an Estimated $53 Billion*. Liberty Mutual Research Institute for Safety 2009. Retrieved November 13, 2010 from http://www.google.ca/search?hl=en&source=hp&q=Liberty+Mutual+Workplace&btnG=Google+Search&meta=&aq=f&aqi=&aql=&oq=&gs_rfai=.

Little, P. L. 2006. The high cost of under-skilled labor. *Industrial Maintenance & Plant Operation*, *67*, 10. Retrieved October 24, 2006 from EBSCOhost database.

Livingston, J. S. 2003. Pygmalion in management. *R0301G, Harvard Business Review, Motivating People*, HBS Press.

Lockwood, N. R. 2007. Leveraging employee engagement for competitive advantage: HR's strategic role. *HR Magazine, 52*(3), 1–11. Retrieved April 26, 2007 from EBSCOhost database.

Love, R. and Dale, B. G. 2007. Benchmarking. In *Managing Quality*, pp. 480–493. USA: Blackwell Publishing.

Lowe, D., Levitt, K. J., and Wilson, T. 2008. Solutions for retaining generation Y employees in the workplace. *Business Renaissance Quarterly*, *3*(3), 43–58. Retrieved November 20, 2010 from EBSCOhost database.

Lutchman, C. 2008. *Leadership impact on turnover among power engineers in the oil sands of Alberta*. Dissertation, University of Phoenix.

Lutchman, C. 2010. *Project Execution: A Practical Approach to Industrial and Commercial Project Management*. Boca Raton, FL: CRC Press.

Lyall, S. 2010. In BP's record, a history of boldness and costly blunders. *New York Times*. July 12, 2010. Retrieved June 20, 2011. Available from http://www.nytimes.com/2010/07/13/business/energy-environment/13bprisk.html?pagewanted=all

Maccoby, M. 2010. The 4Rs of motivation. *Research Technology Management*, *53*(4), 60–61. Retrieved December 26, 2010 from EBSCOhost database.

MacLean, R. 2004. EHS organizational quality: A DuPont case study. *Environmental Quality Management*, *14*(2), 19–27. Retrieved November 21, 2010 from EBSCOhost database.

Maclean, R. 2003. Superior environmental, health and safety performance: What is it? *Environmental Quality Management*, *13*(2), 13–20.

MacLeod, Q. 2010. Managing fireground errors. *Fire Engineering*. *Penn Well Publishing Co*. Retrieved August 2011 from EBSCOhost database.

Madlock, P. E. 2008. The link between leadership style, communicator competence, and employee satisfaction. *Journal of Business Communication*, *45*(1), 61–78. Retrieved July 25, 2008 from EBSCOHost database.

Major Incident Investigation Board (MIIB). 2005. *Initial Report to the Health and Safety Commission and the Environment Agency of the Investigation into the Explosions and Fires at the Buncefield Oil Storage and Transfer Depot, Hemel Hempstead*.

Major Incident Investigation Board (MIIB). 2006. Bunciefield. *HSE (07/06)/CIO, Third Investigation Progress Report*.

Markiewicz, D. 2009. OSHA compliance alone doesn't cut it. An integrated safety and health management system is needed. *Industrial Safety & Hygiene News*, *43*(7), 20. Retrieved July 3, 2010 from EBSCOhost database.

Marsh, T. 2007. A culture phenomena, affective safety culture. *Journal of the Institution of Risk and Safety Management Newsletter*.

Marsh, T., Davis, R., Phillips, R., Duff, R., Robertson, I., Weyman, A., and Copper, D. 1998. The role of management commitment in determining the success of a behavioral safety intervention. *Journal of the Institution of Occupational Safety and Health*, *2*(2).

Marson, G. K. 2001. The value case for investment in occupational health. *Occupational Medicine*, *51*(8), 496–500. Society of Occupational Medicine.

Maxson, R. 2010. Developing safety leaders. *Paperboard Packaging*, *9*(7), 28.

Mayer, R. C., Davis, J. H., and Schoorman, F. D. 1995. An integrative model of organizational trust. *Academy of Management Review, 20*(3), 709–734. Retrieved December 26, 2010 from EBSCOhost database.

McCroskey, J. C. and Teven, J. T. 1999. Goodwill: A reexamination of the construct and its measurement. *Communication Monographs, 66*(1), 90–103. Retrieved December 26, 2010 from EBSCOhost database.

McKayn, L. 2010. Generation green: Why Gen Y and the millennials are greener than you'll ever be. *CRM Magazine, 14*(4), 12–13. Retrieved November 21, 2011 from EBSCOhost database.

Mearns, K., Whitaker, S. M., and Flin, R. 2003. Safety climate, safety management practice and safety performance in offshore environments. *Safety Science-Pergamon, 41*. Retrieved March 10, 2011 from www.elsevier.com.

Medical Benefits. 2002. Liberty mutual workplace safety index. *Medical Benefits, 19*(10), 10. Retrieved November 17, 2010 from EBSCOhost database.

Michalko, M. 2000. Four steps toward creative thinking. *The Futurist,* 4(3), 8–21. Retrieved January 20, 2009 from EBSCOhost database.

Miller, P. 2004. *The Economics of Occupational Health Provision. Health as a Business Management Issue in the 21st Century*. Institute of Petroleum, Energy Institute, ISBN 0-85293-403-3.

Miller, P., Rossiter, P., and Nuttall, D. 2002. Demonstrating the economic value of occupational health services. *Occupational Health Medicine,* 52(8), 477–483. Society of Occupational Medicine.

Ministry of Environment, British Columbia. 2002. *Introduction to the Incident Command System*. Retrieved January 16, 2011 from http://www.env.gov.bc.ca/eemp/resources/icsintro.htm#2.

Mishra, P. K., Samarth, R. M., Pathak, N., Jain, S. K., Banerjee, S., and Maudar, K. K. 2009. Bhopal gas tragedy: Review of clinical and experimental findings after 25 years. *International Journal of Occupational Medicine & Environmental Health,* 22(3), 193–202. Retrieved January 16, 2011 from EBSCOhost database.

Mogford, J. 2005. Fatal accident investigation report; Isomerization unit explosion, *Final Report*, BP Texas Refinery, Texas City.

Mol, T. 2003. *Productive Safety Management: A Strategic, Multi-Disciplinary Management System for Hazardous Industries That Ties Safety and Production Together*. Elsevier, Butterworth-Heinemann.

Mook, B. 2010. Allen Family Foods will fight $1M penalty from Maryland Occupational. *Daily Record, the Baltimore,* June 7th 2010. Retrieved March 13, 2011 from http://findarticles.com/p/articles/mi_qn4183/is_20100607/ai_n54032350/.

Moore, D. A. 2008. Process safety leading and lagging metrics. *American Society of Safety Engineers (ASSE), Middle East Chapter, 8th Professional Development Conference Proceedings*, Kingdom of Bahrain. Ref: ASSE-0208-030.

National Safety Council. 2011. *CEO's who "Get it."* Retrieved March 1, 2011 from http://www.nsc.org/safetyhealth/Pages/211CEOsWhoGetIt.aspx.

Newstorm, J. W. and Davis, K. 1997. *Organizational Behavior: Human Behavior at Work* (10th ed.). International Edition, McGraw-Hill.

Obadia, I. J., Vidal, M.C. R., and Melo, P. F. F. F. 2006. An Adaptive management system for hazardous technology organizations. *Safety Science,* doi:10.1016/j.ssci.2006.07.002.

O'Dea, A. and Flin, R. 2001. Site managers and safety leadership in the offshore oil and gas industry. *Safety Culture Science, 37*, 39–57.

Olive, C., O'Connor, T. M., and Mannan, M. S. 2006. Relationship of safety culture and process safety. *Journal of Hazardous Materials, (130)*. Retrieved March 10, 2011 from www.elsevier.com.

OSHA. 2010. *Process Safety Management Guidelines for Compliance*. Retrieved January 1, 2011 from http://www.osha.gov/Publications/osha3133.html.

Owens, P. L. 2006. One more reason not to cut your training budget: The relationship between training and organizational outcomes. *Public Personnel Management, 35*(2), 163–172. Retrieved April 12, 2007 from EBSCOhost database.

Palmer, S. 2007. Stress, performance, resilience and wellbeing: The fit vs. unfit manager. *Occupational Health & Workplace Wellbeing Session. Institution of Occupational Safety and Health Conference*, Telford, UK.

Parker, D., Lawrie, M., and Hudson, P. 2006. A framework for understanding the development of organizational safety culture. *Safety Science-Pergamon*, (44), Retrieved March 10, 2011 from www.elsevier.com.

Parkes, K. 1993. Human factors, shift work and alertness in the offshore oil industry. *Offshore Technology Report, OTH 92-389*, Oxford University, UK-HSE.

Pennachio, F. 2008. Going beyond limits. *Occupational Hazards, 70*(9), 34–35. Retrieved December 20, 2010 from EBSCOhost database.

Pennachio, F. 2009. The effect of the economy on workplace safety. *EHS Today, 2*(3), 26–27. Retrieved December 20, 2010 from EBSCOhost database.

Peters, M. 2008. Behavior-based motor vehicle safety. *American Society of Safety Engineers (ASSE), Middle East Chapter, 8th Professional Development Conference Proceedings*, Kingdom of Bahrain. Ref: ASSE-0208-032.

Pinker, S. 2010. The real cost of upsetting the work–life balance. *The Globe and Mail.* Monday December 20. Report on Business.

Pisarski, A., Brook, C., Bohle, P., Gallois, C., Watson, B., and Winch, S. 2006. Extending a model of shift work tolerance. *Chronobiology International: The Journal of Biological & Medical Rhythm Research, 23*(6), 1363–1377. Retrieved November 9, 2007 from EBSCOhost database.

Powell, S. M. 2011. Estimating spill may be more law than science. *Houston Chronicle*, January 13, 2011. Retrieved January 16, 2011 from http://www.chron.com/disp/story.mpl/business/energy/7380838.html.

Price, W. H., Kiekbusch, R., and Theis, J. 2007. Causes of employee turnover in sheriff operated jails. *Public Personnel Management, 36*(1), 51–63. Retrieved April 6, 2007 from EBSCOhost database.

Principal Global Indicator. 2010. *IMF Data Mapper.* Real GDP (Annual). Retrieved November 13, 2010 from http://www.principalglobalindicators.org/default.aspx.

Ramlall, S. 2004. A review of employee motivation theories and their implications for employee retention within organizations. *Journal of American Academy of Business, Cambridge, 5*(1/2), 52–63. Retrieved September 30, 2006 from EBSCOhost database.

Ramseur, J. L. 2009. Oil spills in U.S. coastal waters: Background, governance, and issues for congress: RL33705. *Congressional Research Service: Report*, 8/27/2009, pp. 1–36. Retrieved February 6, 2010 from EBSCOHost database.

Reason, J. 2000. Safety paradoxes and safety culture. *Injury Control and Safety Promotion Journal, 7*(1), 3–14.

Results Based Interactions 2004. Leadership skills. *Development Dimensions International, Inc., MCMXCIX*. Pittsburgh, Pennsylvania.

Reifenstahl, E. 2009. The value in developing a company's leadership culture. *Fort Worth Business Press, 25*(39), 24–24. Retrieved July 2, 2010 from EBSCOhost database.

Reiman, T. and Oedewald, P. 2004. *Organizational Culture and Social Construction of Safety in Industrial Organizations*. Pergamon, Oxford.

ReVelle, J.B. 2004. *Quality Essentials: A Reference Guide From A to Z. American Society for Quality (ASQ)*. Quality Press, Milwaukee.

Riordan, C. M., Vandenberg, R. J., and Richardson, H. A. 2005. Employee involvement climate and organizational effectiveness. *Human Resource Management, 44*(4), 471–488. Retrieved April 25, 2007 from EBSCOhost database.

Roberts, M. 2004. *An OH View on Health in Business, Looking Forward. Health as a Business Management Issue in the 21st Century*. Institute of Petroleum, Energy Institute, 2004, ISBN 0-85293-403-3.

Rosen, A. and Callaly, T. 2005. Interdisciplinary teamwork and leadership: Issues for psychiatrists. *Australasian Psychiatry, 13*(3), 234–240. Retrieved November 24, 2007 from EBSCOhost database.

RoSPA. 2006. WSA winners announced. *Occupational Safety & Health Journal, 36*(4), 6. Retrieved July 26, 2008 from EBSCOhost database.

Ruuhielehto, K., Heikkila, J., and Suna, T. 2005. How to help people work safely—Promoting safety culture and safe work practices. *VVT Industrial Systems/ Fortum Oil and Gas Joint Research Paper*. Retrieved March 10, 2011 from www.fortum.com.

Ryan, D. 2010. Are you suffering from audit fatigue? *Industrial Safety & Hygiene News, 44*(10), 101–104. Retrieved March 13, 2011 from EBSCOhost database.

Safety Compliance Letter 2006. *Incorporating Safety Management*. Issue 2461. Aspen Publishers, New York.

Saldana, N. 2004. The Use of Safety Dashboards in a Six Sigma Culture: The Johnson & Johnson Journey for Excellence Continues. Paper presented at the American Society of Safety Engineers (ASSE) Professional Development Conference and Exposition, June 7–10, 2004, Las Vegas, Nevada. Paper # 04-712-1.

Schaechtel, D. 1997. How to build a safety management system. *Professional Safety, 42*(8), 22–25. Retrieved July 4, 2010 from EBSCOhost database.

Sepai, O. and Blain, P. 2000. *A Critical Review of Current Biological Monitoring Techniques for the Assessment of Exposure to Low Levels of Benzene*. Institute of Petroleum, Energy Institute, 2000, ISBN 0-85293-300-2.

Sharbrough, W. C., Simmons, S. A., and Cantrill, D. A., 2006. Motivating language in industry: Its impact on job satisfaction and perceived supervisor effectiveness. *Journal of Business Communication, 43*, 322–343. Retrieved July 26, 2008 from EBSCOHost database.

Silén-Lipponen, M., Tossavainen, K., Turunen, H., and Smith, A. 2004. Theatre nursing: Learning about teamwork in operating room clinical placement. *British Journal of Nursing, 13*(5), 244–297. Retrieved April 22, 2007 from EBSCOHost database.

Skoloff, B. and Wardell, J. 2010. *BP's Oil Spill Costs Grow, Gulf Residents React*. Associated Press. Retrieved November 22, 2010 from http://news.yahoo.com/s/ap/20101102/ap_on_bi_ge/eu_britain_earns_bp.

Slattery, J. and Selvarajan, T. T. R. 2005. Antecedents to temporary employee's turnover intention. *Journal of Leadership & Organizational Studies, 12*(1), 53–66. Retrieved October 1, 2006 from EBSCOhost database.

Smith, S. 2003. America's safest companies: World-class safety. EHS Today. Retrieved December 30, 2010 from http://www.penton.com/Pages/AcceptableUsePolicy.aspx.

Smither, L. 2003. Managing employee life cycles to improve labor retention. *Leadership & Management in Engineering, 3*(1), 19–23. Retrieved November 5, 2006 from EBSCOhost database.

Solanki, A. 2007. Improving line management accountability: A case study. *Best practices in health and safety (2) Session. Institution of Occupational Safety and Health Conference*, Telford, UK.

Spigener, J., 2009. Leaders who get safety: Values and Personality shape personal ethics. *Industrial Safety & Hygiene News*, 43(10), 38–40.

Suncor Energy Inc. 2011. *Sustainable Development.* Retrieved January 9, 2011 from http://www.suncor.com/en/responsible/302.aspx.

Taleb, N. N., Goldstein, B. G., and Spitznagel, M. W. 2009. The six mistakes executives make in risk management. *Harvard Business Review.* Retrieved December 28, 2010 from EBSCOhost database.

Tharaldsan J., Olsen E., and Eikeland, T. 2006. A comparative study of the safety culture. BB offshore system compared with other companies on the Norwegian Continental Shelf. *Norwegian Research Council (HSE Offshore).*

The Ken Blanchard Group of Companies. 2010. *Building Trust: The Critical Link to a High-Involvement, High-Energy Workplace Begins with a Common Language.* Retrieved July 10, 2011 from http://www.kenblanchard.com/img/pub/Blanchard-Building-Trust.pdf

Theron, E., Terblanche, N. S., and Boshoff, C. 2008. The antecedents of relationship commitment in the management of relationships in business-to-business (B2B) financial services. *Journal of Marketing Management, 24*(9/10), 997–1010. Retrieved December 16, 2010 from EBSCOhost database.

Thompson, T. P. 2002. Turnover of licensed nurses in skilled nursing facilities. *Nursing Economics*, 20(2), 66–69. Retrieved April 11, 2007 from EBSCOHost database.

Thornton, S. L. 2001. How communication can aid retention. *Strategic Communication Management, 5*, 24–27. Retrieved November 3, 2006 from EBSCOhost database.

Trevor, C. O. 2001. Interactions among actual ease of movement determinants and job satisfaction in the prediction of voluntary turnover. *Academy of Management Journal, 44*(4), 621–638. Retrieved April 8, 2007 from EBSCOhost database.

Truxillo, D. M., Bennett, S. R., and Collins, M. L. 1998. College education and police job performance: A ten-year study. *Public Personnel Management, 27*(2), 269–281. Retrieved April 24, 2010 from EBSCOHost database.

Turnbull, P. 2010. *Health & Safety Audit.* Retrieved March 13, 2011 from http://www.ehow.com/facts_7431380_health-safety-audit.html#ixzz16Daa9PMi.

U.S. Department of Homeland Security. 2009. *Oil Spills Compendium, 1973–2004.* United States Coastguard. Retrieved February 7, 2010 from http://homeport.uscg.mil/mycg/portal/ep/contentView.do?channelId=-18374&contentId=120051&programId=91343&programPage=%2Fep%2Fprogram%2Feditorial.jsp&pageTypeId=13489&contentType=EDITORIAL&BV_SessionID=@@@@1864989697.1265590118@@@@&BV_EngineID=cccdadejigldhfmcfjgcfgfdffhdghm.0.

U.S. Department of Labor. 2007. *National Census of Fatal Occupational Injuries in 2008.* Retrieved March 10, 2010 from http://www.bls.gov/iif/oshwc/cfoi/cfch0007.pdf.

U.S. Department of Labor. 2009. *Workplace Illnesses and Injuries 2004–2008*. Retrieved March 10, 2010 from http://www.bls.gov/iif/oshsum.htm.

Van Dyke, D. L. (Captain). 2006. *Management Commitment: Cornerstone of Aviation Safety Culture*. Royal Aeronautical Society Montréal Branch, John Molson School of Business, Concordia University, Montreal, Quebec.

Watson, A. 2007. Communicating the benefits of effective health and safety at board level, Communicating Health and Safety Session. *Institution of Occupational Safety and Health Conference*, Telford, UK.

Wesley, J. R., Muthuswamy, P. R., and Darling, S. V. 2009. Gender difference in family participation and family demand in dual career families in India—An empirical study. *The XIMB Journal of Management*, *6*(2), 49–62. Retrieved April 22, 2010 from EBSCOHost database.

WHO, 2008. http://www.who.int/about/history/en/index.html

Wilson, N., Cable, J. R., and Peel, M. J. 1990. Quit rates and the impact of participation, profit-sharing and unionization: Empirical evidence from UK engineering firms. *British Journal of Industrial Relations*, *28*(2), 197–213. Retrieved April 25, 2007 from EBSCOHost database.

Wren, D. A. 1994. *The Evolution of Management Thought* (4th ed.). New York: John Wiley & Sons.

Yueng-Hsiang, H. and Brubaker, S. A. 2006. Safety auditing applying research methodology to validate a safety audit tool. *Professional Safety*, *51*(1), 36–40. Retrieved March 13, 2011 from EBSCOhost database.

Zwetsloot, G. I. J. M. and Ashford, N. A. 2003. The feasibility of encouraging inherently safer production in industrial firms. *Safety Science-Pergamon*, *41*. Retrieved August 14, 2011 from www.elsevier.com.

Index

I

S

For Product Safety Concerns and Information please contact our
EU representative GPSR@taylorandfrancis.com Taylor & Francis
Verlag GmbH, Kaufingerstraße 24, 80331 München, Germany